UNCOUNTABLE

Uncountable

A PHILOSOPHICAL HISTORY OF
NUMBER AND HUMANITY FROM
ANTIQUITY TO THE PRESENT

David Nirenberg
AND
Ricardo L. Nirenberg

The University of Chicago Press CHICAGO AND LONDON

The University of Chicago Press, Chicago 60637
The University of Chicago Press, Ltd., London
© 2021 by David Nirenberg and Ricardo Lida Nirenberg
Published 2021
Paperback edition 2024
Printed and bound by CPI Group (UK) Ltd, Croydon, CR0 4YY

33 32 31 30 29 28 27 26 25 24   1 2 3 4 5

ISBN-13: 978-0-226-64698-5 (cloth)
ISBN-13: 978-0-226-82836-7 (paper)
ISBN-13: 978-0-226-64703-6 (e-book)
DOI: https://doi.org/10.7208/chicago/9780226647036.001.0001

The University of Chicago Press gratefully
acknowledges the generous support of the Divinity School
and the Division of Social Sciences at the University
of Chicago toward the publication of this book.

Library of Congress Cataloging-in-Publication Data

Names: Nirenberg, David, 1964– author. |
Nirenberg, Ricardo L., author.
Title: Uncountable : a philosophical history of number
and humanity from antiquity to the present /
David Nirenberg and Ricardo L. Nirenberg.
Description: Chicago : University of Chicago Press, 2021. |
Includes bibliographical references and index.
Identifiers: LCCN 2021007568 | ISBN 9780226646985 (cloth) |
ISBN 9780226647036 (ebook)
Subjects: LCSH: Mathematics—History. | Mathematics—
Social aspects. | Mathematics—Moral and ethical aspects.
Classification: LCC QA21 .N574 2021 | DDC 510—dc23
LC record available at https://lccn.loc.gov/2021007568

♾ This paper meets the requirements of ANSI/NISO Z39.48-1992
(Permanence of Paper).

For Isabel, and for Sofía

All chance, all love, all logic, you and I,
Exist by grace of the Absurd,
And without conscious artifice we die.

W. H. AUDEN, "In Sickness and in Health"

CONTENTS

CONTENTS

Introduction

PLAYING WITH PEBBLES

The ancient problem of "the one and the many." I suspect that in
but few of you has this problem occasioned sleepless nights. . . .
I myself have come, by long brooding over it, to consider it the most
central of all philosophic problems, central because so pregnant.

WILLIAM JAMES[1]

A remarkable attribute of the species biologists call *Homo sapiens* is
that its members have so often asked themselves about the nature
of their own *sapientia*: the knowledge or wisdom for which they are
named. Equally remarkable is the fact that in answering these questions, humans have been so willing to tear themselves apart. Over
and over they have divided their cognitive capacities into "good"
ones and "bad." They have even imagined that some ways of thinking make humans eternal and godlike, while others lead to mortality, deceit, damnation, even the destruction of the world.

Often enough over these past three thousand years, we humans
have pursued these divisions to the death, clashing over differences
of opinion about what we should know and how we should know
it. We are not talking here only of the many clashes between different religions and cultures of knowledge in the distant past. Even the
two world wars of the twentieth century were understood by many
who lived through them as the consequence of bad choices about
what kinds of knowledge to pursue. World War I, for example, was
explained by leading European and American intellectuals as the
result of mathematics gone bad, an inhuman fusion of arithmetic
and geometry that destroyed Western civilization. (Please contain
your mockery until you have read chapter 1.) Plenty of ideologues
found it easy enough to cast the Cold War as a struggle between two
different theories of knowledge, "Marxism" and "liberalism," "determinism" and "freedom." Perhaps future generations will come

to see the current arguments about the human impact on climate change as yet another chapter in this long history of humanity's division over the nature of knowledge.

Today, mathematical forms of knowledge—computation, artificial intelligence, and machine learning, for example—touch many more aspects of the world than they did in the first half of the twentieth century, or, indeed, in any previous period of this planet's history. Divisions between types of knowledge, such as those between the humanities and the sciences—"the two cultures," as C. P. Snow dubbed them in 1959—are if anything deeper than they have ever been. Yet unlike our predecessors from a century ago, few people today—except perhaps panicked humanities professors who feel their habitat melting away beneath their feet—would consider these divisions deeply threatening. Even fewer would claim that understanding them is in some way essential to humanity.

We are not writing an Apocalypse. Ours is an attempt to understand these millennial divisions so that we might better live with them. How has humanity pitted its various powers of thought so fiercely against itself? And why have the truth claims of numerical relations emerged so powerful from this conflict? Achieving this understanding is a historical task, and the first half of this book (chapters 1-5) set out to provide that history. In chapters ranging from ancient Greek philosophy and the rise of monotheistic religions to the emergence of modern physics and economics, we trace how ideals, practices, and habits of thought formed over millennia have turned number into the foundation stone of human claims to knowledge and certainty. (Readers who have a low tolerance for ancient history, philosophy, or religion may want to skip chapters 2-4.)

Learning to live humanely with these divisions is the goal of the second half of this book (chapters 6-10). These divisions and conflicts of our faculties and our knowledge are not necessary ones. The fragments of our humanity can be brought together in different ways, even in ways that might be truer to basic aspects of the questions we want to ask and the objects we want to know, truer even to our own human being.

This book is therefore not only a history. It is also a philosophical and poetic exhortation for humanity to take responsibility for that history, for the knowledge it has produced, and for the many aspects of the world and of humanity that it ignores or endangers. We seek to convince you that how we humans think about our knowledge has deep consequences for how we live our lives and that we need to become more conscious of the first if we wish to change the second.

But before we can do any of that, we need to be moved by the stakes of the problem. For that, we turn not to history, philosophy, or psychology but to a story.

Blue Tigers

Alexander Craigie is the narrator in one of Jorge Luis Borges's last short stories, "Tigres azules" ("Blue Tigers," published in 1983).[2] Craigie is a Scotsman making a professorial living teaching "occidental logic" in the British colonial city of Lahore (in today's Pakistan) circa 1900. A philosopher and a servant of reason, he is also quite interested in tigers and has been dreaming about them since early childhood. Toward the end of 1904 he reads somewhere the surprising news that a blue variant of the animal has been sighted in the subcontinent. He dismisses the report as impossible. But rumor of their existence continues, and eventually even the tigers in his dreams turn blue. So he sets off to find them.

After some time he arrives at a Hindu village mentioned in some of the reports. When he tells the villagers what he is looking for, their reaction is furtive but helpful. Frequently they come to tell him of a sighting, leading him hurriedly in directions where the beast is said to have just been spotted. Never is it to be found. When after some time he proposes to them that they explore in a direction they seem to have been avoiding, he is met with consternation. That area is sacred, forbidden to man, guarded by magic. Any mortal who walks there might go mad or blind from the sight of divinity. So our narrator sneaks off in the night on the forbidden path.

The ground is sandy and full of channels. Suddenly, in one of the channels, he sees a flash of the same blue he has seen in his dreams. "The channel [is] full of pebbles, identical, circular, very smooth and a few centimeters in diameter." They are so regular as to look artificial, like tokens or counters. Picking up a handful, he places them in his pocket and returns to his hut, where he attempts to remove them from his pocket. He takes some out but feels that two or three handfuls still remain. He feels a tickle, a tremor in his hand, and opening it sees some thirty or forty disks, although he could have sworn that from the channel he had not taken more than ten. He does not need to count them to see that they have multiplied, so he puts them in a pile and tries to count them one by one.

"This simple operation prove[s] impossible." He can stare at any one of them, remove it with thumb and index finger, and as soon as it is alone, it is many. "The obscene miracle" repeats itself over and over. His knees begin to tremble. He closes his eyes, repeats slowly some of Spinoza's axioms of logic, but whatever he does he cannot escape the stones. At first he suspects he is crazy, but with time he realizes that madness would be preferable: for "if three plus one can equal two or fourteen, reason is an insanity."

The professor returns to Lahore. Now it is the pebbles that populate his dreams. He carries out experiments, marking some with crosses, filing others, puncturing a few, attempting to introduce some difference into their sameness by which he might distinguish them. He charts their increase and decrease, "trying to discover a law," but they change their marks and their number, seemingly at random, in no pattern discernible by statistics. He concludes, "The four operations of addition, subtraction, multiplication and division were impossible. The pebbles denied arithmetic and the calculus of probability. . . . Math, I told myself, had its origins and now its end in pebbles. If Pythagoras had worked with these . . . After a month I understood that the chaos was inescapable."

In this story, without theorems or technical notation, Borges set out in narrative a basic precondition for what is habitually called rationality and posed a brilliant thought experiment: what happens when that precondition does not hold? The precondition here is a very simple one of sameness or difference. Can we tell whether something is the same as itself? A blue disk can't be pinned down as identical to itself or as different from others. Hence, it can't be grasped by counting, statistics, or any logical or scientific analysis.

It is not only ungraspable but also terrifying, for the study of blue tigers will drive you mad. In the end the author saves his narrator from that awful fate. After a sleepless night desperately wandering the city, at that hour of dawn when "light has not yet revealed color," Craigie enters the mosque of Wazil Khan. He prays to God for relief. Suddenly he hears a voice asking for alms: a blind beggar has mysteriously appeared before him. At the blind beggar's insistence Craigie gives him the disks, which fall noiseless into his hands, "as if to the bottom of the sea." The beggar's words in return are "I do not yet know the nature of the alms you have given me, but mine to you are terrible. Yours are the days and the nights, sanity, habits, and the world."

Borges has taught us something very important about how we apply sameness and difference to our objects of thought. But notice how starkly the two are divided. On one side the ever-changing, indistinguishable, and uncountable "blue disks" bring unreason, chaos, madness; on the other stable pebbles, countable because unchanging, always the same as themselves, bring reason, science, and sanity. Borges's conclusion seems to imply that we must choose between two types of attention, two forms of life, two kinds of knowledge, each horrifying in its own way. As we shall see, that has been a common teaching across much of human history and philosophy. But our goal in this book is to convince you that resolving the problem in this way is both false and dangerous. There are infinitely many objects of thought in this world that act like well-behaved pebbles, but there are also infinitely many that act like the ones Craigie found on the forbidden path. In later chapters we will discover that with the exception of some very peculiar mathematical objects, every "normal" pebble is also from some perspective a "blue" one. This is true even at the physical foundations of the universe. The great pioneer of quantum physics Erwin Schrödinger—describing the physicist's inability to declare electrons, protons, and other quantum objects as "same" or "different"—sounds much like Craigie trying to count blue disks: "This means *much more* than that the particles or corpuscles are all *alike*. It means that you must not even *imagine* any one of them to be *marked*—'by a red spot' so that you could recognize it later as the *same*."[3] Conversely, as we shall see when studying economics, psychology, and other sciences of self and society, many a "blue" object can be usefully approached as if it were stable, as if it remained the *same*.

If we try to pass off the burden of "blue tigers" onto blind beggars, opting to attend only to the countable, or to live only by laws of reason, we do violence not only to the objects of our thought but also to our own being, just as if we attempt to do the opposite, and tend only to the azure. Yet this is precisely the path often taken. The "occidental logic" Craigie teaches is one example, born from an insistence on submitting the mind to the rule of sameness. Consider the role of sameness in just a few of the fundamental logical principles—we will sometimes call them "laws of thought"—so brazenly flaunted by Borges's blue pebbles.

Of these rules perhaps the most fundamental is the Identity Principle, which declares that for *any* thing, let's call it x, x is the *same* as x, or $x = x$. With certain things in certain circumstances, the Identity Principle works very well: a well-behaved pebble, for example, under moderate temperature and pressure, relatively short spans of time, and unaided human eyes, seems consistent and unchanging and can in good conscience be taken to be equal to itself. (Perhaps this is why pebbles have been used by humans as aids in counting for thousands of years.) For other things under other circumstances—ranging from elementary particles to people's dreams and passions—it does not work so well, if at all.

Another "law of thought" goes by the trade name Principle of Sufficient Reason. This principle was first fully formulated by the German mathematician-philosopher Gottfried Wilhelm Leibniz (1646-1716), but the general idea had been put to work already by the earliest Greek natural philosophers in order to explain why things in the cosmos are as they are. In plain English the principle can be stated as follows: any fact that obtains must occur by virtue of some reason, cause, or ground that makes it happen the way it does and not some other way. The *same* causes must result in the *same* effects.[4]

Yet another is called by modern logicians the Principle of Non-Contradiction. Aristotle, the Greek grandfather of Craigie's "occidental logic," offered this version: "It is impossible for anything at the same time to be and not to be." Indeed he went further: not only can contradiction not exist within a thing, but it cannot exist within a person: "It is impossible for anyone to believe that the same thing is and is not." We cannot hold contradictory thoughts at the same time.[5]

Two things are especially striking about these formulations of logical principles and laws of thought. One is how heavily they depend on "sameness," that is, on some sort of identity or equality. And the other is the willingness with which even the greatest intellects have been willing to apply them, as we have just seen Aristotle do, to the human mind or psyche. "The mind can always intend, and know when it intends, to think the Same." Thus William James, in *The Principles of Psychology* (1890), the founding text of that field in North America. "This *sense of sameness* is the very keel and backbone of our thinking."[6] There is indeed great profit in this submission of the mind to the logic of sameness. But there is also loss, the rejection (or ignoring) of everything in ourselves and in the world that does not conform to those rules, a "kind of dying to oneself," as the Danish philosopher Kierkegaard put it, a half century before James.[7]

Sciences of Sameness

Let's focus for now on profit. The human mind is indeed astonishingly capable of intending (we might also say imposing or projecting) patterns of sameness and difference on the unfathomable and often terrifying complexity of the cosmos. The focus on intending sameness has been productive of systems of knowledge that we can call sciences, systems that have been remarkably powerful in helping humanity cope with, and sometimes understand, or even control that complexity. Consider, for example, three different sciences, built on different kinds of sameness, that provided inhabitants of the ancient world—think Babylon, Egypt, China, in the millennium before the Common Era—with knowledge of the past and the future.

The discovery and mathematization of repetition and periodicity in the movements of the sun, moon, planets, and even of the stars that so densely packed the preindustrial skies gave many ancient societies a sense that they could project the past into the future: a comforting predictive power in a vast and variable cosmos. Long before written record we find the constellation we call Orion painted by Paleolithic hands on the cave walls of Lascaux. And on the cuneiform tablets of the first Mesopotamian scribes, the verb *to count* was applied to the skies in the production of astral omens.[8] Sameness was also put to work in the study of dreams, whose imagery was long thought to provide analogies—seven fat cows = seven abundant years, seven lean cows = seven years of famine, to quote just one biblical example—that could yield predictive knowledge. And finally, magic too achieved its power through perceptions of sameness. The widespread use of dolls, figures, and statues to represent the victim in ancient magic (as in some contemporary practices) is an example of the powers of similarity.[9]

Three different and very ancient forms of predictive knowledge, built on attention to different types of sameness and repetition, each with very different futures. Today *oneirology* (the science of dreams) is scarcely a word. The study of magic is confined to anthropologists, historians, or "the ignorant." Astronomy, however, benefits from billions in annual investment from science foundations and is a monument to the ability of the human mind to make some sense of the structure of the universe.

The point here is not that knowledge has progressed. (When it comes to dreams, it may even be that a certain kind of self-knowledge has been lost by not attending to them.) Our point is rather that the form of attention we today call scientific has been oriented toward certain kinds of sameness—in this case formalizable, axiomatic, mathematical—and not toward others. There are many reasons why this is the case. But one, noted already by the Roman natural philosopher Pliny, writing some two thousand years ago, is that mathematical astronomy provided some seemingly stable powers of prediction about an otherwise overwhelming universe: "O mighty heroes, of greater than mortal nature, who have discovered the law of those great divinities and freed the miserable mind of man from fear. . . . Praised be your intellect, you who interpret the heavens and grasp the facts of nature, discoverers of a theory whereby you have bound both gods and men!"[10]

Pliny thought mathematical astronomers praiseworthy because they derived procedures through which they could predict the movements of the planetary deities, thereby binding "gods." And to the degree that the planets were thought to determine the fate of a person, a science (today we call this astrology, not astronomy) capable of predicting the future position of the planets also offers knowledge about human fate, thereby binding "men."

The Aztecs provide a different astronomical example, instructive because it reminds us of the contingency of certainty and the tenacity of fear. They were sophisticated astronomers, but they believed (as did the ancient Egyptians) that the system that kept the sun appearing regularly in the sky was unstable and that the sun might someday fail to rise if humans did not do their part by making offerings to the gods. In the case of the Aztecs, these offerings involved human sacrifice, a practice that may have contributed to the collapse of the Aztec empire when Hernán Cortés and his handful of Spanish "Conquistadors" arrived in the early sixteenth century. Tired of being slaughtered at sacrificial altars, subject peoples rebelled against the Aztecs and handed victory to the European invaders.[11]

How could anyone think that human action (let alone sacrifice) is necessary to ensure the dawn? What could be more certain from our experience than the sun's rising? And yet the Aztecs were not wrong in worrying about their knowledge of the sunrise. Their refusal to deduce future certainty from previous dawns is defensible in the most sophisticated probabilistic terms.[12] As solar system dynamicists today would tell us, the system *is* unstable. Yet we moderns expend very little of our still considerable religious energy in keeping the solar system going. Again, the point is not that the Aztecs were bad scientists or that we moderns should be more worried about the sunrise. Our point is simply that the desire for certainty can lead us to extend—often inappropriately, sometimes disastrously—

lessons, methods, and sciences of sameness from one domain of knowledge into another where perhaps they do not apply.

The Stars and the Psyche

Another astronomical example, this one from Princeton, New Jersey, Monday evening, February 18, 1952. The art historian Erwin Panofsky and the medieval historian Ernst Kantorowicz are engaged in a discussion about their sense of the Sublime. Both are refugees from the Nazis, and both are permanent members of the Institute for Advanced Study, colleagues of the physicists and mathematicians Albert Einstein, Hermann Weyl, and John von Neumann, all three of whom will appear frequently in this book. As the two conclude their colloquy and step out into a cold and clear New Jersey night, Kantorowicz pronounces: "Looking at the stars, I feel my own futility." Replies Panofsky, "All I feel is the futility of the stars."[13]

The two German-Jewish refugees might have been talking in earnest, manifesting their different metaphysical inclinations. But more probably, given their vast learning and their capacity for irony, they were also playing variations on philosopher Immanuel Kant's famous analogy between the laws of physics that govern galaxies and the laws of the psyche that govern humanity: "Two things fill the mind with ever new and increasing admiration and awe, the more often and steadily we reflect upon them: the starry heavens above me and the moral law within me."[14]

Compressed into this banter under a winter sky are some basic questions about humanity and science. For example, is there a relationship between the forces that govern the starry heavens and those that drive our inner life as human beings? If there is, then what is the relationship between our knowledge of the cosmos and that of our psyche, of physics and psychology, of (multiplying analogies) objective and subjective, law of nature and human freedom?

Kant (1724–1804) lived in a century whose knowledge of the cosmos was being transformed by mathematical discoveries—such as those of Sir Isaac Newton (1643–1727) and Leibniz, two early explorers of calculus—and by the physics and astronomy this new mathematics made possible. Kant and many others wondered whether the phenomena mathematics was useful for might include not only physical but also psychic events: not only the movements of the planets, but also the workings of the mind. He and many others wanted to explore the relationship between the two and to inquire into the possibility of borrowing truth-producing tools that worked well in one domain (such as mathematics, which worked so well in physics) and applying them to the other.

Let's put it in terms of the "laws of thought" we introduced earlier. Over millennia we humans have developed rules for thinking about the world, rules like the Identity Principle, the Principle of Non-Contradiction, the Principle of Sufficient Reason, and others that we will be introducing throughout this book. These strict principles have proven astonishingly successful at discovering certain truths about that world. But do these axioms of reason also apply to our inner selves, to our ethics, for example, or more broadly, to our own feelings and thoughts?

If they do, then there are many questions we will need to ask, among them the implications for what theologians called "free will" and some philosophers came to call simply "freedom."[15] Kant himself thought that this was a most "difficult problem, on which so many centuries have labored in vain," and celebrated his own proposed solution.[16] And if they do not, if these axioms cannot be applied to our inner selves, then the questions are different but just as important. After all, our political and economic sciences have been built (as we will see in chapter 8) by applying the Principle of Non-Contradiction at the level of the psyche and of society. If the human subject is not internally consistent in ways that those theories of the world generally assume, we should certainly want to know, not least because those theories shape all manners of policy, from our wars to our welfare systems.

These questions, too often forgotten, are also venerable, much older than Kantorowicz and Panofsky, older than Kant, indeed so ancient that over the eons, humanity's answers to them have changed and even switched polarity, like the earth's magnetic field. Like magnetism, those answers have also been a force on human thought and action, one that we are seldom aware of, even as it shapes the possibilities for both. A goal of this book is to sharpen our awareness of these basic questions and of what is at stake in how we choose to answer them.

"All That Is, Insofar as It Is, Is Number"

Our contemporary sciences are even more spectacularly successful than those of Kant's day and as capable of breeding strange conceits about our psychic and moral lives. In fact, the stakes to our question may seem even higher in our era of machine learning and quantum computing, of brain imaging and computational neuroscience, and even of cyborgs (human-machine hybrids) in which a failing body part or organ—hand, retina, or tympanum, for example—is replaced by a computer. But long before Charles Babbage's invention of an "Analytical Engine" or computing machine in the nineteenth century, philosophy, science, and (later) science fiction have been fascinated by the boundaries between the human and the calculating machine or computer. Today that fascination has colonized our basic questions about the nature of human consciousness, becoming the common stuff of our imaginings.[17]

Although the question is perennial, at present it feels particularly acute, perhaps even affecting the chances of our short-term survival as a species. We could try to make that dramatic point plausible by discussing some of the many recent books that focus on one symptom or other of our disease. Books, for example, that trace the transformation of rationality in the nuclear age, that criticize "neoliberal" reductions of happiness or the good to an economic calculus, that worry about the increasing extension of algorithm over a human society understood simply as "big data," or that intervene in the current debates around climate science and the Anthropocene.[18]

Our book, however, is not about any of these symptoms. It is rather about the more fundamental problem beneath all of them: the tendency to apply forms of knowledge that are effective in one domain (say logic, or astronomy) to another (say literature, psychology, or anthropology), where the necessary conditions for their application may not apply, without paying sufficient attention to what may be lost in the gap. To quote the ever-quotable Friedrich

Nietzsche, "We have arranged for ourselves a world in which we can live—by positing bodies, lines, planes, causes and effects, motion and rest, form and content: without these articles of faith nobody now could endure life. But that does not prove them. Life is no argument."[19] No, Nietzsche is not quite right. More precisely, they are proved and demonstrated in some domains of knowledge and not in others, but we have extended them to many aspects of our world in which their validity is not, cannot be, demonstrated. Yet we take them as universally true, perhaps because we could not endure to live without the confidence they provide.

We will call this comforting but unexamined extension of our habits of thought in search of illusory certainties "the expansive force of success." We can everywhere find examples of the error and of the false confidence it produces. It is perhaps a universal tendency to confuse, as an old quip has it, the customs of one's tribe for the laws of the universe. But in this book we focus on a particular and quite peculiar set of customs for thinking about sameness and difference: those associated with counting, with number, with logic, and with all the knowledge that flows from them. Peculiar because, although these habits of thought and forms of knowledge are in fact customs in the sense that they are the product of a shared culture and set of assumptions (in this context we might call them axioms), they actually do dictate laws to certain aspects of the universe. It is precisely because of this that there is such a strong temptation to apply these same assumptions to other aspects of that universe, such as our thoughts, emotions, aspirations . . . in short, to every aspect of human life and culture.

Like all habits and temptations, this one has a history. Already in the early sixth century BCE, the Greek sage Pythagoras is said to have maintained that everything can be counted. The statement attributed to him, "all that is, insofar as it is, is number," suggests not only that everything can be counted or measured but perhaps even that knowledge of numerical relations is the only true knowledge, numbers the only true being.[20]

Even the "dark" medieval ages held on to aspects of this conviction. When the twelfth-century scholar Adelard of Bath personified Arithmetic in his treatise *On the Same and the Different* (ca. 1120), he put all things at her command, because "as all visible things are subject to number, they must also be subject to [Arithmetic]. For whatever is, is either one or many. . . . I have no doubt that she should be preferred to all essences, since she takes confusion away from them, and gives them distinction." Considerably closer to our own time and diction, the great logician and philosopher Alfred North Whitehead made a similar statement in *An Introduction to Mathematics* (1911): "Now, the first noticeable fact about arithmetic is that it applies to everything, to tastes and to sounds, to apples and to angels, to the ideas of the mind and to the bones of the body. The nature of the things is perfectly indifferent, of all things it is true that two and two make four."[21]

Perhaps the most important goal of this book is to convince you that it *is not* true of all things that two and two make four. The nature of the things *is not* perfectly indifferent. Counting "ideas of the mind," for example, might require us to treat our thoughts as if they can be arranged as a sequence in time, 1, 2, 3, . . . , where each is the same with itself but different from the others, like pebbles arranged in a row. And then, in order to claim that *any* thus isolated mental state must be caused by a preceding, similarly isolated mental state, we might have to adopt a rigid discipline where one thinks *only* like this: 1, then nothing at all, then 2, then nothing at all, then 3, . . . (already knowing, of course, that after 1 comes 2 and after 2 comes 3 . . .). Often enough our thoughts do not work that way. Nor do our feelings necessarily conform to Aristotle's version of the Principle of Non-Contradiction, any more than our moods need follow Leibniz's Principle of Sufficient Reason rather than the inexplicability celebrated by a Spanish poet: "And suddenly, unannounced, for no reason, here's joy."[22]

We will logically demonstrate the conditions of sameness necessary for two plus two to equal four in chapter 6 and begin there to point out some of the many cases in which those conditions do not hold. But first, we should acknowledge the amazing power of mathematics to impose its principles of identity and sameness on many things that would seem to resist them. For example, when philosophers and poets have wanted to imagine something as removed as possible from sameness and as resistant as possible to the powers of number, they have often had recourse to running water. (In chapter 2 we will dip our toe into Heraclitus's famous river, and the poet Rilke's "swift water course" runs through chapter 7 on physics and poetry.)

But consider this trick discovered by the Bernoullis, founders of the field we today call fluid dynamics. (Like us, they were father and son, but unlike us, they published separately and sued each other for plagiarism.) In *Hydrodynamica* (1738) and *Hydraulica* (1743) they combined the resources of the new physics and calculus, applying Newton's laws of motion to the droplet (*guttula*) or particle (*particula*) of fluid *as if* droplet or particle were mass points or tiny pebbles.[23] From the motions of those small elements they then concluded about the motions of the whole. That is, of course, the whole point of calculus: to simplify an object by splitting it into "infinitesimals," then to put those back together by means of integration. Separation and putting together, analysis and synthesis.

The English word *calculus* comes from Latin for "pebble." The ancients used pebbles to represent numbers as they counted, and the modern English word's etymology serves to remind us that the powers of calculus derive from treating everything it touches as if it were a normal pebble: imperturbably always the same as itself, happily and unproblematically subject to the Identity Principle, remaining constant whether we collect them together or separate them. Since throughout this book we will highlight the importance in the history of thought of this simple, basic property of pebble-like elements, it seems appropriately pretentious to coin a neologism for it. So we will draw on the Greek word *apathes*—imperturbable, impassive—and call elements with this quality *apathic*.[24]

Calculus, like all of mathematics, depends on treating things as apathic. It divides a whole up into small parts—droplets, "infinitesimals"—applies its simplifying magic to the parts, and finally puts the parts together again *with no change* in them or in the whole. (Just as in geometry we can take a polygon, divide it up into triangles, then set them together again, and voilà, the original polygon.)[25] We can apply this wondrous power to problems of enormous complexity, such as the motion of cooling fluid in a nuclear reactor's core, of air over an airplane's wing, or the enchanting vibrations of Stradivarius's violins.

You can already predict the next question. Are there things that do not remain the same, that *cannot* be separated or brought together without difference?[26] We hope to convince you that there are, and will call such objects *pathic*, from the Greek meaning susceptible to change or alteration. If so, then for what kinds of objects, questions, or contexts does it make sense to treat things like pebbles, and for what kinds does it not? What errors do we commit if we apply this marvelous calculating power to objects, questions, or contexts that are not appropriately pebbly? And what kinds of knowledge do we lose if we confine our attention only to those that are?

We stated earlier that every object of thought can be approached, depending on perspective or question asked, as a blue pebble or a normal one (excepting, again, some peculiar mathematical objects). There is choice involved. Mathematical models often proceed by treating their objects as normal pebbles, and there is nothing wrong with that. Indeed the wondrous predictive power of the best mathematical models comes from their power of abstraction, their ability to idealize and simplify, to leave things out. Galileo famously ignored wind currents and viscosity in his model of free fall. If he had focused on turbulence (as James Clerk Maxwell once quipped), modern physics might not have gotten off the ground.

What can be safely left out and what cannot? That is among the most difficult questions confronting the scientist, but it is also a difficult question for all of us who seek to understand something about anything, including ourselves. We suspect that even the great Newton would agree with us on this score if the words of retrospective self-reflection attributed to him are true: "I do not know what I may appear to the world; but to myself I seem to have been only like a boy playing on the seashore, and diverting myself in now and then finding a smoother pebble or a prettier shell than ordinary, whilst the great ocean of truth lay all undiscovered before me."[27] We do not know what truths Newton felt his pebbles had distracted him from. But our own position is that, in particular when it comes to the study of the human, we need to become more aware of the losses our more lithic options require.

Between Pebbles and the Deep Blue Sea

It is one more symptom of the Expansive Power of Success that our educators have often preferred pebbles to oceans. Leonhard Euler (1707–1783), for example, was not only one of the greatest mathematicians of the Enlightenment but also a late representative of a long tradition of scholars involved in the education of princes. Plato played pedagogue to the future politicians of Athens, and Aristotle tutored the young Alexander before he was Great. Descartes exchanged letters with Princess Palatine of Bohemia, while Hobbes taught both the future Charles II and the Duke of Buckingham (a student so disengaged that he reportedly masturbated during his lessons[28]). Euler's tutee was the fifteen-year-old Prussian princess Friederike Charlotte von Brandenburg-Schwedt, and the course took place by correspondence. Beginning in 1760, Euler wrote her some 230 letters, which became best sellers when published with the title *Letters to a German Princess on Diverse Subjects of Physics and Philosophy* after his death.[29]

Euler was an astute critic of many "laws of thought" and especially of Leibniz's Principle of Sufficient Reason. But he, too, had his principles, and the most basic one that he taught the teen princess, indeed the most basic notion in nature, according to him, was that of body. What is body? Euler first dismissed some earlier influential answers to this question before offering his own: *impenetrability* is the defining property of bodies. That word means, Euler explained, that two *different* bodies cannot be in the *same* space at the *same* time. But, says Euler in anticipation of the princess's likely (and chaste) objection, aren't fluids such as water or air penetrable yet bodies nevertheless? No, he replies, they are not penetrable; they only seem to be so because we can plunge our hand in them. But this is because water and air and other fluids are formed of small impenetrable particles with empty space in between, and by moving around, those particles make way for our hand: there is no water in the space our hand occupies. Euler concludes, "This property of all bodies, known by the term *impenetrability*, is, then, not only of the last importance, **relatively to every branch of human knowledge**, but we may consider it as the master-spring which nature sets a-going, in order to produce all her wonders."[30]

The pedagogy Euler offered his royal pupil boils down to one thing: the key to every branch of human knowledge is to treat things as in some sense apathic, as a pebble or a conglomeration of pebbles, so that they can be reduced to the procrustean bed of the Identity Principle, of logic, and of math. We do not reject this pedagogy nor the laws of thought that it teaches. But we insist that it is partial and would offer you some of the things it lacks: a recollection of the ocean's existence, a set of questions that help us to decide whether to think of something as a normal pebble or a "blue" one. Our lesson is simply that we can choose what shifting sands we seek: those of the beaches, the shallows, or the deep.

In many societies, though perhaps in some more than in others, the pathic and apathic aspects of our humanity have often been split apart and pitted against each other. Classical Chinese thought produced many warnings against that strategy even if it could never quite escape it. "So sameness is called the One and differences are called the Way. As each prevails over the other this is thought of as engaging [in struggle], and good fortune and misfortune are spoken of as successes and defeats." Thus *The Book of the Pheasant Cap Master*, written in China ca. 221 BCE.[31] Splitting is not our path. We are not preaching reason or unreason nor urging a systematic choice between stable sameness or endless difference, math and madness. Quite the contrary, we want to hold on to blue tigers, and to logic too. We want to learn from our dreams and our poems and also from our science.

We cannot achieve that simply by creating new systems of logic (as Hegel and others have claimed to do) nor by rejecting logic as a source of understanding about humanity (as Heidegger preferred). The only way to do so, we submit, is to become conscious of the choices we make as we attempt to know something of ourselves and our world and to realize that those choices are not dictated by law but depend on the questions we are asking, on the perspectives and disciplines from which we ask them, on the objects we are studying, on who we are or wish to be, on what we want to know. There are no invariable rules, no laws of thought that can dictate to us what would be the right choice for every question or situation. The choice is ours to make, and in it lies what has often been called human freedom but which we would prefer to call human knowledge.

Law and freedom. In the Christian and Islamic worlds (though not so much in some others, such as the Confucian), this seeming opposition has stood at the heart of thinking about humanity for millennia. Reconciling the two was Kant's "difficult problem." Across those centuries the difficulty has sometimes been exacerbated by exaggerated claims of law, sometimes by those of freedom. This way of imagining the human condition as a struggle between necessity and freedom is certainly not limited to debates about the powers of scientific reason. It has been animated by the many different kinds of systems of thought that claim authority to structure our lives in this world and even in worlds to come. It has been central, for example, in the development of our religions, as when St. Paul characterized the convert to Jesus as "fully freed from the Law" of the Jews (Rom. 7:5–6) or Luther represented his Protestant movement as an emancipation from Catholic legalism. And as the vocabulary itself makes clear, the struggle between "law" and "freedom" is also a political one, a struggle over the rules and norms that govern our communal lives and over who has the power to determine and impose them.

In the chapters that follow we will touch on as many of these spheres of human activity from as many periods, cultures, and disciplines as possible, for they are all related in terms of the choices they present us with—choices between necessity and contingency, certainty and doubt, sameness and difference, eternity and mortality, objectivity and subjectivity, normativity and relativity, among many others. And throughout we will continue to insist both on the value of our sciences and systems, our laws of thought and life, and also on the need to remember that these rules do not plumb the sea of humanity. We will play with pebbles and take swimming lessons as well. We will suggest how we might choose between approaches to a given problem but without establishing any rule, except perhaps the rule that no rule is absolute.

One of the greatest lawyers of the sixteenth century, the "Sheikh of Islam" Ebü-s-Su'ūd, chief jurisconsult of the Ottoman Empire, put it this way when asked about the standing of Sufi mystics in Muslim law: "Knowledge of Divine Truth is a limitless ocean. The sharī'ah [law] is its shore. We [lawyers] are the people of the shore. The great Sufi masters are the divers in that limitless ocean. We do not argue with them."[32] A remarkable statement of epistemic humility from one so powerful and an eloquent formulation of our general point: whatever laws we choose for ourselves to live by, humility about their reach is a prerequisite for the preservation of humanity.

For Professionals: Warranties and Limits of Liability

We have tried to be true to all of the many thinkers and disciplines, periods and places that we touch on in this book. No doubt we have failed. Not every economist will see their view of their field reflected in ours, not every quantum physicist will endorse the stress we place on certain experiments and theories, not every literary critic will sympathize with our reading of their favorite poet. We hope that we have not committed howlers in your particular discipline. Where we have, we hope that you will point them out to us and ask yourselves if whatever errors in detail you may find are merely lamentable or lethal to the more general argument.

When it comes to methodology,

- We have everywhere tried to study—and this we take to be a sine qua non, not only of the historian but of much of the humanities—the sources in their original languages (unless otherwise indicated, all translations are our own) and to attend to what those words might have meant in the context in which they were written. As a corollary, there are many cultures we have not been able to explore because we do not know their languages or their history. We very much wish, for example, that we knew Sanskrit and Chinese.

- Mathematics is also a language (albeit a formalized one) with a history. Here, too, we have tried to honor that history even as we have sometimes translated the mathematics into modern terms (such as those of set theory) that were not used in earlier periods.
- But we have not stayed within the historical contexts of our sources. We have also asked how those sources might resonate with questions being asked by other thinkers, how they might be read by future readers, what relationship they might bear to questions being asked in other times and places, including in our own.
- Historians profess a horror of anachronism. Taken to an extreme, such a horror would make it impossible to speak, since every word we use has changed its meaning over time, and the words available to us today are saturated in our context. Compromise is needed if the past is to be made intelligible and interesting (let alone useful) to the present. We have tried to honor taboos on anachronism while animating history with the needs of the living.
- We are only partly historians (one of us grew up a medievalist, the other a mathematician), and our subject is not only historical but also philosophical, theological, psychological, physical, mathematical, and literary—in short, human. Not all disciplines of humanity have the same relationship to anachronism that historians do: some even consider their truths timeless. We must not, in a book such as this, let the historian dictate laws to the philosopher or the mathematician any more than the other way around.
- In any event, it is our view that the philosophical and the historical cannot be separated. How we can think in a given time and place, what we think we know, and why we think we know it: these are not independent of how we have learned to think from our predecessors and of what our particular past has taught us counts as knowledge.

- Aspects of the thinkable are historical, pathic, and vary across time and place. We must cultivate the ability to recognize difference.
- Our species has faced common problems and brought common tools to bear on those problems. Many of the questions we are asking in this book were also asked, albeit in other languages and with other inflections, in many cultures we have not touched on. Often these cultures have influenced each other even when we do not know enough to perceive the influence.
- Some questions are shared, common, perhaps some are even universal. We must cultivate the capacity to recognize similarity.
- Which to emphasize? There is no rule. It depends on the question one is asking, the perspective one is taking, the "sameness" or "difference" one intends. As you read this book, we ask you to decide whether we made the right choices, and wherever you think we have not, to bring the powers of your own will to bear on these questions that we take to be so vital to our humanity.

1

WORLD WAR CRISIS

*In no historical era has the human being become so much
of a problem to himself as in ours.*

MAX SCHELER[1]

How Mathematics Caused the Decline of the West

According to the philosopher Richard Rorty, no one today believes
that the ways in which we humans think about the nature of our
knowledge is a matter of life and death. "When contemporary philosophers insist" that their "explanations of how objects make sentences true, or of how the mind is related to the brain, or of how free
will and mechanism might be reconciled . . . are 'fundamental' or
'perennial,' nobody takes their claim seriously."[2] Things were clearly
otherwise in 1918, in the final weeks of World War I, when the German historian Oswald Spengler published one of the early twentieth century's best-selling books: *Der Untergang des Abendlandes,*
The Perdition of the Western World, or, as the title is traditionally
and more tamely translated, *The Decline of the West.*[3]

The book swept over mangled Europe like a tidal wave. (A second volume was published in 1922, to less rapturous attention.)
From Mussolini to Thomas Mann, everybody seems to have reviewed, read, or at least debated it. To a world embroiled in industrial war and mourning the deaths of millions, it offered a diagnosis, an explanation of the "last spiritual crisis that will involve all
Europe and America." What was this crisis that was producing the
downfall of the West? Not immigration, globalism, or race, as one
might expect from a book with such a title today. The crisis Spengler wrote of, and that resonated with so many readers, was a crisis
in our form of thought, a crisis provoked by "the tyranny of reason,"

whose "most distinct expression is the cult of the exact sciences, of dialectics, of demonstration, of causality."[4]

Spengler promised his myriad readers that they could overthrow that tyranny if they learned to think anew about the foundations of their knowledge. They should begin, he told them, with what he claimed was his own momentous discovery about number and mathematics, "a fact of decisive importance which has hitherto been hidden from the mathematicians themselves." What is this fact? *There is not, and cannot be, number as such.* There are several number worlds as there are several Cultures," "each type fundamentally peculiar and unique, an expression of a specific world feeling, a symbol having a specific validity . . . which reflects the central essence of one and only one soul, viz., the soul of that particular Culture."[5]

Stripped of the then fashionable jargon of essence, soul, and world feeling, Spengler's teaching amounts to this: number is not the same as itself. Or in the jargon of our own day: mathematics is culturally relative. It is easy to miss the radical flavor of these words today, when it has become a commonplace to talk of science as a "cultural construct." But in his day Spengler could with more justice boast that he was a revolutionary: "So far, no one has dared to assume that the supposed constant structure of the intellect is an illusion."[6] A vast historical account, extending over nearly a thousand pages, purported to demonstrate how cultures and their mathematics rise and fall. Hence the subtitle of his book: "Sketch of a Morphology of World History."[7]

The Decline of the West, in other words, explained to its readers the collapse of their world in terms of questions about sameness, difference, and the nature of knowledge. To a Europe in what seemed its death throes, Spengler offered the following diagnosis: the particular soul of modern European civilization and its powerful mathematics—the Faustian soul, he called it—was a product of the history of thought born in the seventeenth century with the mathematics of Descartes, Galileo, Leibniz, and Pascal. Now the Faustian West and the "mathematic" that had held it in such a powerful grip were dying together, "having exhausted every inward possibility and fulfilled its destiny."[8]

Strange as it may seem today, Spengler was not an anomaly. For much of the first half of the twentieth century many of the world's intellectuals believed that *a choice about how to know the world* had become problematic in a new way and that the convulsive crises they were living and dying through were the result of that choice. Like Leibniz, Kant, and many other philosophers before him, Spengler distinguished between two ways of knowing the world, one organic, the other mechanical; one oriented toward image, the other toward law; one imaginative, the other scheming; one oriented toward the experience of time, the other toward mathematical number; one living, the other dead.[9] Two ways of knowing, one more felicitous than the other. We trust you can tell on which side Spengler thought the danger lay. "Stiff forms are the negation of life, formulas and laws spread rigidity over the face of nature, numbers make dead."[10]

Today it takes a specialized historian to make any sense of this language. But a hundred years ago even the lightly cultured could recognize the vocabulary. They might even hear familiar echoes of Christian scripture in the oppositions between life and death, soul and body, humanity's spiritual capacity for freedom and its slavish subjection to carnal law.[11] And they might have understood that Spengler's splitting of knowledge into two opposing styles was not only an explanation for the shipwreck of their "Faustian" modernity but also a call to conversion from one style to the other.

The legend of Faust: whether in Marlowe's version, Goethe's, or later Thomas Mann's, the protagonist sells his soul to the devil in exchange for a period of great knowledge and power. Spengler and many before him believed that what the "Faustian" West had received in exchange for its soul was a powerful kind of mathematics. Robert Musil parodied precisely this view in his novel *The Man without Qualities*, today considered one of the most acute contemporary meditations on the crisis of modernity in early twentieth-century Europe:

Most of us may not believe in the story of a Devil to whom one can sell one's soul, but those who must know something about the soul (considering that as clergymen, historians, and artists they draw good income from it) all testify that the soul has been destroyed by mathematics and that mathematics is the source of an evil intelligence that while making man the lord of the earth has also made him the slave of his machines.[12]

"When one attacks Spengler," Musil wrote in his 1921 review of Spengler's book, "one is attacking the age from which he springs and which he flatters, for his faults are its faults."[13] Spengler had an uncanny sensibility for the tendencies of his times. His observations about the emergence of mass media, the rise of populist dictatorships, and the persistence of global war would soon come to be hailed as prescient. In due time, and unlike other famous German intellectuals, he would even have the foresight to refuse the Nazi movement the endorsement it avidly sought from him. Looking back after World War II, Theodor Adorno criticized Spengler's philosophy of history (and ignored his mathematics) but marveled at the chilling accuracy of his predictions. We would not be surprised if readers in our own newly populist era rediscover Spengler's volumes and feel that chill anew.[14]

But the "fault" of the age that matters here is this: Spengler thought of the choice between two ways of knowing as a crisis, a choice between life and death, a war being fought on the battlefields of number. He was in excellent company. In a 1971 essay the historian of science Paul Forman even suggested that the popularity of Spengler's ideas created such a hostile environment for German mathematics and physics in this period that it shaped the future of science.[15] We would reverse the optic. Spengler is not cause but symptom — albeit, like fever, a powerful symptom — of an age convinced that its most important problems were caused by how its thinkers thought about knowledge.

"Conflicts of a Philosophical Nature Become Motives for Murder"

The great sociologist Max Weber described the disease in a now famous lecture he gave in Munich to the Free Student Union of Bavaria on November 7, 1917. (A few years later Weber would debate

Spengler in the City Hall of that town.) In "Science as a Vocation," Weber argued that "the increasing intellectualization and rationalization" of human knowledge in modernity means that "there are no mysterious incalculable forces that come into play, but rather . . . one can, in principle, master all things by calculation. This means that the world is disenchanted." The consequences of this "disenchantment" for humanity could not be more profound. Because "civilized man" lives in a world in which knowledge is endlessly progressing, he cannot die fulfilled, "old and satiated with life," as the biblical patriarchs did. The modern subject knows that whatever he has been able to seize in life, "it is always something provisional and not definitive, and therefore death for him is a meaningless occurrence."[16]

We can disagree with Weber about the degree of the world's disenchantment and reject the reading of Tolstoy's "The Death of Ivan Illich" that he offers as evidence of mortality's meaninglessness in modernity. But we could not ask for a more eloquent and sensitive witness to the anti-intellectual revolution Weber perceived in his society. It is as if, he writes, the allegory of Plato's cave in book 7 of the *Republic* had been reversed. In that ancient allegory, human emancipation consisted of learning to see true being in the light of thought rather than attending to the illusions and shadows of life. But "today youth feel rather the reverse: the intellectual constructions of science constitute an unreal realm of artificial abstractions." For them, "genuine reality is pulsating" in lived experience. Intellectual constructions, science, "and the rest are derivatives of life, lifeless ghosts, and nothing else."[17]

Once again, we find forms of knowledge posed against each other in terms of the quick and the dead. Weber understood this revolution as a very dangerous one: hence, large parts of his speech were aimed against the cult of personality, demagoguery, and the call for the politicization of science and a charismatic professoriate, warnings that still resonate today. But he also understood it as part of an eternal war between different aspects of the human, a war he described through the metaphor of polytheism and the unceasing struggle between gods. "The ultimately possible attitudes toward life are irreconcilable, and hence their struggle can never be brought to a final conclusion. Thus it is necessary to make a decisive choice." To his great credit, he was not dogmatic about what, for each individual, that choice should be. He asked only that we make our choices with integrity, with attention to the presuppositions on which they are made, and with each of us attentive to our own orientation, to "the demon who holds the fibers of his very life."[18]

We find Weber remarkable both for his relative lack of dogmatism and for his attention to the roof, so to speak, of the intellectual edifice rather than its basement. What he found so challenging for the modern subject was the limitless horizons of knowledge, the provisional nature of every step and every discovery. Others, like Spengler, focused on the lack of foundations. The aforementioned "crisis" in the foundations of mathematics provides a good example. Spengler jumped into this debate, noting that "even the most 'self-evident' propositions of elementary arithmetic such as $2 \times 2 = 4$ become, when considered analytically, problems, and the solution of these problems was only made possible by deductions from [set theory], and is in many points still unaccomplished."[19]

From Spengler's point of view the fragility of the foundations of mathematics was a point in his favor, but the mathematicians and philosophers whose work he was referring to would have seen it quite differently. For them, the publication in 1884 of Gottlob Frege's *Basic Laws of Arithmetic* seemed to put within reach the dream of building all of mathematics—and thereby much of human knowledge—on one solid foundation of pure logic without paradox or contradiction.

The goal was "to show," as Bertrand Russell put it in a later recollection, "that all pure mathematics follows from purely logical premises and uses only concepts definable in logical terms."[20] In 1902 the young Russell had himself disturbed the dream by pointing out a problematic paradox in Frege's set theory. But a decade of concerted effort by Russell and the mathematician Alfred North Whitehead (whom we met discussing $2 + 2 = 4$ in our introduction) seemed to have put those remaining paradoxes to rest. The publication of their coauthored *Principia mathematica* (whose three volumes appeared in 1910, 1911, and 1913) was received by many as a triumph of mathematical logic, albeit one that was not easy to digest.

(As an example of the indigestibility, you might turn to the proof of the theorem $1 + 1 = 2$, which does not appear until page 362 of volume 2.)

Where critics like Spengler focused on the fragility of these foundations, their builders hoped for strength. For some, these victories of symbolic logic in the realms of mathematics even revived the project of conforming language to logic, thereby establishing the relationship of words to the real and eliminating interpretive ambiguity and error. Frege, for example, discussed the case of "the morning star" and "the evening star." The two refer to the same "real" referent (the planet we call Venus), but their use is not always interchangeable: think, for example, of how a Romeo might experience the difference between the two.

Russell's 1905 theory of description addressed a similar question: "Given that the words used to form the subjects of sentences refer to things, and that a sentence is true if things are as the sentence says they are, how is it that some true sentences containing a referring expression become false if one substitutes another expression that refers to the same thing?" The new logic sought to tame these "problems" of language. (We use scare quotes because a problem from the point of view of a logician may well be very much the opposite for a poet.)[21] These and related efforts—such as those of the Vienna Circle, which picked up the banner of logical empiricism in the years after the Great War—would have a considerable influence on the future of philosophy, producing the schools we nowadays call "analytic."

In short, this was not a one-sided battle. For one school of thought (here represented by Spengler), the tribulations of the present world were caused by an excess of math, logic, reason, and the apathic. For the other school, by a lack of it. For Frege, Russell, and company, the urgent goal was to banish historicism, contingency, psychology, human experience—in short, the pathic—in favor of logic and the eternal. Metaphysics and indeed all philosophy (other than logic) would disappear in this better world of clearer thought, as would the confusions created by our experience of time and change.[22] To put it in Russell's words, "every philosophical problem, when it is subjected to the necessary analysis and puri-

fication, is found either to be not really philosophical, or else to be
. . . logical." Or in another of his lapidary phrases, "To realize the
unimportance of time is the gate of wisdom."[23]

It is difficult for us to grasp the extent to which these rival commitments and convictions about the nature of knowledge became
intertwined with the great political events of the age. When the
mathematical physicist Dr. Friedrich Adler—friend of Einstein and
son of the Austrian politician Viktor Adler—assassinated the prime
minister of Austria in October of 1916, it made sense to him (and to
Trotsky, Lenin, and others who celebrated him) to explain his politics in terms of the new mathematizations of the universe, particularly Einstein's theory of relativity, which he interpreted as legiti-
mating a shift of frames of reference from nation to class. It made
equal sense for his father and his lawyers who were pleading in his
defense to argue that the assassin was not in his right mind, because
he suffered from "an excess of the mathematical."

Adler's acquaintance Otto Neurath also spent time in jail. The
economist and apostle of logical empiricism—and soon thereafter
a founding member of the Vienna Circle—was condemned for his
services as director of economic planning to the short-lived Bavarian
Socialist Republic ("the November Revolution" of 1919). He served
his sentence writing a manifesto aimed at the dangerous ideas he
thought most threatened his world. The title: *Anti-Spengler*.[24]

Even the 1922 centennial celebrations of the Society of German Scientists and Physicians were caught up in this highly polarized swirl of politics, science, and philosophy. The organizers invited Albert Einstein, who had just received the Nobel Prize, to talk
about relativity in physics, and Moritz Schlick—philosopher, friend
of Einstein, and animating spirit of the Vienna Circle—to talk about
the philosophical implications of Einstein's work. But in late June
of that year, the German foreign minister Walter Rathenau was assassinated by the secret ultranationalist "Organization Consul," and
Einstein's name was said to be next on the hit list. Schlick did lecture
on "Relativity in Philosophy," but Einstein's participation was canceled, replaced by the physicist Max von Laue.

Schlick himself would be assassinated in the lecture hall of the
University of Munich in 1936 by one of his former students (whose

PhD dissertation was titled "The Importance of Logic in Empiricism and Positivism"). The murderer claimed in court that the professor's "Jewish" antimetaphysical empiricism had stripped the defendant of the capacity for moral judgment and restraint. The argument met with enormous political sympathy. The killer was convicted but was released shortly after the Nazi annexation of Austria, serving only eighteen months. (Moritz Schlick was not in fact Jewish, except insofar as he was associated with ideas—such as mathematical physics and logical empiricism—that were being attacked as "Jewish" within the anti-Semitic cultural and political discourse of the times.) It was an age, as journalists noted in the Schlick case, in which "conflicts of a philosophical nature could become motives for murder." Mass murder, they might have written with more foresight.[25]

The point of these anecdotes is simply to illustrate our claim that a mere century ago the choice between types of sameness and difference could feel like a matter of life and death. Otto Neurath—as strident an apostle of sameness as one could find—spoke constantly of the "Oneness of Science and Society." But in his times (as perhaps in all others) science, like society, was very much divided, and one of the things it was most divided about was the subject of oneness, or sameness, itself. This division extended all the way down to what most took to be the deepest foundations of science: the foundations of mathematics.

Crisis at the Foundations of Mathematics

It was in this polarized context that in 1920 the mathematician Hermann Weyl published his "On the New Foundation Crisis in Mathematics," focusing attention on a set of questions that had preoccupied mathematicians for a generation or more, certainly long before Spengler put pen to paper. It was a debate between those who came to be called "constructivists" or "intuitionists"—famous names like Henri Poincaré, Leopold Kronecker, and L. E. J. Brouwer—on the one hand and the "logicism," "formalism," or "existentialism" of equally famous mathematicians—like Georg Cantor, Richard Dedekind, and David Hilbert—on the other.[26]

All mathematicians agreed that proofs must be organized logically. But they disagreed on the nature and limits of that logic. There are many examples of such disagreement, but the one that led Weyl to proclaim a "new foundation crisis" involved the Dutch mathematician L. E. J. Brouwer. In 1907 Brouwer began to insist that foundational logical principles are abstracted from our human experience. That experience is, he argued, restricted to the finite. But what about infinity? Since the nineteenth century and the discoveries of Georg Cantor, infinities (it turned out that there is not one but an infinite number of them) were becoming ever more important both in set theory and in mathematics in general. But could our axioms apply to infinite sets of the sort that no human can ever experience?

Brouwer took particular aim at one foundational logical principle, the Principle of the Excluded Middle, which holds that the statement "either p or not p" ($p \lor \neg p$) must always be true, for any proposition p. It is of great importance to mathematics, not least because, if you accept it as universally true, it allows proof by contradiction: you can prove p by disproving not p even if you cannot actually construct any other type of proof for p.

As early as 1907 Brouwer had argued against such proofs. In 1918 and 1919 he published two papers on "intuitionistic set theory" that began to systematize his view of a mathematics independent from the customary axioms of logic, treating mathematics as a mental construct rather than an object already existing: something produced temporally, inside the mathematician's mind at a given place, time, and state of knowledge. He used his method to construct a new explanation of the continuum, one that abandoned the ancient idea that it was made up of points or real numbers. It was this construction that converted Hermann Weyl (whose own book on the continuum had appeared in 1918) to Brouwer's intuitionism and led him to proclaim the "new foundation crisis." By 1928 that crisis had become so personal that it provoked the departure of Brouwer, Einstein, and other important mathematicians from the board of the greatest mathematical journal of the day, *Mathematische Annalen*, edited by David Hilbert.[27]

Why did the debate about questions of what constitutes a proper foundation for mathematical proof become so vociferous? The mere existence of questions about foundations cannot explain that. After all, the problem of foundations long predated the "new crisis." And postdate it as well: today, Brouwer's challenge remains largely unanswered, yet these questions exercise few mathematicians, and Hilbert's "logicism" prevails as math's way of going about its business. The problem of foundations eventually ceased to exercise minds not because it was solved but because it lost the resonance that made the question so good to think with, the resonance with other vital concerns of the age.[28]

If mathematics was ground zero for anxieties about sameness and difference in the early twentieth century, this was in part because the application of its laws and methods had such extraordinary success in providing the foundations for other domains of knowledge. Physics was perhaps the most important of these: it was in its realm that mathematics seemed most powerfully to dictate the conditions of possibility for being. As the American philosopher John Dewey put it in his *The Quest for Certainty* of 1929: "[Starting with Galileo] the 'laws' of the natural world had that fixed character which in the older scheme had belonged only to rational and ideal forms. A mathematical science of nature couched in mechanistic terms claimed to be the only sound natural philosophy." [29]

The marriage of physics and mathematics proved so capable of accounting for cause and effect in the physical world that it had become a model for those aspiring to certainty, or even determinism, in human affairs. Here, too, Spengler portentously opined, pronouncing Western physics, like mathematics, to have reached its end. And again, choices of number and sameness were at issue: "in the rapidly increasing use of enumerative and statistical methods, which aim only at the probability of the results, and forgo in advance the absolute exactitude of the laws of nature," Spengler saw "deep and utterly unconscious" skepticism opening an abyss beneath physics.[30]

Why skepticism? Like many people in his day, Spengler understood statistics as chance, as the probability rather than the certainty of something occurring at a given point in time. This he opposed to mathematics and "mathematical law," in which he saw certainty rather than chance, determinism rather than probability.[31]

He did not hesitate to map this imagined dualism onto the existential categories of his age. "Statistics," he claimed, "belong, like chronology, to the domain of the organic, to fluctuating life, to destiny and incident and not to the world of exact laws and timeless eternal mechanics." Again, the staging of a choice between sameness and difference, between on the one hand mathematics, causality, deadly law, and on the other statistics, chance, freedom, and life.[32]

Causality vs. Chance, Determinism vs. Freedom

Spengler was tapping into a debate that was raging within physics itself. It is easy to find lectures by physicists of the day with titles like "On the Present Crisis in Mechanics," "The Present Crisis in German Physics," "Concerning the Crisis of the Causality Concept," and "On the Present Crisis in Theoretical Physics," this last given in Tokyo by Albert Einstein in 1921. Einstein had apparently read Spengler's book: "Sometimes in the evening," he wrote in early 1920 to the mathematical physicist Max Born (who would receive the Nobel Prize in 1954), "one likes to entertain one of his propositions, and in the morning smiles about it."[33] But the crisis that concerned the great physicist was provoked, like the one in mathematics, by rival commitments to the pathic and the apathic at the foundations of his discipline.

"The goal of theoretical physics," Einstein declared in "On the Present Crisis," "is to create a logical conceptual system, resting upon the smallest possible number of mutually independent hypotheses, which allows one to comprehend causally the entire complex of physical processes." Einstein is putting his chips on the table, and firmly on the side of "causality." What does that mean? Here is the explanation offered in 1920 by Moritz Schlick: "The principle of causality is . . . the general expression of the fact that everything which happens in nature is subjected to laws which hold without exception."[34] Given the *same* cause and conditions, always the *same* result: hence the synonyms mechanical, deterministic.

But new discoveries were mining the ground beneath causality and its laws. Already in the late nineteenth century, for example,

Ludwig Boltzmann had interpreted the Second Law of Thermo-dynamics as a statistical law (briefly put, heat flowing from a colder body to a warmer one is an extremely unlikely event, but not impossible). In what sense could it then be called a law? There were many who seized on these triumphs of statistics to emphasize indeterminacy and absolute chance in nature.[35] In 1914 the physicist Max Planck—soon to receive the Nobel Prize—chose to address the issue of "this dualism" between causal and statistical laws. Marking the end of his term as rector of the University of Berlin, in an address titled "Dynamic and Statistical Lawfulness," he criticized attempts to deny causal laws and regard all regularity as statistical: "the concept of an absolute necessity would be lifted from physics entirely. Such a view must . . . be an error as disastrous as it is short-sighted."[36]

This "shortsighted" view was already shared by many physicists by the time that Spengler retailed it to a broader public four years later. One of the reasons for its appeal is that it helped those physicists engage with some of the most exciting questions of their discipline, questions raised by "the investigation of the structure of atoms and the radiation emanating from them" (to quote from Niels Bohr's Nobel citation of 1922). Those questions had been transforming physics ever since Lord Rutherford, Marie and Pierre Curie, and others observed that radioactivity emanated from within atoms, not from interaction between them. (The Curies coined the word *radioactivity*, and Marie's researches in it earned her two Nobel Prizes, in physics in 1903, the first Nobel given to a woman, shared with her husband, and in 1911 in chemistry.) That discovery alone exploded an ancient fantasy of sameness: the belief that atoms were stable and indivisible particles at the foundations of the universe.

In the following decades a number of the greatest names in physics, including just about every figure mentioned thus far—oriented their thought toward understanding the nature of atoms and their parts. What is striking about this group, apart from their talent, is their awareness of the bearing that their theorizing about physics had on the foundations of thought and the strength of their disagreements about the philosophical implications of their discoveries. That disagreement is the reason why we found Einstein writing to Max Born about Spengler in 1920.

At that time Einstein, like Max Planck, maintained a strong view of causality. "God does not play dice with the universe" has become one of his more famous quips. Max Born, on the other hand, was already a convert to acausality and a leading advocate, along with Werner Heisenberg (Nobel 1932) and others, for the statistical foundations of quantum mechanics. The young Wolfgang Pauli (Nobel 1945) put the outlines of the debate with rare evenhandedness in his 1924–1925 article "Quantentheorie" (Quantum Theory): the moment of transition of a single excited atom "appears, according to the present state of our knowledge, to be determined solely by chance. It is a much discussed but still undecided question whether we have to regard this as a fundamental failure of the causal description of nature, or only as a temporary incompleteness of the theoretical formulation."[37]

The discussion of this question over a handful of years produced research that transformed our understanding of sameness and difference in the universe. We will discuss the physics in more detail later. But to put the problem in terms of our introduction, the world's leading physicists were debating whether elementary particles behave as pebbles, or as blue tigers, or both. And the answer seemed to be that it depends. A troubling answer, because it suggests that there is no apathic and stable "sameness" at the quantum-mechanical foundations of the universe.

Some of this group seem more dogmatic than others. Born and Heisenberg, for example, pursued their strong wager for acausality in the German press: "quantum mechanics establishes definitively the fact that the law of causality is not valid."[38] Erwin Schrödinger, on the other hand, repeatedly changed his mind. His 1922 inaugural lecture as professor of physics at Zurich was much influenced by his mathematician colleague there, Hermann Weyl. Titled "What Is a Law of Nature," it was a broadside against causality—understood as the postulate "that every natural process or event is absolutely and quantitatively determined at least through the totality of circumstances or physical conditions that accompany its appearance." Discovery of the mysteries of the atom, Schrödinger told his audience, requires "liberation from the rooted prejudice of absolute causality."[39] But by 1925 he was making a very different choice, elaborating his wave mechanics as a causal description of atomic processes in space-time explicitly in opposition to Heisenberg and Born's acausal matrix mechanics and explicitly in dialogue with what he took to be the nature of human knowledge. As he put it in the second paper of his wave mechanics trilogy, "we really cannot change the forms of thought, and what cannot be understood within them cannot be understood at all. There are such things, but I do not believe that the structure of the atom is one of them."[40]

"Forms of Thought" is a phrase we more expect in a philosophical essay than a paper on mathematical physics that won its author a Nobel Prize (awarded in 1933). But that is the point. For the participants in these early twentieth-century debates whom we've discussed thus far, questions about math or physics were also questions about politics, philosophy, psychology, and anthropology: that is, questions about humanity and its place in the cosmos. Perhaps even more remarkable, these physicists and mathematicians were often aware that they occupied a place in a long history of thought and that the questions that felt so critical to them had felt critical to earlier thinkers as well: "reminiscent of . . . the foundation crisis in ancient Greece," as the mathematician Max Dehn put it before the assembled faculty of the University of Frankfurt in 1928.[41]

The Crisis: A Poet's View

This shared conviction is remarkable. Educated Europeans — whether mathematicians or physicists, novelists or historians — were perfectly willing to believe that the vast convulsion in which their civilization found itself in the first half of the twentieth century was the product of choices about sameness and difference, choices whose history they often understood as beginning in ancient Greece. Poets and philosophers, too, shared this conviction, perhaps none more than the French poet Paul Valéry (1871–1945).

Valéry was deeply engaged with the scientific questions of the day as well as the philosophical. He was also among the many we've met in this chapter who saw in the catastrophe of the Great War the collapse of a civilization. In 1919 he described a world that "calls progress its tendency toward a fatal precision," a world that exiles the "phantoms" of the spirit, until it creates "the miracle of an animal society, a perfect and definitive colony of ants."[42] A year later Valéry published what would become perhaps his most famous poem, "Le cimetière marin" (The Cemetery by the Sea).

In a country mourning the loss of millions of its young men in the Great War, then just ended, the cemetery is overdetermined. But the sea is the Mediterranean, and the poem a cenotaph for the civilization carried on its waves, the thought world we call Western. Even its meter is a memorial: the hendecasyllabic line of medieval Occitan, Italian, and Spanish poetry.[43]

> Ce toit tranquille, où marchent des colombes,
> Entre les pins palpite, entre les tombes . . .
> Midi le juste y compose de feux
> La mer, la mer toujours recommencée!
> Ô récompense après une pensée
> Qu'un long regard sur le calme des dieux!

> This quiet roof, where dove-sails saunter by,
> Between the pines, the tombs, throbs visibly.
> Impartial noon patterns the sea in flame—
> That sea forever starting and re-starting.
> When thought has had its hour, oh how rewarding
> Are the long vistas of celestial calm!

Already in the first stanza we can feel the marmoreal density of a funeral monument, its words tightly joined to one another by allusion, etymology, history, sense, and sound.[44] Here and throughout the poem we encounter the constant movement between sameness and difference; the changing and the unchanging ("I am what's changing secretly," "the dead lie easy"); eternity of the soul and mortality, the quick and the dead. Then suddenly, unexpectedly, this charnel reverie bursts out in a stanza of poetic reproach against a Greek thinker from the distant past:

> Zeno, Zeno, cruel philosopher Zeno,
> Have you then pierced me with your feathered arrow
> That hums and flies, yet does not fly! The sounding
> Shaft gives me life, the arrow kills.

Zeno's logical paradoxes, with their demolition of change and human experience, demolitions achieved through the enunciation of powerful principles articulated in pursuit of eternal unity: these are the life and death-dealing laws of thought against which the poem's concluding verses explode:

> The wind is rising! . . . We must try to live!
>
> .   .   .   .
>
> Break, waves! Break up with your rejoicing surges
> This quiet roof where sails like doves were pecking.

Valéry's work would be admired by many — T. S. Eliot, W. H. Auden, Wallace Stevens, Jorge Guillén, Rainer Maria Rilke, to name only poets. It would also be criticized, not least for its attempt to take seriously the vast history of ideas. Borges combined admiration and critique, calling "The Cemetery by the Sea" an ambitious poem, but an unwieldy, airless large machine.[45] What else can one expect of an attempt to capture in 144 lines of verse both the tragedy and the glory of a three-thousand-year history of human feeling and rational thought?

In the poem the struggle over sameness and difference that Valéry places at the "origins" of philosophy and science is sometimes referred to openly, sometimes in more occult harmonies, not separate from sensations and seascapes. This is as it should be. In poetry, ideas are not to be treated in the same way as they are in prose.[46] Some twenty years later (ca. 1940), writing in another exceedingly dark hour, Valéry provided a starker statement of the problem as he had come to see it, now in less than ten lines of prose:

Finally the sage said to me, after having spoken to me for more than thirty hours running and instructed me in everything that one needs to know:

"I will summarize the doctrine for you. It consists of two precepts:

> 'All things different are identical.
> All things identical are different.'

Come and go between these two propositions in your spirit, and you will first come to see that they are not contradictory; then, that thought can form only one or the other, and can only move from one to the other. There is a time for the one and a time for the other, and who thinks one, thinks the other. That's all."[47]

We find this an admirably undogmatic statement of the choice that so animated the thinkers of the age, the choice between sameness or difference. That lack of dogmatism is all the more admirable in that it was so rare. Many of the thinkers who understood the crisis of their age in terms of choices in the development of knowledge were convinced that they knew where things had gone wrong and what choices had been badly made. We have already seen some dogmatic positions staged among mathematicians and physicists. Let us conclude now with philosophers, since these see themselves as specifically charged with thinking about the nature of thought.

The Crisis: Philosophers' Views

Edmund Husserl (1859–1938) provides a preeminent example of a philosopher deeply engaged in interrogating the relationship between human experience and the rules of reason. His work began with *The Concept of Number* (his 1887 dissertation) and *The Principles of Arithmetic* (his first book, 1891). Increasingly he sought to create a "science of consciousness" capable of integrating reason with experience, the fundamental concepts of logic and mathematics with the "lifeworld." It was through his work that some of the mathematicians and physicists we've encountered arrived at their own questions and choices. Herman Weyl, for example, encountered Husserl and his phenomenology early in his career and acknowledged its deep influence on his mathematics already in 1917, a few years before his 1920 proclamation of mathematical "crisis."

Like all of our protagonists in this chapter, Husserl had lived through World War I (in which he lost a son) and through the turbulent politics of the Weimar Republic. He would witness the Nazis' rise to power, and in 1933 he was barred from professional life and publishing because of his Jewish birth, though he had converted to Protestantism nearly fifty years before. Thereafter he was shunned by many of his former colleagues, including most notoriously by his own student and successor in his professorship, Martin Heidegger. Husserl had always resisted bringing philosophy to bear explicitly on the political, but in this last decade of his life he sought to intervene in the calamity he was living through, and he did so in a way that should now be familiar: by asking what wrong turn in the nature of knowledge could have produced this collapse of humanity.

His answer can be found in *The Crisis of European Sciences and Transcendental Phenomenology*, published posthumously in 1954.[48] He conceded in his introduction that "universal philosophy, together with all the special sciences, makes up only a partial manifestation of European culture." But for Husserl this rational part is "the functioning brain, so to speak, on whose normal function the genuine, healthy European spiritual life depends." He was convinced that it was because something had gone wrong with that part of the brain that the world was now plunged in "the radical life crisis of European humanity," and his goal was to diagnose the mental illness by charting its history.

Husserl's key question should by now sound familiar: "Can reason and that-which-is be separated, where reason, as knowing, determines what is?" In less Germanic words, can reason provide stable foundations for being? According to him (and many others), since Galileo the modern sciences have answered this question with an ever more imperious "yes." Their mathematical victories in explaining the cosmos and making it predictable promised us the certainty of universal truths.[49] But each of their victories, while seeming to conquer new territories, in fact receded ever farther away from "universal philosophy" and "humanity."

Why was this so? One could imagine a number of approaches to this question. Husserl might have questioned the possibility of any meaningfully *universal* philosophical truths about humanity (except perhaps the universal truth that every philosophical truth about humanity is contingent). Or he might have attempted to distinguish between aspects of our humanity that are subject to the rules of reason and others that are not. But Husserl did not choose the path of skepticism, contingency, or anthropological psychology. Instead he divided mathematics itself into two: one that is grounded in our experience of being, and one that is abstract and alienating.

Among the Greeks, according to Husserl, geometry had proceeded not through calculation but through measure, so that the pure intuitions of the Greek geometers had been derived from their embodied experience of space and time. But Descartes and his modern successors had translated geometry into algebra, an "arithmetization of geometry." What had once been rooted in our human experience of space and time was now translated into calculation and formula. Because this new approach to mathematics (which he calls "logistic" or "technique," with negative implications) had extraordinary predictive value, "some were misled into taking these formulae and their formula-meaning for the true being of nature itself."[50]

Once modern mathematics created this split between reason and "that-which-is," it could no longer give humanity access to its "life-world," but only to a "scientific world," and this "scientific world," for all its predictive power, could yield only superficial truths about humanity. Hence, "European humanity" drifted into a crisis that did not lessen the practical successes of the sciences but shook "to its foundation the whole meaning of their truth," shattering "faith in 'absolute' reason, through which the world has its meaning, the faith in the meaning of history, of humanity, the faith in man's freedom."[51]

How then to repair the damage to humanity? Husserl's plea was that we need to rediscover the intuitive mathematics on which the faith had first been built, a mathematics now concealed beneath the "inauthentic" axioms of an excessively abstract modernity.[52] If an authentic mathematics could be recovered, the grounding power of reason would once more be reconciled with human freedom, the powerful but naively "objectivist" and materialist claims of the natural sciences could be delimited, crisis could be averted, and the first true "science of the spirit" could be achieved at last.[53]

There is little we agree with in these words, but there are two lessons we want to take from them. The first is that, to this famous philosopher aware of calamity unfolding about him, the most urgent battle was an appraisal of the history of how humanity had learned to think about its own knowledge. As he put it in his Vienna lecture of 1935, "Philosophy and the Crisis of European Man": "What is to be said, then, if the whole mode of thought that reveals itself in this presentation rests on fatal prejudices and is in its results partly responsible for Europe's sickness? I am convinced that this is the case." We could not sympathize more with Husserl's conviction that our inherited ideas and concepts (prejudgments) about the nature of knowledge can be prejudicial to humanity.

And the second lesson: it is easy for diagnosis to become dogmatic. Note how, in Husserl's case, the disease is located in one particular way of knowing, namely, science and mathematics. The elder Husserl was increasingly absorbed in describing what he took to be a long history of conflict between the "natural" and the "humanistic" sciences, but the responsibility he assigned for that conflict was vastly asymmetrical. Again from the Vienna lecture: it is because the "modern scientist" was in the grips of these prejudices that he rejected "the possibility of grounding a purely self-contained and universal science of the spirit."[54] Throughout this book we will argue, to the contrary, that the "modern scientist," like the modern humanist, acts wisely in cultivating consciousness of the contingent conditions of her knowledge. The fact that Husserl believed his own approach to be no less than the first true "science of the spirit" is sufficient evidence that he himself remained in the grips of a number of the "fatal prejudices" he was warning about.

Both of those lessons emerge with vivid clarity from the debate staged at the Swiss sanatorium and resort town of Davos between two great German philosophers, each of them convinced in their own way that clarifying the powers of the "sameness" that mathematics offers was among the most critical tasks confronting humanity. The series of debates staged at Davos in those years could serve us as an allegory for the intellectual struggles of the age, as they did Thomas Mann in his novel *The Magic Mountain* (1924). In 1928 Albert Einstein had lectured there on his mathematical physics (and also entertained the crowd by playing the violin). How appropriate, then, that Ernst Cassirer and Martin Heidegger should take the stage a year later to debate the question "what is it to be a human being?"

We have already seen how the discoveries of mathematical physics were being put to the work of interrogating—through concepts like causality, determinism, law, and freedom—the status of the human and indeed of every object in the cosmos, from the smallest (elementary particles) to the largest (the universe itself). In Einstein's universe, as Cassirer put it elsewhere, "no sort of things are truly invariant, but always only certain fundamental relations and functional dependencies retained in the symbolic language of our mathematics and physics, in certain equations." Space, time, matter: these could now be divorced entirely from human experience and thought of as ideas of relation conceived of and expressed in the formal and rule-bound language of mathematics rather than as metaphysically real substances or containers. Everything, even "thingness" and matter itself, was invariant only to the extent that it was expressed in terms of mathematical reason.[55]

Cassirer and Heidegger had very different attitudes toward these latest victories of reason. For Cassirer, what was distinctive about the human was precisely this potential of our finite reason to reach for the infinite, the "necessarily universal," the eternal. The important question, as he rephrased it to Heidegger, was, "How does this finite creature come to a determination of objects that as such are not bound to finitude?" From Cassirer's point of view, the mathematization of matter in the new physics was one more glorious manifestation of a basic attribute of the human condition that, following Kant, he called "spontaneity." By this he meant that it is the spontaneous projections of the human mind that shape the world so that it appears as ordered or objective. Cassirer called these projections "symbolic form," and he considered the capacity for such form as the defining feature of the human being. The power to create the world through the spontaneous projection of symbolic form (of which mathematics was only one: myth, poetry, language itself were others): this was for Cassirer the source and expression of human freedom.[56]

For Heidegger, by contrast, humans are creatures whose experience is built on an abyss of nothingness, and their fundamental attribute is finitude. As he wrote in "What Is Metaphysics?," delivered shortly after the Davos debates, "We are so finite that we cannot even bring ourselves originally before the nothing through our own decision and will. . . . Our most proper and deepest finitude refuses to yield to our freedom." This "nothing" had nothing to do with logic; nothing to do, for example, with the null or empty set on which a year or two earlier John von Neumann had refounded mathematics. In fact whereas for Cassirer mathematics was one of the most notable expressions of human freedom, for Heidegger it was quite the contrary. In his writings of this period, Heidegger frequently attacks "Logistik" and mathematics, denying that they could provide a basis for *any* knowledge about Being, and insisting that, to the contrary, they block our path to knowledge of Being.

Modernity, Heidegger suggests, has grabbed the wrong end of the Greek "mathematical."[57] Like Spengler and Husserl, albeit for different reason, he saw in Descartes an important pioneer in this inversion of priorities. "Mathematical knowledge," he tells us in *Being and Time*, "is regarded by Descartes as the one manner of apprehending entities which can always give assurance that their Being has been securely grasped. . . . Such entities are those *which always are what they are*." To put this critical observation in our terms, Heidegger is saying that mathematics is an appropriate tool for apathic objects, objects that remain always the same, that adhere to the mathematical Principle of Identity. Descartes's error is not that he likes math but that he dogmatically treats all Being as if it met these conditions of sameness.[58]

Heidegger rightly insists that philosophers should ask whether or not their tools of thought are appropriate to the questions to which they apply them. This is especially true of the questions of "nothingness" that interest him above all. He knows that the proofs of logicians, for example, depend a great deal on the "not": "Ontology and logic, to be sure, have exacted a great deal from the 'not,' and have thus made its possibilities visible in a piecemeal fashion." But, he went on, "has anyone ever made a problem of the *ontological source* of notness, or, *prior to that*, even sought the mere *conditions* on the basis of which the problem of the 'not' and its notness and the possibility of that notness can be raised?"[59]

This would seem to be at a position very much like the one we have outlined as our own in the introduction: the conditions of notness, like the conditions of sameness, cannot be taken for granted and must be explored. (We go further, and maintain that in both cases the conditions are contingent rather than necessary.) But it was not enough for Heidegger to urge a critical self-consciousness on the claims of logic or mathematics, or to differentiate his "nothing" from logical negation, zero, or the null set. He wanted to put Being entirely outside reason's reach. Heidegger knew, of course, that mathematics provided a powerful tool with which to formalize "systems of Relations." But, he insisted, "one must note that in such formalizations the phenomena get levelled off so much that their real phenomenal content may be lost." The phenomena that interest Heidegger "resist any sort of mathematical functionalization; nor are they merely something thought."[60]

For Heidegger the problem is not just that mathematics and reason give us access only to a small part of Being (the part he characterizes as "pure substantiality"). Worse—and it is here that Heidegger becomes dogmatic—the problem is that mathematics and reason actively block our path to all the rest, that is, to everything that is most important about Being in the World.

According to Heidegger, all previous philosophers had repressed this disturbing truth, preferring fantasies about achieving freedom from finitude, fantasies built out of the powers of logic and rational thought. Hence, it was necessary to reread the entire history of philosophy, violently if need be. The result, he promised the audience at Davos, would be nothing less than the "destruction of the former foundation of Western metaphysics (spirit, logos, reason)."[61] There were plenty in that age of crisis eager to accept his argument, forgetting that Heidegger's banishment of reason from the grounds of Being is just as tyrannical a dogmatism as reason's rule.

In restaging this encounter between Heidegger and Cassirer, our goal is not to anoint one or the other as a saint of our particular devotion. It is simply to point out that throughout their encounter in 1929 both of these great philosophers were convinced that our choices about the powers and limits of reason determine crucial aspects of our humanity. Both agreed that mathematics posed a particularly important example for their debates because its Principle of Identity was so powerfully capable of producing certainties about the universe. Both shared the general conviction that those choices could alienate humanity from something crucial in itself, producing crisis, and that humanity was at that present moment living through such a crisis. What they disagreed about on the stage at Davos was which choices were right and which were wrong.

An Aside on Race as a Principle of Identity

There is another Principle of Identity that they might have disagreed about, one that was everywhere in the European air in the 20s and 30s even if it was not spoken of onstage during their Davos debate. It is something of an irony that one can be a stringent critic of certain claims of identity (in this case mathematical) and yet be a strident apostle of a different one. Sameness of "race" is also an "identity," for some even a "law of nature," and one that many believed powerfully shaped the possibilities of thought. Heidegger was certainly among these, and wrote often, both about the forms of thought to which the "Jewish race" was prone and about the importance of Germanness, *Volk*, blood, and soil as conditions for thought. We can set aside questions of politics—such as his enthusiastic embrace of the Nazi party—and still ask what it means about the foundations of his philosophy that these categories of sameness remained uninterrogated across his career.

Mathematical sameness and racial sameness were interrelated in these years. In the 1920s mathematical reason and its application to the physical world was frequently and negatively associated with "Judaism," and the tendency only grew stronger with time. The famous physics textbooks of Phillip Lenard (Nobel Prize in physics 1905) even warned physics students not to study too much mathematics, because it carried Jewish intellectual influences that killed "feelings for natural scientific research."[62] (Lenard was, as you might expect, a particularly bitter opponent of Einstein's physics.) Or as Johannes Stark (Nobel in physics 1919) wrote in the pages of *Nature*, "we must establish and recognize the fact that the natural inclination to dogmatic thought appears with especial frequency in people of Jewish origin." Leading mathematicians were also involved in the struggle, founding journals such as *Deutsche Mathematik* (German mathematics) in order to further a less abstract and more human (understood explicitly in *Deutsche Mathematik*'s case as a "less Jewish") mathematics.[63] The anti-Jewish flavor of these efforts may tempt us to dismiss them as marginal, but they were, like those of our other examples thus far, attempts to address the crucial challenges that mathematics seemed to pose to a humanity seemingly defined by a duality of sensation and reason and a simultaneous yearning for certainty and for freedom. Not even the seemingly "objectivist" sciences were immune from the lure of these racial fantasies of sameness.[64]

We have not come close to exhausting the long list of "intellectuals" for whom the key question of the first half of the twentieth century was that of the relationship between forms of knowledge and possibilities for life. And though we have focused here on German-speaking lands, we could just as well have turned to the French of Henri Bergson, or to José Ortega y Gasset's 1933 Spanish lectures on Galileo, published in 1942 as *Esquema de las crisis* (*Man and Crisis*, 1958). We should also explore the formation of other disciplines in the period, such as the new "science of the soul" (psychology, literally) whose founder, Sigmund Freud, was very much aware of the tensions inherent within such a phrase and alive to the roles of those tensions in the crisis of modernity.

American Variations:
John Dewey on the Quest for Certainty

But piles make for tedious prose, and historical examples, however mountainous, are never enough for certainty.[65] So we will conclude this chapter with just one more, chosen to correct any possible misimpression that, because we have thus far mentioned only European thinkers, the United States experienced no crisis or was exceptional in being exempt from the effects of this long history of thought. We have already met the American philosopher, psychologist, and educator John Dewey and his 1929 book *The Quest for Certainty: A Study of the Relation of Knowledge and Action*. There, and with his customary combination of clarity and depth, he outlined another version of our choice.

According to Dewey, the dominant preference throughout (implicitly Western) history had been for the eternal. The great thinkers of the past had been engaged in "the quest for a certainty which shall be absolute and unshakeable." They had been convinced that "the distinctive characteristic of practical activity, one which is so inherent that it cannot be eliminated, is the uncertainty which attends it." Hence, they had reached for the certainties of reason and slighted the finitude of the sensible. Our "practical activity" as humans in the world had been dismissed as a source of knowledge in favor of more eternal truths. This error was the root of a vicious vine that had implacably overgrown the long history of thought and threatened to strangle our humanity.[66]

Dewey provided a brief sketch of that history, pointing to multiple paths that had (falsely) promised these immortal certitudes. Religion was an important but, to Dewey's mind, outmoded one whose replacement by reason had been more or less achieved already in ancient Greece. "For deliverance by means of rites and cults, [Greece] substituted deliverance through reason." And at the beginnings of this fantasy of salvation through reason he placed the discovery of geometry, which "seemed to disclose a world of ideal (or non-sensible) forms which were connected with one another by eternal and necessary relations which reason alone could trace. This discovery was generalized by philosophy into the doctrine of a realm of fixed Being which, when grasped by thought, formed a complete system of immutable and necessary truth."[67]

In chapter 4 we will reject the too easy opposition between religion and reason, two paths toward certainty that have always been and continue to be intimately related, coconstitutive rather than substitutive. But let us leave quibbles aside and simply notice the shape of Dewey's argument, which you will by now find familiar: across history the human yearning for eternal certainties, propelled by the powers of mathematics, has produced a tyranny of reason that alienates humanity from the sensible, finite, and transient in itself. Like Spengler, Husserl, Heidegger, and many others of his day, Dewey offered his own revolutionary solution to this tyranny: the distinctive truths of what has come to be called American "pragmatism."

The trajectory Dewey describes—yearning for certainty, false promises of reason, emancipatory revolution—is one we have encountered over and over again in this chapter. We focused on the first half of the twentieth century in order to make plausible the possibility that our own age may itself participate in some stage of this ongoing struggle between certainty and crisis despite its refusal to take seriously such questions or their history. We should also note what we did *not* do: we did not choose a side: the pathic over the apathic, or vice versa. We did not try to elevate logic over experience or the sensible over the nonsensible. We did not (like Heidegger) accuse reason of blocking the way to thought, or conversely, stigmatize and attempt to overcome the contradictions and ambivalences of our senses, emotions, or language. Nor did we attempt to separate any of these into authentic or inauthentic, liberating or alienating, as nearly every one of the thinkers we have encountered in this chapter tried in some way or other to do. On the contrary, we tried to emphasize throughout that such choices are themselves the product of a history of thought about the human that has us in its grips. They are symptoms of our disease, not antidote or cure.

Why do we think about our human nature and knowledge in the way that we do? Why is our anthropology, our way of thinking about the human, split in the ways that it is? It is both fashionable and true to point out that every historical context is unique unto itself, but it is also true (albeit less fashionable) that certain problems abide, and ours is one of them. "And here," to quote once more Dewey's *Quest for Certainty*, "is a justification for going back to something as remote in time as Greek philosophy."

2

THE GREEKS

A Protohistory of Theory

A *protohistory of theory* cannot replace the *protohistory of theory.*
It can only recall what has eluded us.

HANS BLUMENBERG[1]

Why start with the Greeks? The Babylonians, for example, were far more precocious and also far more adept at applying mathematics to problems in the world, from tax collection to astronomy. The Egyptians, too, were practiced at calculation and measurement (indeed the Greek Herodotus attributed the invention of geometry to Egyptian efforts to survey the ever-shifting banks of the Nile). We can find early and important inquiries written in Sanskrit and Chinese. And if we knew more about the ancient civilizations of Central and South America, we would presumably find their achievements fascinating.[2]

But we are writing the history of a way of thinking about knowledge, not a history of number or of mathematics. In the previous chapter, we saw how many of the European and American protagonists understood the bitter debates in which they were enmeshed as originating in ancient Greece, distinctive to a tradition they sometimes called "Western." A distinctive way of proving mathematical propositions did indeed originate in ancient Greece. Those proofs used assertions of sameness and difference in order to produce seemingly incontrovertible conclusions within the rules and methods the (very small) community of mathematicians had set for themselves.

We will retell that story across the next three chapters, but not in the heroic key in which it has so often been told. For what makes Greece so important in our history of humanity is not its specific contributions to arithmetic or geometry or even its development of specific kinds of demonstrations, such as deductive proofs. What matters for our history is the ways in which the purposeful choices about sameness and difference that produced the necessary truths of mathematics in the Greek world came to be imagined in that world as necessary across many interrelated domains of thought, including those we today call physics, psychology, philosophy, and religion.

In this chapter we will describe a construction project at the foundations of Greek thought, one in which sameness and difference produced a new type of "certainty." We will suggest that this new certainty called many other certainties into question and thereby provoked a crisis similar to the one we vicariously lived through in the previous chapter, albeit this time spurring no world war but only a philosophical polemic between the partisans of sameness and of difference. Then we will explore in the following chapter how that polemic was systematized by Plato and Aristotle into a framework that came to seem necessary to Western thought for almost two thousand years.

Sameness and Difference in Ionia

Roughly three thousand years ago, a necklace of Greek-speaking cities was strung along the Ionian coast of Asia Minor (modern-day Turkey). By the late seventh century BCE, some of these were flourishing trading ports joined in loose federation and in frequent contact with the great powers to the south and east: with Egypt, with Babylon, and beyond. The city of Miletus, for example, had a trading outpost, Naucratis, in Egypt as well as numerous colonies on the Black Sea and in Southern Italy. Phocaea, the northernmost Ionian city, colonized what was to become Marseille in France.

Beginning with the ancient Greeks themselves, tradition has assigned a great deal of invention to the inhabitants of these Ionian colonies. Thales of Miletus (ca. 624 — ca. 546 BCE), for example, was placed among the Seven Sages and given the title of first philosopher and first mathematician. A number of geometric theorems were also ascribed to him by this "heroic" tradition:[3]

1. A circle is bisected by a diameter.
2. The angles at the base of an isosceles triangle are equal.
3. If two straight lines intersect, opposite angles are equal.
4. If two triangles have one side and two angles equal, the two triangles are equal.
5. The angle inscribed in a half circle is a right angle. (This was also attributed to Pythagoras.)

The propositions on this list are various. Perhaps (or even probably) they were not discovered by Thales. But they do share a striking and seldom-noted feature: in all of them, sameness plays a peculiar role. We can make this more obvious by stating the propositions more fully. Take, for example, number 4: given two triangles, if one side of the first triangle is the *same* (i.e., the same length) as a side of the second, and if two angles of the first triangle are the *same* as two angles of the second triangle, then the other two sides of the first triangle are the *same* as the other two sides, respectively, of the second triangle, and the third angle of the first triangle is the *same* as the third angle of the second triangle.

There is considerable debate about whether the productiveness of exploring the sameness and difference of angles was a Greek discovery or whether it was also a feature of more ancient Egyptian and Babylonian geometries. The most extreme champions of Greek primacy claim that "angles first became objects of measurement" among the Greeks. Others find plenty of Near Eastern predecessors.[4] Disputes over precedence are, apparently, as old as mathematics. Since the historical record allows us to say virtually nothing with any certainty about Thales or his contemporaries, priority is beside the point. What matters here is simply that the insights about sameness attributed by the later Greek tradition to Thales constitute an example of a type of *theoretical* assertion that *changed the cognitive work that could be done by the distinction between sameness and difference.*

We mean assertions of this sort: if among *a, b, c,* etc., there hold certain relations of sameness, then among *x, y, z,* etc., there must hold certain relations of sameness. In other words, from the sameness of certain measurable, or more generally, determinable objects, we can deduce with certainty that other determinable objects must also be the same. "Equivalence relations are both the raw material and the machines in the factory of Greek proofs," a historian of Greek mathematics has put it.[5] Our more general point is that *same-to-same assertions* of this type will play an enormous role in the development not only of mathematics, but of many types of Greek thought. But sticking for now with geometry, such assertions are not possible in plane geometries with only points and line segments.[6] They become possible when angles become measurable, that is, capable of being "the same" or "different."

The figure of Thales—again, we can speak only of the figure created by later tradition—is not only a mathematician but also as a student of nature. "Traditionally," wrote Simplicius of Cilicia in the sixth century CE, "Thales is held to be the first to reveal the investigation of nature to the Greeks."[7] Thales the "physicist" was interested in the origins of the cosmos. Why is there something rather than nothing? And what is the first and most elementary form of that something?

According to Aristotle, Thales held that water is this first indestructible form, from which all things arise and (perhaps) also into which all things again dissolve, "the substance persisting but changing in its qualities." "Because of this" (Aristotle continued) "they say there is no absolute coming to be or destruction, but its nature is always preserved." And hence Thales, "the originator of this sort of philosophy" (still Aristotle's words), "declared the earth to be upon water."[8]

From such cosmologies Thales and his students were said to have provided explanations of certain natural phenomena. Earthquakes, for example, were produced by agitation of the waters on which the earth floated.[9] Some scholars stress the nonmythological, "naturalistic" aspects of these explanations, which they see the Milesian school as having pioneered. (Homer, by comparison, attributes earthquakes to the god of the seas, Poseidon "the earth shaker.")[10] We will again avoid claims of priority. But Thales and his students do seem to have sought explanations that abstained from invoking the manifestly noncausal, unpredictable character of the mind or psyche of a god or human. They maintained (according to later tradition) that "necessity is strongest, for it exercises power over everything."[11]

The vocabulary of necessity points toward what we might call laws and general principles of causality. But although these general principles were not hoisted on the shoulders of gods, they did rest on indefinitely multiple and mostly implicit assumptions of sameness. A beautiful example can be found in the traditions surrounding Anaximander of Miletus, considered to be Thales's disciple. Anaximander, too, was credited with many feats of physics by later tradition, including the prediction of earthquakes and determining the proportions of the cosmos.[12]

According to Aristotle's *On the Heavens*, Anaximander deserves the credit for placing earth at the center of the cosmos. The claim is worth citing in full.

> But there are some who say that it (the earth) stays where it is because of *sameness*, such as among the ancients Anaximander. For that which is situated in the center and is related *equally* to the extremes, has no inclination whatsoever to move up rather than down or sideways; and since it is impossible to move in opposite directions at once [i.e., at the same time], it necessarily stays where it is.[13]

Aristotle's Anaximander assumed radial symmetry for both the earth and an outer cosmic shell. Given this sort of symmetry or sameness the earth has no reason to move in any direction rather than its opposite direction; so it stays put at the same spot.

We have italicized the words "equally" and "sameness" to show that Anaximander's assertion about the earth, like the geometrical propositions of Thales, is of the same-to-same type. Some physicists today hail this idea as having the same originality and import as Newton's. According to this triumphalist narrative, where others imagined the earth supported by columns, heroes, elephants, or tortoises, and Thales had proposed an earth floating on water, his disciple now asserts an earth held in place only by the samenesses of symmetry.[14]

This argument attributed to Anaximander is often characterized as the first formal (although not explicitly formulated) application of the Principle of Sufficient Reason, a principle that would become a "law of thought" for the sciences of the European Enlightenment. Some have seen something intensely modern in Anaximander's derivation of scientific argument from symmetry.[15] Let's not be carried away by precocious modernity, for there is something archaic, too, in these ideas. The Ionians did not hesitate to understand their own society (*polis*) and the cosmos as in some sense the same and to extend concepts like "center" and "sameness" from one to the other.[16] If we wished, instead of stressing scientific precocity, we might say that Anaximander's symmetrical cosmos was born out of an extreme form of ethnocentrism, a confusion of the customs of one's tribe with the laws of the universe.

Already in this very early example, we can see that sameness claims have great power. With that power comes great risk. Aristotle's treatment of Anaximander in *On the Heavens* provides us with some examples of the dangers (beyond the one just given of ethnocentrism) inherent in arguments from sameness. One danger is that they often encourage us to ignore potential sources of difference and even to assume all the "samenesses" necessary for the argument we want to make. To paraphrase Aristotle's own statement of a different law of thought, the Principle of Non-Contradiction, "one must add all further such samenesses as may be necessary to meet logical objections."[17] In other words, the sameness specifications necessary for sameness reasoning can be open ended, unspeakable, inexhaustible, indeed often unknowable, at least when we refuse to take refuge in utopia, dogma, or myth. (We will frequently stress this point, not least in our treatment of modern economics.)

The samenesses attributed to the universe by Aristotle's Anaximander, for example, are insufficient for the conclusions drawn. In addition to radial symmetry one need also assume that the outer shell itself is perfectly *homogeneous* (of the *same* nature), since otherwise (to gloss it in the language of Newtonian gravitation) it might happen that, for example, one portion of the shell is denser than the diametrically opposite portion, and hence exerts stronger gravitational force, attracting the earth toward it. And we must also assume that the intervening space is *isotropic*, that is, that all directions, while *not* being the *same* direction, are nevertheless the *same* in many respects: that they all present the *same* resistance (if any) to motion, and that they all transmit the influence of the surrounding shell in *equal* measure, etc. We should also assume that the earth behaves in the *same* way toward all directions. And so on and on.

Aristotle's engagement with Anaximander introduces us to another common danger involving same-to-same assertions: that of extending sameness from one domain to another, where it may not apply in the same way. Aristotle himself fell into this trap in opposing Anaximander's theory. "This argument is clever, but it is not true: for according to it whatever is placed at the center must remain there, even fire, for the resting property is not peculiar to earth." Aristotle then provides a couple of analogies meant to show just how ridiculous Anaximander's argument is. If he were correct, then a hair would never break under tension, however great, if that tension were evenly distributed; and an equally hungry and thirsty man, if he were equidistant from food and drink, would be bound to remain unmoving where he is (a paradox that later became known as "Buridan's ass," with man replaced by donkey).[18]

Note the slippage involved in these analogies: Aristotle moves from a situation where a same-to-same assertion is applied to planet earth to two different situations in which a same-to-same assertion is applied to (a) a hair and (b) a man's will. In case (a), one may try to uncover the hidden sameness assumptions that lie behind Aristotle's statement, among them his reliance on the continuity of matter and his rejection of the hair as a chain of cells or Democritean atoms. In case (b), however, in addition to questionable statements of sameness, like feeling hunger and thirst "in equal measure," Aristotle is assuming that if Anaximander's principle applies to matter, it should apply to states of the soul as well. We are facing an (unexamined?) analogy, a choice to perceive similarity between realms where one might also have chosen to perceive difference and to extend rules from the one to the other, in this case from physics to psychology.[19]

We will have frequent occasion to ponder this fateful choice, since our book is in large part an exhortation to interrogate more

self-consciously the consequences of extending laws of thought derived from one domain to others where the necessary conditions of
"sameness" may not apply.[20] In this respect Anaximander himself
can provide us with an admirable example of caution and epistemic humility. In what is perhaps the most enigmatic fragment to
have reached us from ancient Greece, he is said to have confronted
the impossibility of identifying a sufficient reason for why there is
something rather than nothing.

> Anaximander said that the origin and first principle of all exist-
> ing things is the *ápeiron*. From it all things are born, and in it the
> same things are dissolved because of their guilt; for they must pay
> to one another the penalty incurred by their injustice, according
> to the order of time.[21]

What is this *ápeiron* at the origins of being? The Greek word literally means "no boundaries." It has often been interpreted and translated as "the infinite," from a Latin word that also means having no
boundaries at all, neither outer nor inner.[22] It has also been likened
to a mixture of all sorts of different things, a kind of primordial
chaos.[23] But the notion of the *ápeiron* as a mixture, no matter how
thorough or homogenized, is not logically coherent.[24] It would be
more coherent to consider it utterly formless, a sort of alchemist's
menstruum universalis.[25]

All beings, all things, Anaximander said, originate in the indis-
tinction of the *ápeiron*; then, in due time, they have to pay to one another the penalty incurred by their injustice. But what is the nature
of the injustice that needs to be compensated? What cosmic law
did each thing violate when it separated from the *ápeiron*? Anaximander didn't say, but to us it seems that the law of causality itself
is at stake. For in the perfect sameness and symmetry of the *ápeiron*
there can be no cause, no "sufficient reason" allowing any definite
thing to separate from it.

Why is there something rather than nothing? When our physicists today deal with the problem, they often place some necessary
asymmetries (i.e., lacks of sameness) in the early universe in order
to explain the causes and effects of the big bang or why matter and

antimatter did not cancel each other out. To Anaximander's great merit, he faced squarely the possibility that there is no ascertainable *because*. Having discovered a Principle of Reason, he did not insist on building the entire universe on it. Instead, he left room for the criminal and the uncaused at the beginning and end of all differentiated being.[26] Or as the great poet Auden put it two and a half millennia later, "All chance, all love, all logic, you and I, / Exist by grace of the Absurd."[27]

Pythagoras and Company

Miletus fell on hard times, and Milesian thinkers dispersed on Mediterranean waves. Among the exiles, according to later tradition, was Pythagoras (ca. 570–ca. 490 BCE), who is said to have left the Ionian island of Samos for southern Italy circa 530 BCE.[28] As with all the figures in this story, we have no contemporaneous historical records to rely on: what we know about Pythagoras and the early Pythagoreans comes mostly from sources either slightly later (Heraclitus, Plato, Aristotle) or much later (Porphyry of Tyre, Diogenes Laertius). Many of those stories attributed fundamental mathematical discoveries—such as the so-called Pythagorean theorem, or the mathematical relation between musical intervals—to Pythagoras and even imagined the moment of discovery. For example, Pythagoras is said to have wandered by a blacksmith's shop and noticed that hammers of different size produced different tones when they struck metal.[29]

We should doubt all these stories, and indeed, many of the leading historians of ancient mathematics refuse to credit Pythagoras (or indeed virtually any pre-Socratic, i.e., Greek philosopher living before Socrates) with any mathematics at all.[30] But there is no doubt that within a century after Pythagoras's death, the Pythagoreans had come to be associated by the Greek tradition with a highly distinctive numerical approach to the cosmos. Aristotle, for example, wrote that "the so-called Pythagoreans, applied themselves to mathematics. . . . And through studying it they came to believe that its principles were the principles of everything."[31]

Aristotle's description certainly makes some sense of the thought of Philolaus of Croton (ca. 470–ca. 385 BCE), the first Pythagorean of whose own writings we have some fragments. In one of those Philolaus proclaims, "All things which are known have number. Without this, it is not possible for anything at all to be understood or known."[32] Scholars still argue about whether Pythagoras himself or any of his earliest followers might have held such views, and what, if any, their specific mathematical achievements could have been. But what matters to us here requires no such resolution. We want only to notice that the first principles or powers (*archaí*) attributed to the early Pythagoreans suggest the combination of the study of shapes with that of number, or of what would come to be called geometry, with arithmetic.[33]

The Pythagorean first principles differ from those of a Thales or an Anaximander in telling ways, not the least of which is that they are given as pairs of opposites. Two slightly different lists of such pairs have reached us from Aristotle and from Plutarch, but the two share so much in common that we have consolidated them here for ease of reference. In each pair, the Pythagoreans are said to have deemed the first term better than its opposite:[34]

1. Bounded (limited)/unbounded (unlimited)
2. Odd/even
3. One/many (in Plutarch, dyad)
4. Right/left
5. Male/female (not in Plutarch)
6. At rest/in motion
7. Straight/curved
8. Light/dark
9. Good/bad (not in Plutarch)
10. Square/oblong (unequal sides)

Look at the first pair. For Anaximander and "the Ionians," the unbounded had been the first principle. The "Pythagoreans" seized the scepter in the name of the bounded in order to inaugurate the reign of number.[35] Among numbers, too, there is distinction, namely, the odd and the even. Why are odd numbers superior? Once again, the human habit of mapping our customs of thought onto the cosmos:

in the Pythagorean counting system odd numbers were thought to be male, even numbers female. Notice, too, the abundance of declarations in favor of absolute sameness, such as rest better than motion and unity better than plurality.

Entire books could be written about this curiously mixed bag of opposites whose extension proved surprisingly productive. The opposition between even and odd numbers, for example, enabled intellectual feats of the first magnitude in the domain of logic and math.[36] This expansive tendency is also evident in the eventual and fateful adoption—attributed retrospectively to Pythagoras and the early Pythagoreans—of number and numerical proportion as real, eternal, and immutable being, the being of all other beings, and the only absolutely certain, solid ground. In Aristotle's formulation of their teaching, "All is number."[37]

We will look at just one of the feats made possible by the application of this new science of the odd and the even: the famous "Pythagorean theorem," with its discovery that the diagonal of the square of unit side was incommensurable. Was the discoverer Pythagoras or an early Pythagorean? The point has been endlessly disputed, and our story requires no correct answer to it. What matters is simply that it *was* discovered, if not by an early Pythagorean, then by a near contemporary.[38]

What were those tools? To *prove* the Pythagorean theorem in the particular case of the isosceles triangle, nothing was needed other than the results attributed to Thales that we listed above. And to *prove* the incommensurability (the "irrationality of $\sqrt{2}$") they only had to ascertain that the product of an even number times any number is always even and that the product of two odd numbers is always odd.

They required as well two additional logical principles, "laws of thought" whose history should not be taken for granted: the Principle of Non-Contradiction and the Principle of Excluded Third. The Principle of the Excluded Third is necessary in order to ensure that a number must be either even or odd (no third alternative), while the Principle of Non-Contradiction guarantees that no number can be both even and odd (except perhaps for 1, which the ancient philosophers considered either both or not a number).

Here is a later version of the proof, which we adapt from Euclid:

Suppose p and q are two natural numbers such that $p^2 = 2q^2$ (*). We may assume that p and q are not both even, for if they were, by "simplification" we could find smaller numbers, not both even, satisfying (*). Now p^2 is even, as (*) shows. Therefore, p itself must be even, for otherwise it would be odd (Principle of the Excluded Third), but odd times odd is odd, not even, so (Principle of Non-Contradiction) p is even. Hence, p is a multiple of 2: $p = 2r$ for some number r. But then, multiplying p times p and substituting in (*), we get $2r \times 2r = 2q^2$. Dividing by 2, we get $2r^2 = q^2$, which means that q^2 is even. This, as before, implies that q itself is even; hence, both p and q are even, contrary to our assumption that p and q are not both even. By the Principle of Non-Contradiction, that means that there are no numbers p and q fulfilling (*).

Adapted from Euclid, *Elements*, bk. 10, prop. 117 (possibly an interpolation). See also *The Thirteen Books of Euclid's Elements*, trans. Thomas L. Heath (Mineola, NY: Dover, 1956), 3:2.

Regardless of who in fact discovered it, it did not take more than a century for the theorem and its consequences to become widely known. Already in Plato's *Laws* those who are unaware of incommensurables are dismissively compared to "suckling pigs."[39] The discovery was notorious because it was momentous, bringing with it the many paradoxes of infinity and the problem of the continuum, problems and paradoxes that have occupied sharp minds ever since. And it was momentous at least in part because it brought together two different techniques—number and geometry—for achieving certainty in our knowledge based on two different kinds of sameness, and in the process it called that certainty radically into question.

There is an irony here, one deep enough to call for explication. The discovery of the "Pythagorean" theorem that we've been describing arose from the application of a new tool of certainty, a

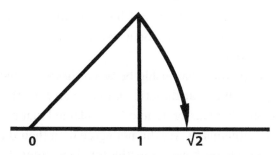

FIGURE 1. Under the assumption that rotations conserve lengths, then on any line there are line segments that "are not": here the line segment [0, √2].

novel form of indirect proof propelled by the merger of geometry and arithmetic: in this case, a proof by contradiction driven by the difference between even and odd. Mathematical proofs like these, whether direct or indirect, had (and still have) a formidable power of persuasion. In this sense the "Pythagorean" turn vastly extended the terrains of certainty.

But the application of this new hybrid tool for achieving certainty revealed an abyss in at least two senses. First, the discovery of surds—irrational numbers like √2—filled what had once seemed firm foundations of number with innumerable holes. Second, this discovery undermined confidence in the reality of geometry.

Recall some of the grounding assumptions attributed to the Pythagoreans. Assumption 1: what we call natural numbers (1, 2, 3, 4, . . .) exist, and beings coincide with having number. Assumption 2: geometry describes actual space, the space of our experience, which means, among other things, that sameness (of lengths, of angles, of areas, etc.) is tantamount to *A* and *B* being applicable or fittable one onto the other by means of translations, rotations, and reflections. In the relationship between these two assumptions lurk the crises that could be provoked by the proof that the hypotenuse of the isosceles right triangle is not commeasurable with the side (in our language, that √2 is not rational) (fig. 1).

By rotating the hypotenuse 45° until it coincides with an extended side, this implies that on any line on which we have marked a unit of length, that is, a segment [0, 1], we must have a line segment [0, √2] not commeasurable with the unit or, in other words, which *does not have number*, and hence (within the conceptual structures of our ancient interlocutors), which "is not": it is not a being, it does not exist.

We are calling the relationship between these two assumptions the marriage between arithmetic and geometry. In the previous chapter we saw that Husserl (and others) blamed that marriage for the crisis of mathematics. We did not agree with their diagnosis of alienation, but we do have sympathy for their point. Imagine an alternative history in which the "Pythagoreans" had only proven (as they also were said to have done) that there cannot be two integers p and q, different from zero, such that $p^2 = 2q^2$. Such a proof would not have opened the earth beneath Greek feet. There was nothing in arithmetic per se that demanded that one must have two integers p and q such that $p^2 = 2q^2$. But because arithmetic had been wedded to geometry, the lack of a number (a fraction) corresponding to our $\sqrt{2}$ corresponded to the lack of a point on the geometric line where a unit of measurement had been established. That line could now menace: "are you implying that I lack a point?"

It seems intuitive that if one travels along the line from say, zero to three, somewhere, sometime, one will have to touch $\sqrt{2}$. But how can one pass through a point that isn't there? One might very reasonably decree, "let some points on the line correspond to numbers, while others, like $\sqrt{2}$, do not. That way we avoid the idiocy of saying that some points exist, but other don't." But then geometry would lose one of its prized possessions, its intuitiveness.

Worse, mathematically inclined contemporaries could deduce that it is not just *one* point in the line that suffers from this type of not being but an infinite number. A simple indirect proof would have revealed to them that any rational multiple of $\sqrt{2}$, such as $\frac{3}{5}\sqrt{2}$, is affected by the same infirmity. Those points that "are not" are (as a modern mathematician might say) everywhere dense, meaning that between any pair of points P and Q you can always find any number of infirm points lying in between P and Q. In short, thanks to the deployment of powerful techniques for achieving (mathematical) certainty, the whole space of what had seemed the most solid ground was suddenly flooded by not being: a crisis both for the "being" of number, and for the "reality" of geometry. That crisis had been caused by new applications of sameness and difference. It would be met in the same way.

Elea: The Logical Reconstruction of a World

Parmenides of Elea and Heraclitus of Ephesus are most notable among those who responded to the crisis. They were both capable of simultaneously perceiving the power of these new forms of knowledge and the perils of the pit that these had opened up beneath their feet. Their very different responses to the crisis will serve to sketch two distinct paths into the future of thought, staked (if you will forgive a shift in metaphor) by two distinct wagers on sameness or difference.

We will not attempt to reflect the many difficulties and obscurities of each figure's teachings and legacy. Like the caricaturist, throughout this book our goal is deliberately reductive: to make prominent a salient feature in each visage so that each may more memorably stand for different types of attitudes toward the nature of our knowledge. We will speak first of Parmenides and his disciples, the Eleatics, not because of chronological order (the dates for both figures are murky) but because they were much more sympathetic than Heraclitus to the "sameness" assumptions that underlay the logical principles and indirect proofs we have just been discussing and did not hesitate to put those methods to metaphysical work.

The precise nature of Parmenides's relations with the Pythagoreans or other mathematical schools is a matter of ever-shifting scholarly debate. Among the very few things that the ancients reported as fact about Parmenides's life "is that he was at some time a Pythagorean," though some modern specialists have preferred to stress the independence of Parmenides's teachings, and a few have even maintained that Parmenides and Zeno, and not the Ionians or Pythagoreans, were the true creators of Greek mathematics.[40] We need not settle these debates in order to imagine that the mathematical discoveries of the age were among the provocations that prodded Parmenides and his followers into an extreme commitment to sameness.

Only fragments of Parmenides's teaching have reached us, and those are in the form of a poem that is written in dactylic hexameters, the meter of epic poetry. This philosophical epic begins with a supernatural chariot voyage beyond the portals of night and day to the abode of an anonymous goddess who reveals all truth to the voyager. Says the goddess,

It is needful that you learn all things:
both the unshakeable heart of very convincing Truth
and the opinions of mortals in which there is no true belief.[41]

This revelation is sharply two sided—*alētheia*, or Truth, on the one hand, and on the other *doxai*, or opinions. Its Januslike faces are uncompromisingly opposed. Imagine, for example, that *A* expresses the (Aristotelian) opinion that a stone falls because it is made of earth, which moves down unless held, while *B*, more up to date (and Newtonian), says that the truth is that the earth and the stone exert on each other a gravitational force, which, and so forth. We might want to conclude that at least *A* and *B* agree that the stone does fall. But Parmenides's goddess rejects the possibility of partial agreement between Truth and opinion, between being and appearance.

Nor is she concerned with "saving the phenomena." Her Truth is designed for disembodied minds, not for the senses, whose data she leaves to "the opinions of mortals in which there is no true belief." Here, too, we might want to see the influence of Pythagoras, who (according to Porphyry) thought that the mind needed to be freed through philosophy from "the impediments and fetters within which it is confined" by "the unsoundness in the operation of sense. Pythagoras thought that mind alone sees and hears, while all the rest are blind and deaf."[42] Sharp indeed is this split between the truth of the mind and the falsehood of the senses, between freedom and slavery.[43]

The "unshakeable heart" of Parmenides's Truth rests on such splits and their corollaries. Fragment 2 articulates one of the most important of these Herculean pillars:

Come now, I will tell you—listen to my saying and take it with you, / the only two ways of thought that can be conceived of: / the first, that *It is*, and that it is impossible for it not to be, / is the way of Conviction, for Truth is its companion. / The other, that *It is not*, and that it must needs not be, / that, I tell you, is a wholly untrustworthy path. For you cannot know what is not (for it is impossible) / nor utter it.[44]

There is no alternative to the difference here between *is* and *is not*: no common, no higher ground. Parmenides does not articulate his logical principles, but we seem to be standing on the Principle of Non-Contradiction (it is impossible that a thing can be and not be at the same time, as it would soon be expressed) and the Principle of the Excluded Third (either being or not being, but no other possibility). With these same tools we have just seen the Pythagoreans open inexhaustible veins of logical/mathematical problems— namely, how to link the continuous and the discrete. But now they are being put to work to close and seal those veins so as to bind us with a new logical prohibition—to abstain from thinking or saying, "what is not," and "what must needs not be."

Parmenides tells us a bit more about this "what must needs not be" that we are prohibited from speaking about: "It is the same thing that can be thought and that can be."[45] We take this to mean that something that (logically?) cannot be is something that (logically?) cannot be thought. But what is it that cannot be thought? Different periods and places have offered different candidates, often involving nothingness in its many guises: the void or empty space of the atomists, a total erasure of space and time, a cosmos without god(s).

The problem with this test of thinkability is that things considered unthinkable from one time, place, or perspective can turn out to be perfectly thinkable in another. Indeed, among the more striking shifts in historical horizons of thinkability is that of the relationship between nothingness and being. Whereas in antiquity, *ex nihilo nihil fit* (nothing is created out of nothing) was the hegemonic doctrine, with the expansion of Christianity. creation *ex nihilo* became not only thinkable but orthodox, and the alternative was blasphemy.

But sticking here with mathematical examples, irrational numbers (such as $\sqrt{2}$), finite space-time with no boundaries, counting infinities, a two-dimensional sphere not attached to a three-dimensional ball—these and many others were once considered unthinkable. Thinkability is both deeply contingent and highly subject to historical change.

Are the apparent holes of "nothingness" created by the Pythagorean proof that some line segments are not commensurable with the unit of measurement thinkable or unthinkable? When facing such a decision, there are choices to be made. Here are just a few possibilities. (1) Relativize or revise your logical principles (such as the Principle of Non-Contradiction, the Principle of the Excluded Third) and the claims of indirect proofs. (2) Revise the notion of geometric sameness mentioned above. (But there is no apparent way to do so that does not at the same time make geometrical discourse enormously more difficult.) (3) Reject the notion that geometry is a faithful description of reality. (But this would have clashed with the identity of thought and being proclaimed.) The first of these is the way of Heraclitus. The third would be adopted first by mathematicians and then by physicists in the nineteenth and twentieth centuries.

Parmenides chose none of the above, taking instead the path of absolute sameness. Since not being is forbidden, and since Pythagorean ontology (numbers exist, and being is the same as having number) leads logically to the *absurdum* of a space-time flooded by not being, we must reject Pythagorean ontology. All is one, all is the same, and all numbers other than 1 are not true being but merely opinion. Motion or change are not real, since all points are one point, and anyway it would make no sense to think of being moving through not being.

It seems plausible that Parmenides's confrontation with the "theoretical crisis" posed by the discovery of incommensurables influenced the extreme insistence on unchanging sameness that is characteristic of his philosophy.[46]

That what is, is uncreated and indestructible,
alone, complete, immovable and without end.

Nor was it ever, nor will it be; for now it is, all at once,
a unity, continuous.

Such was the Eleatic doctrine, achieved by putting the decision between thinkable being and not thinkable not being at the center of critical thought. Indeed the Greek word "crisis," in the sense of a major theoretical decision, appears for the first time in this same fragment: "the crisis about these things lies in this: / it is or it is not."[47]

We cannot abandon Parmenides without pointing out *another* theoretical decision on his part, one that is contained in all of these fragments we have cited and that by itself was sufficient to cause the clash of his Truth with human life: namely, his *choice* to listen more closely to the strictures of logic than to the data of our senses and the experience of thinking. It was his privileging of logical principles over life experience, perhaps even more than his preference for being over not being, which proved so fateful and influential. Parmenides' "true" world, the decision between "is" or "is not," was constructed exclusively out of the types of sameness and difference that logic recognizes.[48]

Ionia Strikes Back: Heraclitus's Demolition Service

Heraclitus of Ephesus made different choices. Among other things he despised Pythagoras, of whom he wrote that "much learning does not teach intelligence." In Pythagoras' case, according to Heraclitus, it could even produce "evil arts." He called Pythagoras "chief swindler," "prime impostor," "bullshitter extraordinaire."[49] Moreover Heraclitus never mentions number (*arithmón*), an omission all the more remarkable given that he often mentions measure (*métron*).[50] We might deduce that he was highly skeptical of the Pythagoreans' science of even and odd, with its strict either/or.

The tenets Heraclitus chose for his philosophy have often been adumbrated in two phrases: "everything flows" and "unity of opposites." Neither phrase is found in the extant Heraclitean fragments, but already in Plato's *Cratylus*, Socrates attributes to Heraclitus the doctrine that "all beings move and none remains still" and then links the phrase to others concerning flowing rivers.[51] So, even though the words "everything flows" may not come from Heraclitus's own *kalamos*, they can plausibly be taken to express his view, which a modern physicist might reformulate as all things physical move and change even if some things do move more fluidly or more visibly than others.

As for the unity of opposites, Heraclitus never uses the word "opposites" (ἐναντία) in the teachings attributed to him.[52] The omission may be deliberate given the importance of that concept for the Pythagoreans. So we should probably *not* attribute to him the view that all opposites are one and the same. But we should take him to be very much concerned with demonstrating that the "opposites" and related logical principles that the Pythagoreans were applying to the world with such astounding results were far from universal. Hence, he spoke neither of number nor of opposites.

Heraclitus focused instead on physical things, among which he included some very important things that other thinkers in other times and places might exclude from the physical (things such as the soul and the mind; and geometry, conceived of as a description of physical space). In dealing with these physical things, Heraclitus's point is repeatedly that the laws of thought derived from Pythagorean oppositions need to be relativized or revised. In other words, his teaching is not that opposites are the same, but that they are contingent on the question we are asking, the perspective we are taking, and the kind of object we are inquiring about.[53]

Fragment B 60 provides a nice example: "The way up and the way down are one and the same." That this is not about the unity of opposites but rather about the relativity of the categories of opposition becomes clearer if we refer it to the sun and the other stars. In Anaximander's space, for example, where the earth is in the middle of a sphere and the stars go around in circular paths, the way up and down are indeed the same, and the Pythagorean opposition of right and left becomes as relative as the opposition of rest and motion, for the right direction becomes the left at the antipode. Another fragment: "In [the screw of] the carding-comb the straight and the curved paths are one and the same." This one seems directed against the Pythagorean opposition straight/curved. Over and over again

Heraclitus's fragments relativize the principles of thought on which the Pythagoreans built their oppositions.[54]

Heraclitus's demolition here is mathematical: when you are on a two-dimensional plane, a circle is curved, and a straight line is straight. But go up one dimension into three-dimensional space and you can combine curved and not curved, a circle on the x-y plane, and a straight line on the z axis: a three-dimensional spiral "synthesis" of seeming "opposites." In geometry, it is generally the case that what requires a reflection in any dimension can be done by a rotation in one dimension higher. Heraclitus puts this rule to work relativizing the distinctions and oppositions of his more categorical rivals. An important lesson, but one too easily forgotten, even by the likes of Immanuel Kant.[55]

Heraclitus attacks Pythagorean morality as well as mathematics. Fragment 61 relativizes the Pythagorean opposition good/bad: "The sea is the purest and most polluted water. Fish can drink it, and it is good for them; to men it is undrinkable and destructive."[56] We could go on, adding to these examples others refuting each of the oppositions on the Pythagorean list, except for the original one, even and odd: Heraclitus never spoke of numbers. Perhaps he was allergic.

He did, however, speak often of water. We have mentioned the sea but not yet dipped into the famous river fragments. The most famous of all is, "For and on the same people who step into the same rivers, other and other waters flow." Our translation may sound awkward (compare the vulgar version, "you can't step into the same river twice"), but it is true to Heraclitus's language, which is poetic, polyvalent, dense. Its syntax and its meaning repeatedly and deliberately violate those very laws of thought, such as noncontradiction, whose sovereignty he is challenging.[57]

Does Heraclitus contradict himself or not? The question has concerned philosophers since Aristotle, but it misses the point. The fragment is concerned with metaphysical notions even more fundamental than the Principle of Non-Contradiction: namely, the notions of sameness and difference on which that principle was built. *The important question is not whether Heraclitus contradicts himself but how he draws attention to the contingent nature of our choices between sameness and difference and thereby provides a radical critique of noncontradiction itself.* (By critique we do not mean a dismissal of the principle but rather a becoming conscious of the conditions of its possibility.)[58]

Let's start with the waters, for it seems easy enough to agree that the waters flowing *on* us are others and others, ever different. That is to say, *for* all of us the waters are ever different. But what does it mean to say, as Heraclitus does, that the river is the same? What is the river if not the waters? It might mean that the banks are the same and that we agree in calling a river "the same" when its banks are the same. But strictly speaking, riverbanks do *not* stay the same, nor do riverbeds. The Nile provided the ancient world with an egregious example whose fickle banks motivated the first geometers, according to Herodotus. Perhaps we could agree that banks and beds do change, but not as rapidly, not as visibly, as the waters; and so for convenience we call them "the same," while we call the waters "different." In sum, Heraclitus is trying to convey that the sameness or difference of the waters and the river are both subjective and conventional. Difference and sameness *sono cose mentale*, they are mental, as Leonardo da Vinci said of painting: they properly pertain to the psyche.

A similar argument applies to people, the waders. What do we mean when we say that a person is "the same"? Heraclitus pushes us to perceive that this emphasis of sameness over difference is also a matter of perspective. His point is not so much that of modern physiology, that all the cells in each of our bodies are constantly changing and replacing themselves (like the planks of Plutarch's famous ship). It is rather a more psychological one, that our own sameness or difference are subject to our psyche, or if you prefer, to our memory and attention. Sameness and difference in physical affairs are, for Heraclitus, relative, subjective, indeed psychological. (We hasten to add that distinctions between the physical and the psychological are themselves deeply historical: those of Heraclitus and his age will not be the same as ours.)

We might almost say that the term *psychological* was coined by Heraclitus himself. In another fragment he tells us that "were you to go in search of the limits of *psychē*, you would never find them, even if you tried every path: so deep is *psychē*'s lógos."[59] So far as we know, this is the first occurrence of the word *psychē* in conjunction with the word *lógos*: a meaningful moment in the history of psychology. The fragment speaks of the boundaries and limits (πείρατα) of the *psychē* and says that they are unreachable, impossible to find: we hear echoes of Anaximander's *ápeiron*, that formless, infinite womb and tomb.

According to Heraclitus, just as there cannot be a definite, sufficient cause for any determinate thing to arise from the *ápeiron*, there cannot be a *lógos*—a definite cause or a complete account—of the states of our soul/mind, our feelings and thoughts. The lack of boundaries of the *ápeiron* and the impossibility to find boundaries to/in the *psychē* suggest a similar fluidity in both. Should we speak in modern terms of "stream of consciousness"? We do not know, but we suspect that, were Heraclitus alive today, he would treat the gurus of artificial intelligence much the way he treated Pythagoras and his followers.

In the light of these fragments (B 12, B 45), it appears that for Heraclitus physical sameness and difference have an unfathomable *logos*, since they pertain to *psychē*, whose limits cannot be found. Another fragment pulls yet more solidity from beneath our feet. "*Psychē* has a *logos* that increases by itself."[60] Not only are the boundaries of the *psychē* indeterminable and changeable: so, too, are the boundaries of its unfathomable *logos*.

We could not be further from the Pythagorean, who dwells only in the delimited; nor from Parmenides, with his extreme sameness; nor for that matter (and looking forward a century or two) from an Aristotelian anthropology, with its rational man incapable of self-contradiction. This is presumably why some 2,500 years later Nietzsche would call Heraclitus "the soft breath of Spring" blowing through the permafrost of Western philosophy. One need not be so partisan in order to perceive that here, at what so many have called the beginnings of our "scientific" traditions of thinking about the mind and world, we can already speak roughly of two schools poised in polemical antithesis.

We can characterize the antithesis with a pun. *Psephos* in Greek meant "pebble," which the Latins called "calculus." For the ancients (and the not so ancients: think of the abacus) pebbles were important aids in counting: hence our English word "calculate." The Pythagoreans had even used them in their efforts to define the properties of even and odd.[61] Like integers, pebbles have distinct, nonfluid boundaries, and when thrown together they do not lose any of their difference or individuality, difference that we in any event ignore in reducing them to number. If, in the spirit of Heraclitus, we were allowed to indulge in serious wordplay, we would say that where he offered us psychology, the Pythagoreans gave us psephology; the *logos* of the *psychē* and the *logos* of pebbles. At different times and places, those two have sometimes been taken to be the same. We want to maintain that they are different.

3

PLATO, ARISTOTLE, AND THE FUTURE OF WESTERN THOUGHT

If I believe that someone else is capable of discerning a single thing that is also by nature capable of encompassing many, I follow "straight behind, in his tracks, as if he were a god."

SOCRATES, as reported by Plato

The teachings of the "founding fathers" introduced in the previous chapter reach us only because their great debate between sameness and difference came to seem to later generations a necessary one for interpreting the world. Choosing between these positions—often while attempting to reconcile them—became one of the primary tasks of Greek philosophy and science, and thereby, of all the many later cultures that would learn from Greek teachers and their texts (including Judaism, Christianity, and Islam, as we will see in the next chapter). Of all those Greek teachers, perhaps none are more important to the continuing history of divisions between sameness and difference than Plato and his student Aristotle. For it was Plato—often writing in the character of his teacher Socrates—who attempted in the fifth century BCE to reconcile sameness and difference, truth and appearance, certainty and crisis. Plato, too, was tempted by number into building his philosophy on foundations of sameness, foundations that Aristotle would criticize and partly tear down but also rebuild and refine. It is through their collected teachings that Greek fantasies of sameness proved influential enough to serve as synecdoche for much of the future of Western thought.

Dialectics: Division into
Difference or Unification into Sameness?

Here is Plato's Socrates describing to his young student Phaedrus the two most basic principles that he thinks characterize philosophical thought:

> The first consists in seeing together things that are scattered about everywhere and collecting them into one kind, so that by defining each thing we can make clear the subject. . . . [The second] is to be able to cut up each kind according to its species along its natural joints, and to try not to splinter any part, as a bad butcher might do. . . . I am myself a lover of these divisions and collections, so that I may be able to think and to speak; and if I believe that someone else is capable of discerning a single thing that is also by nature capable of encompassing many, I follow "straight behind, in his tracks, as if he were a god." God knows whether this is the right name for those who can correctly do this or not, but so far I have always called them "dialecticians."[1]

Division into difference or unification into sameness: the choice is not only the basis of all knowledge but also so difficult that the person who can make it correctly deserves to be called godlike. As for the activity of these godlike people, Socrates assigns to it the honorable name that he elsewhere associates with his own, namely, dialectics.

We agree: the choice is a difficult and important one, so difficult and important that thinkers across the centuries have by and large refused to fully recognize the radical freedom we have in making it and the contingency of the truths the choice produces. Instead, they have often preferred to restrict that freedom through dogma. That preference has produced massive amounts of immensely valuable knowledge, as we shall see. But it has also blinded us to a great deal, including much knowledge of the nature of human knowledge and knowledge of ourselves as humans.

We can already see symptoms of a squint in Socrates's words, especially if we compare them with a fragment from Heraclitus:

"Conjoinings: wholes and not wholes, converging and diverging, harmonious dissonant; and out of all things one and out of one thing all." No clearly cuttable joints here. Socrates is reputed to have said of Heraclitus's work that "it needs a Delian diver to get to the bottom" of it, and indeed Heraclitus's diction is obscure and resonant with contradiction. Socrates seems not to want to go so deep. His rather brutal metaphor of cognition as butchery asserts that no matter how fundamental the choice, there are "natural joints," unified, essential, stable, indivisible categories of "sameness" that we are forbidden from carving into difference.[2]

Over and over again Plato's dialogues explicitly posit and explore a fundamental conflict between opposing categories of sameness and difference and assign to that conflict a primary and perpetual role in the generation of all philosophy. And yet at the very same moment that these dialogues present themselves and the discipline of philosophy as fully committed to the critical exploration of the possibilities for both, their yearning for stability at the foundations of being is such that they stack the deck in favor of sameness. We might say that Plato wants to allow for our felt "Heraclitean" experience in a world of flux while at the same time holding onto "Parmenidean" criteria of being in which existence excludes change. This seemingly impossible task—to find an eternal and unchanging foundation while making sense of an endlessly changing cosmos—became the grail for many future philosophical, scientific, and religious quests (a sharp distinction between the three is anachronistic).

Plato's dialogues pioneered the critical methods and the dogmatic shortcuts adopted by later generations of knights-errant. We can see both the critical exploration and the sleight of hand at work in a dialogue—*Parmenides*—that Plato deliberately staged as a direct engagement with the logic of pure sameness emanating from Elea. *Parmenides* presents itself (improbably) as the record—committed to memory by an audience member—of a philosophical discussion held a half century before between Plato's philosophical forefathers: an elderly Parmenides, a middle-aged Zeno, and a stripling Socrates.

Sameness and difference are, not surprisingly, at the heart of the discussion. The stage is set by a reading of Zeno's treatises, starting with the argument that "if existences are many, they must be both like and unlike, which is impossible." Socrates sets out to attack this thesis (128e–129a), which he presents as a version of Parmenides's thesis that "existence is one." He does so by making the distinction that Plato's work put at the center of so much philosophy to come, a distinction between things and ideas. It is not surprising, he says, that all things participate both in the idea of sameness or likeness and in the idea of difference. Surprising, indeed truly disconcerting, would be if someone could show that an abstract idea, such as absolute sameness or sameness as such, participated in its opposite, becoming difference.[3]

Note Socrates's strategy: he is taking refuge in the fortress of the absolute. In the sensible world of things, difference and sameness may be mixed, but not in the world of pure ideas. The realm of the absolute does not produce contradictions of the sort on which (according to Socrates) Zeno had built his theses. Absolute sameness can never be disrupted by difference. (Perhaps this is a good place to point out that on these grounds our own position—that there is no absolute sameness but only different types of sameness—would likely have struck the Socrates of *Parmenides* as either a real wonder or as utter nonsense.) Socrates is here proposing a sketch of an idea with a very long future, the idea that what we have come to know as Plato's theory of Forms, according to which for every predicate or property of things in the world there corresponds a single, non-sensible, eternal, absolute, and unchanging Form.

The wily old Parmenides will have none of this. Instead he maneuvers the young Socrates into halfheartedly conceding that if there are unchanging ideas (Forms) of sameness and of difference, there should also be ideas or Forms of all sorts of things, including hair, mud, and dust. But how could such humble relics of our finitude be eternized as Forms? Young Socrates admits discomfort at the possibility.[4] And now Parmenides proceeds to wrap Socrates in a tissue of sophistry involving the possibility of *different* things participating in the *same* idea. Can one and the same idea reside in different things at once? Or does one thing participate in one part of the idea and a different thing in a different part, so that the idea would then have parts? But then the idea of the small, or absolute smallness, would have parts, and a part would be still smaller than the absolute small, wouldn't it? And so on.

How delicious to find a character in a work attacking one of his author's most famous teachings. It is not surprising that centuries of readers have been attracted to these pages, many (like Aristotle) seeking to find in them arguments either for or against Plato's theory of Forms (Aristotle voted against).[5] Our goal is quite different. We are not trying to decide between the rival claims of the characters of Socrates and Parmenides. We would rather extract from these pages a sense of the shared philosophical moves by means of which all parties in the dialogue are grappling with conundrums of sameness and difference, for these moves would for millennia continue to serve those who want to wrestle with a changing world while holding on to an ideal of sameness.

Whatever their disagreements, *Parmenides*'s protagonists all commit to a set of (largely unstated) "axioms" about the relationship between things and Forms (which we will abbreviate F) even as each participant in the dialogue draws from those commitments quite different conclusions.[6] What are those implicit commitments? Scholars disagree, but here, for the sake of illustration, is a list.[7]

- Causality: Things that are F (other than the F) are F by virtue of partaking of the F.
- Separation: The F is itself by itself, at least in the sense of being separate from, and hence not identical with, the things that partake of it.
- Impurity-S: Sensible things are *impure* inasmuch as they can (and, in fact, often do) have contrary properties.
- Purity-F: Forms cannot have contrary properties.
- Uniqueness: For any property F, there is exactly one form of F-ness.
- Self-Predication: For any property F, the F is F.
- Oneness: Each Form is one.

What is striking about this list of commitments underpinning Plato's theory is the degree to which not only ideas (Forms) but also the participation of sensible things in those ideas are all subjected to the rules and principles of thought—such as identity and noncontradiction—that govern number. Through axioms like these, Plato sought to bring the mutable world of flux under the rule of sameness.[8]

Number is everywhere in *Parmenides*, which is perhaps not surprising given that the dialogue is dedicated to exploring the question of whether the one or the many does or does not exist (128d–e, 137b). The one and the many is a version of the question of sameness and difference that translates with particular ease into the language of number, and indeed, the dialogue quickly becomes a numbers game as the interlocutors focus on the relationship between one and its being in order to generate first two, then three, then the infinite series of counting numbers. If one exists, then number exists, they agree, "but if number exists there must be many, indeed an infinite multitude, of existences."[9]

On and on they go, using the series of numbers (1, 1 + 1, 2 + 1, 3 + 1 . . .) in order to generate contradictions ("The one is, then, only one, and there can be no two"; 149d) through which to argue about the existence of the one or the many. Given the premises and the procedure, it is not surprising that the dialogue arrives at the conclusion that if the one did not exist, all thought would be impossible. All would be unlimited, so that even if you try to take from a thought "what seems to be the smallest bit, it suddenly changes, like something in a dream; that which seemed to be one is seen to be many, and instead of very small it is seen to be very great in comparison with the minute fractions of it" (164d). What could be worse than thoughts that change shape when you try to grasp them, like Borges's blue pebbles? Far better to assimilate ideas to numbers and adopt the dialogue's conclusion, "that if the one is not, then nothing is" (166c).

Plato's Pact with Elea

Parmenides is in some ways an idiosyncratic dialogue, but in its yearning for a foundation of absolute sameness beneath the apparent permanent flux of the world, it is fully representative of the Platonic corpus. Representative, and aptly named, for in it we see the beginnings of the Athenian Academy's pact with Elea, an alliance offensive and defensive between Plato's school and the long-dead venerable figures of Parmenides and Zeno through which Plato manages to embrace the Eleatic and Pythagorean logico-mathematical absolute certainties while at the same time claiming to bridge the gulf that separated those certainties from the phenomenal world of change and eternal truth from mere opinion.

It is difficult to think of a Platonic dialogue that does not show at least some trace of this alliance, but some, such as the *Sophist*, are more explicit than others about the legacy. The *Sophist* begins with a character named Theodorus of Cyrene introducing a "Stranger from Elea" to Socrates. The Stranger's provenance is obviously significant, but so is the choice of intermediary: Theodorus was a figure in the history of mathematics credited with having proven the "irrationality" of the square roots of all nonsquare numbers from 3 through 17 (the famous case of 2 having been already proved). All the more portentous, then, is his opening proclamation that philosophers are divine.[10]

A problem immediately arises in the discussion: there are those who are widely considered philosophers but who only pretend to be—namely, the sophists. How to define the sophist so as to distinguish him from the true philosopher? Immediately we can see that we are in territory of sameness and difference and that the territory is dangerous indeed, for the Stranger does not hesitate to describe the problem in frightful terms: "a wolf is very like a dog, the wildest like the tamest of animals. But the cautious man must be especially on his guard in the matter of resemblances, for they are very slippery things" (231a). The dialogue will eventually come to the conclusion that—much as Socrates put it in the *Phaedrus*—getting things right in the matter of sameness and difference is the very definition of philosophy: "Shall we not say [says the Stranger] that the division of things by classes and the avoidance of the belief that the same class is another, or another the same, belongs to the science of dialectic?" (253d). The sophist will be found to fail precisely in this. Hence, this dialogue that purports to be about the definition of the sophist must explore sameness, difference, and other Platonic principles of being.

As interlocutors so often do in Plato's oeuvre, the discussants are quick to agree that "number must exist, if anything does" (238b). They agree with even greater alacrity that one large part of human knowledge is especially easy to disassociate from truth: the part we call imitation, representation, image, mimesis, art, and indeed physical reproduction in general. The first victim in the Stranger's hunt for the sophist is the artist, because he "professes to be able, by a single form of skill, to produce all things, [so] that when he creates with his pencil representations bearing the same name as real things, he will be able to deceive the innocent minds of children, if he shows them his drawings at a distance, into thinking that he is capable of creating, in full reality, anything he chooses to make" (234b).

Sculpture is slightly more difficult than drawing to dismiss from the world of sameness-dependent truth. "Creating a copy that conforms to the proportions of the original in all three dimensions and giving moreover the proper color to every part" produces things that do seem to deserve the name of likenesses. But the case is exceptional: the instant sculptors depart from those exact proportions in order (for example) to account for the perspective of the viewer, they cease to produce likenesses and produce instead semblances (*phantasma*). The illustrative example the Stranger gives is that of colossal statues, in which "artists, leaving the truth to take care of itself, do in fact put into the images they make, not the real proportions, but those that will appear beautiful" from the perspective of the viewer (236a).[11]

Art here serves the Stranger with what seems like a simple starting point for "the extremely difficult question" of how "one may say or think that falsehoods [like art objects] have a real existence without being caught in contradiction." (236e–237a) It is at precisely this point in the argument that the Stranger quotes Parmenides. "'Never let this thought prevail,' says he, 'that not-being is; but keep your mind from this way of investigation.'" He then deploys examples of art in order to produce the contradiction that will move the argument forward. The image (*eidolon*), says the Stranger, is not true, but "it has a being of a kind" (240b).

This is a nice example of the ways in which—for the sake of his argument—Plato stacked the deck against certain kinds of sameness and in favor of others. It also reminds us that the tricks he utilized could have a long future: in this case a future as long as the (still ongoing) history of Western art. Platonic explorations of the onto-logical status of art objects would eventually intersect with "Judeo-Christian" criticisms of idolatry—such as St. Paul's locution that "an idol is nothing in the world" (1 Cor. 8:4)—raising new questions about art's relation to sameness, being, and eternal truth. Some early Christians, such as Tertullian (writing in Latin ca 200 CE) in his *On Idolatry*, looked back at Greek philosophy in order to widen Paul's condemnation: "*Eidos* is the Greek term for *form*; similarly, its diminutive *eidolon* gives us our word *formula*. Therefore, both form and formula always really mean *idol* (*idolum*)." Others, such as Origen (writing in Greek at more or less the same time) tried to de-ploy Platonism in order to open a space for Christian art by differen-tiating between types of likeness or sameness, in this case between a likeness, which is, and Paul's idol, "that which is not."[12]

In the end Origen condemned both likenesses and idols. And yet he continued to put images to work in order to make rationally in-telligible a crucial theological problem: the "sameness" that identi-fies God with Christ and the "difference" that distinguishes the two. The Father, according to Origen, is a statue so large as to hold the entire earth, while the Son is another statue perfectly alike in aspect and every other respect, except without the measureless immen-sity. How near and how far we are from the critique of perspective in the Stranger's colossus. As for art (rather than theology), we can hear the Stranger's words continue to echo among theorists of art thousands of years after the penning of the *Sophist*. According to the sixteenth-century artist, architect, and "founding father" of art history Giorgio Vasari, sculptors of his day maintained "that sculp-ture imitates the true form, and shows its aspects, as you go around it, to all vantage points. . . . And many of them do not hesitate to say that sculpture is as far superior to painting as the truth is to a lie."[13]

Forgive the excursus. We felt compelled to provide at least some evidence for our claim that these moves shaped the possibilities for thought in many domains of human activity. Art is only one of those domains, and we cannot explore it further, though we yearn to com-pare the artifacts of worlds innocent of Plato, such as China, with those produced in the "Western" aesthetics that he so thoroughly saturated.[14]

Let us return to the *Sophist*, where we might say that Plato sacrificed art out of contingent philosophical motives, though doubtless he felt them to be necessary: privileging the intellectual and eternal (i.e., sameness); demoting the perspectival, the relative, or the contingent (i.e., difference); and elevating noncontradiction to a law of thought, all in the service of denying the possibility of nonbeing.[15]

How Difference Mingled with Being

Images are the first step in the dialogue, not the last. Next and more difficult, judging from Plato's pedagogically self-conscious ordering, comes language (238a–241d). The Stranger does not stigmatize language in the same way that he does images, perhaps because—as he will later say—without interweaving categories of shared meaning there can be no philosophical communication.[16] But language does not provide an easy way out of the problem. How can one remain loyal to Parmenides and yet make any statement about the nonexistent without falling into contradiction? For what true statement can be said about something that is not?[17] It is at this point that the Stranger worries that he might be repeating the sin of Oedipus (himself a stranger in Thebes): "Do not assume I am becoming some sort of parricide—for in defending ourselves we will have to examine and test the doctrine of my father Parmenides. . . . We must dare to lay un-filial hands on that pronouncement" (241d–242a).

What follows is an impious exposure of the father's purported flaws. The Stranger drives the Eleatics into self-contradiction by exploiting the gap between the implicit axioms that govern forms and the language with which we speak of them. You say that "being is one"; are then "being" and "one" the same? And if so, how can you, who assert that nothing exists but unity, give two names to the same thing (244b)? (Centuries later questions like this one will torment Christians and Muslims trying to theorize the radical unity of God.) Then he ridicules the father's mathematical analogy. Parmenides had said that the whole of being is like a ball (presumably three-dimensional) that is the same throughout. But a ball has a center and extremes (a two-dimensional sphere), and those are different parts of the ball, aren't they? Now, how can a whole that is one have multiple parts? We can hear, between the lines, the snickers and the sneers (244e).

The climax of this shaming is what seems to be a repudiation of the father: "I say that anything that has any capacity whatsoever either to effect a change [τὸ ποιεῖν ἕτερον, "to make other, different"] in anything of any nature, or to be affected [τὸ παθεῖν] no matter how minimally, or by the vilest or slightest thing, even though it be only once, has real being (247e). Where Parmenides had identified being with sameness and immunity to change (the apathic), the Stranger now identifies it with difference and susceptibility to change (the pathic). Do we dare to hope that at last we are in a position to reject both Parmenides and Heraclitus? Can we turn a deaf ear both to "the champions either of the one or of the many forms of the doctrine that all reality is changeless, and [also to] . . . the other party who represents reality as everywhere changing"? "Like a child begging for both," must we now "declare that reality or the sum of things is both at once—all that is unchangeable and all that is in change" (249d)?

Anyone familiar with the broad impetus of Plato's teachings knows that the dialogue cannot end here, as indeed it does not. Instead, the Stranger turns discussion toward the most basic principles of the universe, the *archaí*, so to speak, out of which all else is produced. What are these? We have already encountered different opinions on the question, but the Stranger chooses to begin with two, rest and motion. These have the advantage of mapping onto the two schools—Parmenidean changelessness or Heraclitean change—while also translating easily into physics.

But the opposites rest and motion cannot by themselves be the whole caboodle. If they were, says the Stranger, then saying that rest and motion *are* would be saying that they are either at rest or in motion, but motion cannot be at rest or rest in motion. (One wonders why can't rest be at rest and motion in motion.) The Stranger therefore posits a third principle, namely, being, or existence (249e–250c). He treats these three principles—existence, rest, and motion—as if they were three different things, *A*, *B*, and *C*, each unchanged by the presence of the others, just as in geometry we can take three points without affecting any of the three.[18]

And yet some commingling there must be, for if motion and rest had no share in being, then motion and rest would not be (251d–252a). But if there is some commingling of the three, some form of participation of the one in the other, then the bottomless problem that has thus far led the dialogue to contradiction threatens us once again. Some more absolutist sleight of hand is necessary: not *all* principles can commingle. Rest and motion, says Theaetetus with the Stranger's approval, cannot mingle without destroying each other, for rest cannot be in motion or motion be at rest (252d).

We call this "sleight of hand" because the prohibition can easily be shown to be arbitrary, a sophistic matter of definitions. Why, for example, can we not say that when some moving object slows down, this is a commingling of rest with motion? And although the prohibition claims to be grounded on an analogy with physics, we need not for that reason accept it as true, even in the physical realm. Following in the steps of the scoffer, we could easily object that in a rotating ball some diameter always remains at rest, or that a mariner is at rest on her boat but not on the water.

But no scoffer emerges at this point in the dialogue. And it is now, after (arbitrarily) declaring that rest cannot mix with motion, that the Stranger uses terms very similar to those we encountered in the *Phaedrus* in order to define the philosopher-dialectician as he who knows what ideas or classes or genera can be mixed, "which kinds are consonant, and which are incompatible with one another," where there are connections and where there are divisions, "dividing according to kinds, not taking the same form for a different one or a different one for the same." (253b–d). This art (says Plato) furnishes a solution to the problems of participation posed without relief in *Parmenides*.

But in order to reach the solution, we need to add sameness and difference to our three principles of being, motion, and rest. Otherwise we can't count those fundamental principles, for in order to count them we must "prove" that each is the *same* as itself and *different* from the others (254d–255b). With these five categories in hand, the Stranger reveals himself not to be a parricide at all. For now he can deny that not being is "the opposite" of being (of which, re-member, "father" Parmenides forbade us to think). What happens, he explains, is rather that difference mingles with being. So "*x* is not *y*" means "*x* is other, or different from, *y*." In other words, we should still avoid speaking of not being as predicate or existential quantifier: we should never say, "*x* is not." Instead we should take not being exclusively in its determinate, relational (or copula) sense and say, "*x* is not *p*." A pious pact with Parmenides about the same-ness of being has been achieved while at the same time allowing for difference in the world.

But note *how* the pact was achieved: by positing a fivefold edi-fice, all of which rests on an assertion of absolute sameness, namely, the assertion that motion and rest are exclusive, that is, they cannot coexist in any being. The possibility of change is derived out of logi-cal first principles of sameness, and those principles will regulate the ways in which difference or change occurs, in the guise of fate, necessity, or law. That is the gist of Plato's pact with Elea.

Creationism: From Physics to Psychology

Plato's physics provides perhaps the best example of the power of this pact. (In fact judging from the importance of motion and rest among Plato's first principles, we might well think that the prin-ciples followed from the physics.) The *Timaeus* sets forth that physics in the form of an account of the creation of the universe and everything in it. Timaeus begins his account with a question that splits the world into the same debate we have been following: "What is that which always is and has no becoming, and what is that which is always becoming and never is?" (27d).

He then proceeds to state all the Parmenidean pieties we have en-countered in our previous dialogues: "that which is apprehended by the intellect and reason is always in the same state, but that which is conceived by opinion with the help of sensation and without reason is always in the process of becoming and perishing and never really is." From this it follows that a divine craftsman (a demiurge; 28a) must have patterned creation on the unchangeable rather than (like an artist) making a copy of something already created.[19] Notice what Plato has achieved by formulating the creation story this way: the choice between sameness and difference is turned into a prerequisite of creation—the creator god must choose—rather than a contingent possibility. Once framed as a necessary choice, an either/or of sameness and difference, Timaeus's task is to explain how, from the creator's unmovable and eternal beginning, this entire world of seeming change was created.

He will achieve that task using nothing but the first principles we have already encountered: being, motion, rest, same, and different, plus all the resources of number that derive from those principles. What resources of number? To begin with, proportion:

> The fairest bond . . . which makes the most complete fusion of itself and the things with which it combines. . . . For whenever in any three numbers, whether cube or square, there is a mean, which is to the last term what the first term is to it, and again, when the mean is to the first term as the last term is to the mean— then the mean becoming first and last, and the first and last both becoming means, they will all of necessity come to be the same, and having become the same with one another will all be one.[20]

Out of proportions God derived water, air, fire, and earth, creating the heaven and the world and "harmonizing them in proportion."

He created (we paraphrase 33–37) the cosmos as a smooth ball, the most self-similar (i.e., symmetrical) shape, made of the four elements, which are by nature in geometric proportion, and made the ball rotate uniformly on the same spot. Then He put the Soul at the center of the cosmic ball, but somehow, even before time was created, He made the Soul prior in birth. This Soul He made of a blend of the Same and the Other or Different, using the force of Being and of geometric proportion to blend them. And in the Soul he fashioned two circular bands, perpendicular to each other: on one band moved the Same and on the other the Different.[21] And He gave sovereignty (*krátos*) to the Same over the Different. Note that the elements of the Soul are the five categories we encountered before, in the *Sophist*: Being, Sameness, Difference, Motion, and Rest. All this creation, thus far, was eternal, since time had still not been created; yet there was uniform motion in it (hence a precise clock). Then He created time, in Timaeus's famous words, as "a moving image of eternity": "and when he set in order the heaven, he made this image eternal but moving according to number, while eternity itself rests in unity, and this image we call time." Once again we find ourselves reading the articles of treaty with Parmenides and Elea (37c–38a). The triumph is total. Time, which may seem to us the ultimate source and example of change and flux, is now cast via mathematical astronomy into the opposite, an example or copy of sameness and eternity.[22]

The Craftsman now generates all that is through number, proportion, and geometric forms. And then the gods made by the Creator get into the game. With the cooperation of Intellect and Necessity, they construct the soul and body of the human being. Unfortunately, incarnation comes with a catch: encasement in a body disturbs the proportional movements of the soul. Hence, the human is born in tumult, irrational, because the appropriate ratios and proportions "were twisted by them in all sorts of ways, and the circles were broken and disordered in every possible manner." This is why human souls are "at first without intelligence." But if the revolutions of the circles are corrected, if their proportions are restored to their natural form, then the individual becomes "a rational being." The goal of education is this restoration of the proper numerical ratios to the human soul, for only by aligning the motions of the soul with those of the universe can human well-being be achieved (43b–44d).

On this account, not only the human body and the human soul but also human virtue, wisdom, and happiness all derive from the physical proportions of the universe expressed in number.[23] Our psychology—and that of the Craftsman as well—is the product of mathematical physics.[24] How seriously did Plato take this mathematics of the soul? Very seriously indeed, judging from the educational programs he advocated in book 7 of the *Republic*, which emphasized "those disciplines that are purely concerned with number" and urged legislation requiring that "those who will share the highest offices in our city to turn to arithmetic . . . until they reach by pure thought the contemplation of the nature of numbers" (525a–c). Math is "the bond between this world and the ideal plane" (527b–c).[25]

In the *Republic* Plato was relatively cautious. He presented math as the link or bond between the world of appearances and that of truth, but he did not equate doing math with the highest participation in the Forms or with the Forms themselves.[26] Elsewhere Plato proclaimed the pedagogical powers of math with more abandon. In book 7 of the *Laws*, for example, Plato's mouthpiece presents the study of mathematics—arithmetic, mensuration, and astronomy—as the indispensable first element of an education without which one is more like a pig than human. "We simply cannot dispense with its character of necessity." "Even God is never to be seen contending against necessity."[27] Hence, the *Laws* propose teaching young children games that make out of math "a good deal of fun and amusement," making Plato perhaps the first advocate of early childhood mathematics education (819b).[28]

The *Epinomis* goes even further. The authenticity of this once popular but now rarely cited dialogue has been debated, in part on the grounds of its extreme enthusiasm for number.[29] Could the great philosopher be so dogmatic, or should we consider this the work of vulgar epigones? The dialogue is indeed dogmatic, but the dogmatism is, as we have seen, not alien to Plato's thought. The opening question is one basic to Plato's writings, indeed, to all philosophy: "what are the studies that will lead a mortal man to wisdom?" The answer on offer from the Athenian (the same interlocutor as in the *Laws*) is quite Platonic, albeit unusually stark: neither the arts nor any techniques of production, but only "knowledge of number," can lead man to wisdom (976e).

In the *Epinomis* we might see an attempt to combine the *Timaeus*'s physics with the psychological anthropology of the *Republic* and the *Laws*. Here number is the source of all knowledge, the difference between man and animal, and the greatest gift of the gods. It was Uranus (a.k.a. Cosmos or Olympus) who, wheeling his stars through the heavens, gave us "the gift of the whole number series, and [with it] he gives us likewise the rest of understanding and all other good things." "This is the greatest boon of all." Through this number within us we can align our soul with the cosmos.

There is a moral theology lurking here, and the *Epinomis* does not hesitate to make it explicit: number "is the source of all good things" and "all that partakes of evil is destitute of all number" (978a–b). The declaration is quite extraordinary. Evil is the absence of number, good its presence and contemplation. And the prize for grasping this teaching is even more extraordinary, nothing less than a wisdom that lives beyond death. He who fully grasps the gift of number "will have but one allotted portion, even as he has reduced the manifold within himself to unity, and in it will be happy, wise, blessed, all in one" (992b). If we moderns are tempted to deny that the *Epinomis* could have flowed from Plato's pen, it is perhaps because here more than in any other dialogue, Plato's theories of sameness explicitly embrace the yearning for the eternal that animates them.[30]

∗   ∗   ∗

We began this chapter by describing the challenge Plato took on: the challenge of maintaining the unchanging and eternal at the foundation of being while at the same time accounting for our experience of constant change. Repeatedly we have heard him suggest that one part of creation, namely, the human, has a choice: either to clamber back toward immortality along ladders of number (Plato's preference) or to fall away in pursuit of images, appearance, poetry, and other deceitful things. The future of every one of these moves is still with us. In saying this, we obviously do not mean that Plato's ideas determine ours or that they were transmitted without change across time. On the contrary, just as Plato arrived at his own ideas about number and knowledge, sameness and difference, in an effort to work through the challenges posed by the arguments of his predecessors, the same is true of Plato's successors, and above all of his most famous student, Aristotle.

Aristotle Wrestles with Plato

Students, if they are great ones, criticize the teachers they have learned most from. Aristotle not only did so but elevated his criticism into an ethical maxim that became famous: "it would seem to be obligatory, especially for the philosopher, to sacrifice even one's closest personal ties in defense of the truth." No less a personage than Don Quixote would adopt the maxim (in its Latin version) to express his own quixotic calling.[31]

Aristotle wrestles with his teacher throughout his *Metaphysics*, a book that begins with a history of Greek knowledge from its origins to his own day. In that history, Plato is presented as a follower of Pythagoras, albeit a follower with his own vocabulary, "for the Pythagoreans say that things exist by imitation of number, and Plato says they exist by participation, changing the name." From Forms to the objects of mathematics to sensible things, Aristotle expounds his teacher's system. But it quickly becomes clear that he does not agree with that system: not with the theory of the Forms or with the role Plato and others assigned to mathematics. Aristotle saw that role as exaggerated: "mathematics has come to be the whole of philosophy for modern thinkers, though they say it should be studied for the sake of other things."[32]

His attitude was not due to math phobia: Aristotle was a well-trained mathematician, and his work shows a much greater awareness of the advanced mathematical techniques of his time than that of his teacher. But Aristotle did not think that either mathematical objects or Plato's Forms could provide the foundations—the principles, or first things (*archaí*) of knowledge—for which both he and Plato were searching.[33]

In order to explain why they could not, Aristotle engaged in extended critiques of both aspects of Plato's teachings, insisting on the limits of mathematics and even of deductive reasoning more generally.[34] As he put it at the opening of his *Nicomachean Ethics*, just as there are many different "arts and sciences, their ends are also many," and they require different methodologies. In political science, for example, one can only speak "roughly and in outline," not least because people do not always react to the same goods in the same way. "It is the mark of an educated person to look in each area for only that degree of accuracy that the nature of the subject permits." "It is evidently equally foolish to accept probable reasoning from a mathematician and to demand from a rhetorician demonstrative proofs."[35]

Such statements suggest skepticism about the unity of knowledge and the search for any foundation of sameness. And yet the yearning for sameness keeps creeping in through the cellar door, so to speak, of Aristotle's thought. We could even accuse him, if we wished, of the foolishness of demanding demonstrative proofs from a rhetorician. That is more or less what he does in his *Rhetoric*, where Aristotle has plenty to say about the use of syllogism and deductive reasoning in rhetorical demonstrations and offers nothing less than a rhetorical theory of *apodeixis* (demonstration). That word is the same one he uses to set the most rigorous standards for formal demonstration in science in the *Posterior Analytics*, as well as (less frequently) in his writings on physics and biology.[36]

It is easy enough to show that his own demonstrations in his writings on rhetoric, physics, and biology do not meet the strict standards for syllogistic and deductive demonstration he sets up in his theory of science. How could they? But that is precisely our point. For Aristotle, the very idea of what demonstration is was already permeated by the procedures that gave mathematical demonstrations their overwhelming power: enunciate starting points considered either axiomatic or undebatable, proceed from those by assertions of sameness or difference and principles of noncontradiction or excluded third, and arrive at a conclusion or discovery that asserts itself as necessary. Aristotle was certainly suspicious of mathematics as ontology, but as methodology, it was powerfully attractive. The debt is already evident in the opening lines of his *Posterior Analytics*: "All teaching and all intellectual learning come about from already existing knowledge. This is evident if we consider it in every case; for the mathematical sciences are acquired in this fashion, and so is each of the other arts. And similarly too with arguments—both deductive and inductive arguments proceed in this way."[37]

But consider this basic question: what constitutes the necessary and indubitable starting point for such demonstrations? That question proved difficult enough to answer for mathematics, let alone for physics, physiology, rhetoric, and all the other domains in which Aristotle worked.[38] To answer it one needs to know what is so certain that it does not require any demonstration and what remains in some way uncertain. Of course it could be that there *are no* necessary starting points: that every possible starting point is itself contingent, based on choices and assumptions. But Aristotle did not entertain that possibility. He thought that there were indemonstrables (we are tempted to call them *archaí*)—primary truths, self-evident, unconditional, and necessary—that could serve as starting points. Moreover, he thought that those starting points were so clear that all educated people should agree about them. As he put it in the *Metaphysics*, "not to know of what things one may demand demonstration, and of what one may not, argues simply want of education. For it is impossible that there should be demonstration of absolutely everything; there would be an infinite regress, so that there would still be no demonstration."[39]

And here is where we find sameness being smuggled back to provide the necessary foundations. We could think of many examples, including Aristotle's theory of the immutable *aether* out of which he posited that heavenly bodies are made: a theory of celestial sameness and sublunary difference that would retain its grip on physics until Newton.[40] But we will limit ourselves to the one he offers here to demonstrate the necessary agreement of the educated, "the most certain principle of all things," "regarding which it is impossible to be mistaken": "it is impossible for anything at the same time to be and not to be." Anyone who doubts this is not only uneducated. They are not human or even one of Plato's "pigs." They are "no better than a mere plant."[41]

In our introduction we called this the Principle of Non-Contradiction, and Aristotle provides a slightly fuller version of that principle here: "The *same* attribute cannot at the *same* time belong and not belong to the *same* subject in the *same* respect; we must presuppose, in the face of dialectical objections, any further qualifications [that is, *samenesses*] which might be added."[42] We've added italics to highlight the sea of sameness, sameness everywhere. It is remarkable that Aristotle, in order to eliminate the possibility of infinite regress, is willing to call for potentially infinite further assumptions of sameness. Just as remarkable is the easy consensus he assumes. Socrates had held that only the godlike know how to choose well between sameness and difference, but for Aristotle it is the common hallmark of every educated man. Even more remarkable is how Aristotle treats those who don't agree with him. We know at whom he is aiming, for he tells us so himself: "it is impossible for anyone to believe the same thing to be and not to be, as some think Heraclitus says; for what a man says he does not necessarily believe." Critique of noncontradiction can count only as hypocrisy, not as philosophy.[43]

There are other "sameness" truths that Aristotle thinks of as before demonstration. These include many that we've already encountered, laws of thought like the Principle of Identity and the Principle of Non-Contradiction or the impenetrability of solid bodies (the impossibility for two solid things to be in the same place).[44] In later chapters we will see why none of these can provide the firm foundations for knowledge, the "necessity" that Plato and Aristotle both were looking for, each in their own ways. But we don't need to explore them further here in order to make our point. That point is not, of course, the silly one that Plato and Aristotle were wrong in their physics, their psychology, their mathematics, or their philosophy. Our point is rather that both shared an impulse within the Greek tradition, an impulse to think about the relationship between the cosmos and the individual in terms of necessary choices, often imagined in mathematical terms, between sameness and difference.

This impulse came to be shared across many domains of human thought—from philosophy and logic to myth, cult, and religion—and across many types of people, from the most learned to the most ignorant, throughout many parts of the world. How did it happen that an idiosyncratic and highly academic approach to knowledge came to be so widely shared across so much of the ancient, medieval, and modern world? Part of the answer lies in the conquests of Alexander the Great—a student of Aristotle's according to tradition—whose armies marched Greek ideas into far-flung provinces of the ancient world. But the arts and sciences have always been specialized activities, then as now. In order to understand how so much of the world came to imagine its fate in terms of a necessary choice between sameness and difference, we need to turn our attention from philosophers and their schools to even more zealous missionaries: the teachers and preachers of monotheism.

4

MONOTHEISM'S
MATH PROBLEM

God keeps strict count of all things.

Qur'an 72:28

It is common nowadays to think of science and religion as opposed. To the contrary, faith and reason are twins born of sameness and difference. They are fraternal rather than identical, at times fratricidal, but twins nevertheless. This was already evident in the previous chapter, which described the challenge of maintaining the unchanging and eternal at the foundation of being while at the same time accounting for our experience of constant change. In its pages we watched Plato try to meet this challenge using rules of logic derived from thinking about numbers in order to generate contradictions that could drive forward his arguments about the nature of being. We saw him use mathematics to separate the world into different strata or levels of reality, from the unchanging and uncaused to the varied flux of our embodied experience. We traced how he related the levels to one another, insisting that even the most mutable things participate in some way—often imagined mathematically—in the eternal. Repeatedly we heard him suggest that one part of creation, namely, the human, has a choice: either to clamber (at least part way) on mathematical ladders toward immortality or to fall away in pursuit of images, appearance, poetry, and other deceitful things.

The future of every one of these moves is still with us, in part because they were not Plato's alone. On the contrary, according to an influential narrative in the history of ideas, he was, like all of the

Greek figures we have encountered thus far, part of a vast movement that stretched across Eurasia and the Mediterranean world
and included teachers like Parshvanatha and Zarathustra; Siddhar-
tha the Buddha and Ezekiel; Confucius and Lao Tzu; Isaiah and
Deutero-Isaiah; Jesus and his apostles; Muhammad and his believers. Some have called this movement "The Great Leap of Being,"
others "The Rise of Transcendental Visions," or "The Age of Criticism," still others "The Axial Age."[1] In the twentieth century the
German philosopher Karl Jaspers, who coined that last term, saw
in this movement "the spiritual foundations of humanity . . . the
foundations upon which humanity still subsists today."[2]

We do not need to agree with particular details of this history of
ideas in order to perceive two teachings that all of these teachers
had in common, two teachings that might seem to be in paradoxical relation to each other. The first is skepticism. We have just seen
thinkers like Socrates and Plato emphasize the immense difficulty
of knowing anything true about the world or about ourselves. They
taught that the things of this world as we perceive them in the flesh,
no matter how solid or certain they seem, provide us at best with
partial, relative, transient glimpses of truths; at worst with deeply
deceiving corruptions. In his *Concluding Unscientific Postscript*, the
modern philosopher Søren Kierkegaard summed up well the lessons of this "Greek skepticism": "There one learns thoroughly . . .
that sensate certainty, to say nothing of historical certainty, is uncertainty, is only an approximation, and that the positive and an immediate relation to it are the negative."[3]

This skepticism toward what we know through the senses is one
of the founding commitments of the transcendental turn. But skepticism alone does not transcendence make; on the contrary, skepticism is as capable of corroding the claims of transcendence as it is of
creating them. Here we come to the second commitment: many of
these teachers subscribed to the further proposition that although
there are no eternal and unchanging truths to be found in physical
things, such truths—Plato's Forms, to stick with an example we've
become familiar with—do exist on a different plane of being, one of
eternal, unchanging truth. Our world merely partakes in, or in some
way imitates, or (more optimistically?) points us toward, that other

transcendent "world" of eternal sameness and unchanging truth.[4] Our task as humans — so these teachers would claim — should be to organize our lives so that we come to participate as much in the eternal as possible. (We will not address a third commitment implied in this "should" and shared by many of these teachers: the missionary commitment to spread their transcendent truths.) In the later words of Jesus, one of the most influential of all teachers, "Do not store up for yourselves treasures on earth, where moths and vermin destroy, and where thieves break in and steal. But store up for yourselves treasures in heaven, where moths and vermin do not destroy, and where thieves do not break in and steal" (Matt. 6:19–21).

The Eternal Sameness of the Soul

It is not at all easy to reconcile these two intimately related commitments. Plato himself had pointed out the "most frightful consequence" of associating absolute good with unchanging sameness: if this is the case then God (who is absolute good and full knowledge) can know nothing of human things, and humans can know nothing about God (*Parmenides*, 134c–135c) The later history of Greek philosophy, Christianity, Islam, and Judaism is to some extent the history of wrestling with that frightful consequence. As Christianity, Islam, and Judaism would later do, Plato's own grappling at times extended to treating the realm of truth as another world, or as a heaven or afterlife. In the *Phaedo* (107a–108c), for example, Plato tells of the treatment of individual souls after death (their reward would be in accord with the level of philosophical knowledge of the truth that they had achieved in their past life). And in the last book of the *Republic*, he offers the myth of Er as an account of reincarnation, explaining that the reason most of us do not remember our previous lives is that only the wisest, the best endowed with *sophrosyne* and self-control, are able to abstain, upon arrival at Hades, from quenching their thirst on the waters of the Lethe, the river of oblivion.

We don't know whether Plato meant these different stories as literal descriptions of the afterlife, though some later readers took them that way. The important point is that he does seem to have

presented the soul (the word he uses in Greek is *psychē*) as participating in some way in eternity and as being in some essential way
ontologically distinct from and less subject to change than the body,
with its senses, passions, and appetites.[5] In other words, in this philosophy it is not just "truth" that is split between seeming and being,
between this physical world and the transcendent. It is also the human who is split, divided between body and soul, between the transient and the eternal, between difference and sameness.

In a moment we will study the consequences of these splits for
religions to come, especially Christianity and Islam. But one could
formulate some of the more fateful consequences without any recourse to god or gods, as Plato's most famous (but rebellious) student did not long after his teacher's death. In book 1 of his *On the
Soul* (*De anima*), Aristotle reminds us of the ancient principle "like
is known by like," which, he says, was the doctrine of Empedocles
and of Plato in *Timaeus*.[6] In keeping with this principle, apathic objects (e.g., mathematical objects) must be thought and known by an
apathic part of the soul or mind, pathic objects by a pathic part. The
apathic part must be always the same as itself and therefore different and isolated from other parts of the soul or mind that are pathic,
that is, that are acted on, changed, or influenced by affects, that feel
or perceive. In a famous but obscure passage, Aristotle offered a
model of the mind (Gk. *nous*) that met these demands by separating
it into two parts, one mutable, the other "separable, apathic and unmixed," "immortal and everlasting."[7] The passage has absorbed the
sages of many centuries and sects. We will not join in that endless
labor of explication except to point out the obvious: these distinctions and divisions are all in the service of determining some part of
the human that participates in the eternal.

Aristotle was looking for an immortal foundation of sameness
within the human. He found it in what he calls the "apathic" intellect. No gods or rivers of oblivion here. Unlike his teacher Plato with
the story of Er, Aristotle refuses explicit assistance from the mythical in his quest for immortality, claiming only the armor of logical
objectivity. But this armor, too, is mythical, built out of many of the
same fantasies we encountered in Plato's *Timaeus* and other dia-

logues and all the more dangerous for having repressed its origins. Chief among those fantasies is that the soul can be understood in the same way, using the same "samenesses," as number and geometry. As Aristotle put it earlier in his treatise, "the facts about the soul are just about the same as those concerning geometrical figures."[8] Polygons can all be put together with triangles or split apart into triangles. As it is with polygons, so it is with the soul, according to the Philosopher, all with the goal of discovering its eternal parts and isolating them from the mutable, the mortal.[9]

From Plato to Moses

The prophets of this transcendent age encountered the same challenges as its philosophers: that of maintaining the unchanging and eternal at the foundation of being while at the same time accounting for our experience of constant change. Often enough they met these challenges in similar ways, and often enough in "conversation" with each other. "What is Plato if not Moses speaking in Attic Greek?" This quip, attributed to an obscure Pythagorean/Platonist philosopher from the second century CE called Numenius, was reported by the great early Christian teacher Clement of Alexandria (ca. 150– ca. 215 CE) and taken up and made famous as a pithy adumbration of the ties between Greek philosophy and Judeo-Christianity by bishop Eusebius of Caesarea (ca. 260–340) in the *Preparation for the Gospel*, his monumental exploration of the relationship of Christ's teaching to the teachings that had come before.[10]

Clement and Eusebius were claiming that the sages of these earlier traditions were working on a common problem, a problem for which Christianity would provide the true answer. As Clement had put it in his own compendium of Greek philosophy and Christian teaching,

Plato the philosopher posits as the goal of life "well-being," and says that this is "becoming like unto God to the extent possible," in this either coinciding somehow or other with the doctrine of the Law [of Moses] (for great natures who are devoid of passions somehow or other hit on the truth, as says the Pythagorean Philo when expounding the works of Moses), or because, as one who was always thirsting for learning, he had been taught it by learned men then living.[11]

How extraordinary to find one of the great early Christian teachers convinced that whatever their place in time, space, or culture, "great natures . . . devoid of passions" (apathic intellects) all reach the same truths. How striking: Plato congruent with Moses. But who is this Philo the Pythagorean expounding the works of Moses in whom Clement finds the same conviction as his own?

"Philo the Pythagorean," better known to history as "Philo the Jew" (Philo Judaeus, ca. 25 BCE–ca. 50 CE), was a prominent citizen of the Egyptian city of Alexandria with connections to the Hasmonean and Herodean dynasties and to the priesthood in Judea. His activities were many, but his great surviving project, stretching some twelve volumes in its bilingual Loeb edition, was to read the Hebrew Bible alongside Greek philosophy.[12] Or more accurately, Philo read the Hebrew Bible in its Greek translation, called the *Septuagint* after the seventy scholars said to have translated it in the third century BCE, under Ptolemy II. The translation was necessary because many Jews living in the Eastern Mediterranean lands conquered by Alexander the Great had adopted the conqueror's Greek as their language. Already here, Philo faced an important problem of sameness and difference: can God's words be translated? Are the revelations of God's teaching in translations of scripture exactly the same as the ones contained in the original or different? After all, there are many ways of translating a given word and many possible ambiguities. Hellenistic Jews traditionally solved the problem by imagining the *Septuagint* to be a miraculous translation. Philo shared that view but expressed the miracle mathematically, invoking the strict sameness available in arithmetic and geometry: "For just as I suppose the things that are proved in geometry and logic do not admit any variety of explanation but the proposition that was set forth from the beginning remains unaltered, in like manner I conceive did these men find words precisely and literally corresponding to the things . . . that it was desired to reveal."[13]

Problems of translation aside, Philo believed that divine scripture and Greek philosophy taught related truths about the nature of God, the human, the world, and the types of "sameness" that mediated between them. And among the most important of these "samenesses" for Philo was mathematics: hence, the sobriquet "Pythagorean." Not that Philo had read Pythagoras, whose ascribed works (if they had ever existed) were nowhere to be found in his day, not even in Alexandria's famous library, except as fragments and alleged citations in the teachings of others. Clement called Philo a Pythagorean because Philo borrowed from Plato's mathematically inflected philosophy, especially as taught in the *Timaeus*, and Clement believed (like many others in the ancient world) that Plato had learned that philosophy from Pythagoras's school.

Philo the Jew was indeed deeply influenced by the *Timaeus* and drew frequently on it and other Platonic works in order to make new sense of the Bible's teachings about the relationship between God, humanity, and the world.[14] In his worries about this relationship, Philo was in good and diverse company. One did not need to be an "Abrahamic" monotheist in order to wonder how primordial sameness could give rise to multiplicity and difference, or how suffering and mortal humans might participate in the eternal.[15] Such worries were widely disseminated among ancient intellectuals of many different stripes and sects, and they were often met with some variant of a by now familiar view. Alexander Polyhistor, for example, sounds a great deal like the *Timaeus* in his *Succession of the Philosophers* (ca. first century BCE):

> The principle [*archē*] of all things is a monad. And from the monad the indefinite dyad exists, like matter for the monad, its cause. From the monad and the indefinite dyad come the numbers, and from the numbers the points. From them come lines, out of which come planar figures. From planes come solid figures, and from these, sense perceptible bodies, from which come the four elements—fire, water, earth, air.[16]

But Greek-speaking believers in the Hebrew Bible had special reason to worry, because although the scriptures do describe God's

act of creation and sometimes attempt to account for man's participation in divinity, those accounts rarely aligned explicitly with the
ideals of Greek philosophy.[17] Philo's genius lay in applying Platonic
solutions to the sharp questions that philosophy was posing to Hebrew Scripture's accounts of God and creation. How to reconcile
the Hebrew idea of a God who creates and intervenes in the world
with the philosophical ideal of the divine as unchanging? And conversely, how can the human soul be created and yet participate in
eternity? In books with titles like *On the Creation* and *Questions and
Answers on Genesis*, Philo deployed many of the Platonic moves we
have already encountered in order to guarantee both the unicity of
God and the incorruptibility of at least part of the human soul. Unfortunately, the treatise *On Numbers* that we know he composed did
not survive antiquity. Had it reached us, we would expect to find it
deploying emanations of number and geometry in order to secure
the philosophical integrity of the God of Abraham, one of the many
ways that math could come to the service of monotheism.[18]

Again, that monotheism did not need to be Abrahamic. In fact,
perhaps the most influential Hellenistic writer on these subjects
was yet another inhabitant of Alexandria, the "neo-Pythagorean"
Nicomachus of Gerasa (ca. 70–150 CE), author of the (lost, except
for fragments) *Theology of Arithmetic* and a (surviving) *Harmonic
Manual* and *Introduction to Arithmetic*. The last of these alone was
translated twice into Latin (first by Apuleius and then by Boethius),
into Syriac in the eight century, twice into Arabic in the ninth, and
into Hebrew in the fourteenth. Nicomachus was a follower of philosophy, not prophecy, and above all of the philosophy of Plato.[19]
But perhaps more explicitly and insistently than Plato ever had,
Nicomachus identified forms and numbers, the divine first principle
and the monad, and made number "the necessary point of departure for every philosophy of the One." As he put it in the *Theology*,

> God and the monad can be compared or assimilated to each
> other, being all things in nature in a seminal way.... [The monad]
> generates itself and is generated from itself, is self-ending, with-
> out beginning, without end, and appears to be the cause of en-
> during.[20]

The Divine Mind and Word, too, are compared to the monad by Nicomachus in his effort to preserve primordial unity in the plurality of creation. Just as the monad remains itself through multiplication, so God remains One and transcendent through creation. *Compared* is perhaps too soft a word. In the *Introduction* we see that he more or less identifies number and the mind of God so that creation becomes a divine exercise in arithmetic:

> The pattern was fixed, like a preliminary sketch, by the domination of number pre-existent in the mind of the world-creating god, number conceptual only and immaterial in every way, but at the same time the true and the eternal essence, so that with reference to it, as to an artistic plan, should be created all these things, time, motion, the heavens, the stars.[21]

From such a point of view, the human study of mathematics is not simply a tool for understanding the world. It is also a ladder between the human and the One, that is, between the multiplicity of creation and the unity of God, between human and divinity.

Christianity: From One to Three

Nicomachus was far from the first representative of this "pagan" philosophical quest for "the One." *Henology* is what this quest is called in the trade, from the Greek word *hen*, "one."[22] It is easy enough to see how henologies like this one might intersect with the monotheistic musings of Jews like Philo or with those of another Jew teaching at more or less the same time as Philo but on the other side of the Sinai. There is no doubt that as they walked the dusty roads from town to town spreading their gospel, Jesus and his followers blazed many new paths into the interpretation of scripture and discovered many new ways of thinking about the world and its creator. But there is also no doubt that some of the same problems that bothered a Greek-speaking reader of Hebrew Scripture like Philo would also bother a Greek-speaking reader of Hebrew Scripture like the apostle Paul, whose writings are the earliest we have from a follower of Jesus.[23]

In those writings God is one: "One lord, one father, one baptism, one God, father of all" (Eph. 4:5–6). He is, as Paul writes in the first chapter of his letter to the Romans, eternal, everlasting. And yet he is certainly also in some form of relation to the mutable world. He has, for example, designated a Son and brought forth creation. Moreover, though God is invisible, man is capable of deducing his existence and his everlasting power through "the mind's understanding of created things." It is only because man has chosen to ignore the creator and prefer the creation "in the likeness of an image" (*en omoiômati eikonos*) that "uncomprehending minds became darkened" (Rom. 1:3–4, 19–25).

In our earlier discussion of Plato's *Parmenides*, we already noted that its explorations of the relationship of images to sameness, being, and eternal truth would eventually intersect with Paul's criticisms of idolatry.[24] Our point here is more general: Paul's formulation of the problem of how to mediate between man and the essential sameness of God has a familiar philosophical feel. We would be foolish to declare Paul a card-carrying member of any particular philosophical school of his age. His surviving writings do not admit such systematic categorization. But neither should we deny that the fortifying flavor of Hellenistic philosophy has, like fluoride in the water, seeped into the concepts with which Paul's letters articulate his theology, anthropology, and cosmology.[25] At the very least we can say with certainty that whatever the specifics of Paul's own commitments, his readers immediately started applying to his and other early Christian teachings many of the philosophical tools for the production of sameness that we have encountered in our previous pages, and not least that of number.

It is in fact rather astonishing how quickly number colonized the new movement. By the time the first catalogs of error were being produced by early Christians like Irenaeus of Lyon (who wrote his influential *Against Heresies* between 170–200 CE), their pages were already densely peopled by the "false" (in Irenaeus's opinion) apostles of arithmetic. Sometimes names are given: Ptolemy, "Epiphanes," and Valentinus were apparently among the more infamous of those who used number in order to explain how a world of multiplicity and change emerged from One absolute and unchanging God and how we mortal creatures could, through number, regain that absolute unity and achieve salvation.[26] Some of the writings of these mathematically inclined followers of Jesus achieved gospel status before being repressed by an emerging Catholic consensus in the fourth century. One, titled *The Gospel of Truth,* reemerged in 1945 among a trove of manuscripts known as the Gnostic gospels, uncovered by Egyptian laborers digging for fertilizer near the town of Nag Hammadi in upper Egypt.[27]

We could wander into these texts and get lost in the elaborate systems (similar to those of the *Timaeus*) that they developed in order to get from "monad" to "tetrad" to "ogdoad" and "pleroma" and on to all created things. But we do not need to stray beyond bastions of Christian orthodoxy in order to find number being deployed to unite the pathic human to an apathic God. On the contrary, we could probably demonstrate, had we world enough and time, that each significant development in early Christian theology was capable of stimulating a numerical defense. The sharpening fourth century debate over the Trinity provides a good example. Defending the triune God of the Nicene Creed against the Arians circa 360 CE, the elderly Marius Victorinus—translator of Plotinus and late convert to Christianity—did not hesitate to invoke the by now familiar logic of the monad: "God is the One that generates the monad from himself and reflects love in his unity. So too also in the Many: each and every unity has its own number since it is reflected by others beyond difference." Thus does the One become three without compromise of oneness, a position that would be taken up some thirty years later (in 397) by Aurelius Augustinus, known to posterity as St. Augustine of Hippo, in the explication of the Trinity he undertook in book 1 of *On Christian Doctrine.*[28]

We will linger a few pages on Saint Augustine (354–430 CE) because he provides an excellent example of the virtues and perils of number for the Christian. Crowned in his own lifetime a champion of the Catholic consensus, he began as something of an "arithmologist." One might even label "Pythagorean" his early treatises *On Order, On the Immortality of the Soul* (387), *On the Quantity of the*

Soul (387–88), and *On Music* (391). Even the slightly later *On the Free Choice of the Will* (391–95) does not shrink from invoking human numeracy as proof for the existence of God.[29]

The logic goes something like this. The intelligible structure (*ratio*) of number imposes itself on our minds and to all minds in exactly the same way independent of the senses and of any individual perspective. These truths are common to all, immovable, unchangeable, and eternal. "Seven and three are ten not only at the moment, but always; it never was and never will be the case that seven and three are not ten. I therefore declare that this incorruptible numerical truth is common to me and to any reasoning being."[30] However, he proceeds, like all things in the world our minds are impermanent, always in motion, and incapable of comprehending anything eternal by themselves. Ideas of number must therefore be implanted or facilitated within them by a being that is as eternal and unchangeable as number. "Now you had conceded," Augustine concludes triumphantly, "that if I were to show you something above our minds you would admit it to be God." Number exists and is eternal, we are mutable and yet capable of thinking eternal number, therefore something eternal must have imparted it to our minds, and therefore eternal God exists.[31]

Reading this treatise we might almost confuse Augustine for a "monadic monotheist" like Nicomachus. He insists that although true oneness cannot be found in any *physical* object, all *numbers* derive from the one. He sees number and wisdom as linked and is even willing to claim that they are the same:

> Enough of wisdom's being found inferior in comparison with number! They are the same.... Even if we cannot be clear whether number is in wisdom or derives from wisdom, or whether wisdom itself derives from number or is in number, or whether each can be shown to be the name of a single thing, it is certainly evident that each is true, and unchangeably true. (2.11.32)[32]

Should we speak then of "Augustine the Pythagorean"? Augustine was aware of the danger. In his *Soliloquies* he posed the question explicitly: how is the knowledge of geometric objects—intelligibles that are, like God, the product of a priori knowledge and not of the senses—different from the knowledge of God? The answer he provided there was not very convincing: knowledge of geometry is like knowledge of the earth, knowledge of God is like knowledge of the sky, and the latter ranks higher than the former.[33] As he did over other difficult questions, over the course of his writings Augustine changed his mind about the place of number. Just a few years after equating wisdom and number in *On the Free Choice of the Will*, in his *Confessions* of 397–401 CE he thanks God for having delivered him from the fraudulent divinations of the "*mathematicians*," with their overweening confidence in reason.[34]

Augustine seems now to see the risk of a competition between number and Christ, between *arithmós* and "Logos," as mediators between the eternal and the created.[35] Confronting that choice, he prefers Word over number. But this does not mean he escapes the dichotomization of sameness and difference he has imbibed along with so much philosophy. Even in his *Confessions* he cannot resist availing himself of the certainties of number, convinced that there is none greater available to mortal souls. If only humans could reach the same certainties about God and the unseen as they can about $7 + 3 = 10$.[36] And he remained convinced that the "being" of number is other than that of things in the world. Like God, number is not changing or "becoming"; number truly *is*.[37]

Like God. And therein lies a world of theological difficulty. For the more we treat number as similar to God, the more number threatens to limit or constrain the omnipotence of divinity. An older Augustine confronted the problem in his attempt to explain the biblical account of creation in his great explication of the "literal" meaning of Genesis, *De genesi ad litteram* (written from 401 to 415). Like Philo, Augustine felt the pull of number theory when talking of the creation. Why six days? Because six is the first perfect number. How do we get from the one to the many? The monad's mediation. And so forth. But then the problem rears up: does all this application of mathematical knowledge to the divine mean that number (as well as measure and weight) were in God before creation? Does God have number (and weight and measure)? Do these tell us something about God's nature and thereby limit that nature?

Augustine looks for compromises—God is *numerus* but also *sine numero*, "number without number," in the felicitous formulation of one modern commentator—but in the end attempts replacement. Where in book 4 of the work he was still deploying neo-Pythagorean theology to link God to creation, by book 6 he has adopted the language of "eternal reasons," *rationes aeternae*, rather than arithmetic. We might call this an evasion, rather than a resolution, of the competition between number and God. But it was an evasion that seems to have satisfied St. Augustine and many (though far from all) of the Christian theologians of the future.[38]

"He Is God, the One"

Not many of those future Christian thinkers would come from Augustine's Hippo, because in the seventh century, North Africa would fall to the followers of another prophet of the One God, Muhammad of Arabia. According to his traditional biographers, Muhammad received through the archangel Gabriel the first of the revelations that came to be called the Qur'an, circa the year 610 CE, while living in the city of Mecca. Those revelations present themselves as very much in continuity with the teachings of earlier Jewish and Christian prophets (as well as earlier prophets to the Arabs), all of whom make numerous appearances in the Qur'an and in other early Islamic teachings and traditions. There is a great deal to be said about the similarities (and the differences) between these various prophetic teachings, but for our purposes in this chapter, two are especially important.

First, Muhammad's God—the God of Jesus, Mary, Joseph, Moses, and Abraham—was very much a creator God. In fact, that very first revelation from Gabriel to Muhammad was, according to tradition, about God's creation of man:

> Proclaim! In the name of your Lord and Cherisher, who created—
> Created man out of a leech-like clot of congealed blood.
> (Qur'an 96:1–2)

And second, the Qur'an's God was emphatically and resolutely one:

1. *qul huwa llāhu aḥadun* Say: He is God, one,
2. *allāhu al-ṣamadu* God, the Self-sufficient.
3. *lam yalid wa-lam yūlad* He neither begets nor is begotten,
4. *wa-lam yakun lahu* And there is no one that is a likeness
    kufuwan aḥadun to him.[39]

Thus runs in its entirety the shortest of the Qur'an's chapters, sūra 112, titled *al-Ikhlāṣ* (The Purity), a chapter understood by many Muslims past and present as a statement of complete and unadulterated devotion. It would be difficult to find in any of the scriptures of the Abrahamic faiths a more crystalline statement of divine sameness, of oneness admitting no difference or change. Small wonder that one future Muslim monarch, exasperated by the theological disagreements of the learned in his court, is said to have cut short all debate by violently proclaiming, "the foundations of the faith are Chapter 112 of the Qur'an and this sword!"[40]

We do not know if the earliest community of believers felt that these two teachings—of an active God and an absolutely same one—needed in any way to be reconciled. But we do know that as Islam rapidly expanded beyond the Arabian Peninsula and conquered vast parts of the Greco-Roman and Persian world, many Muslims took up the question we've encountered in the previous pages and in the process transformed the possibilities of thought not only for Muslims but also for Christians, Jews, and indeed, for the future of philosophy. To find such questions we need look no further than the commentaries on this lapidary sūra, which for all its seeming clarity posed many questions.

Especially problematic was the word *ṣamad*, translated above as "self-sufficient," a unique occurrence of the word in the Qur'an. This unique epithet eventually "came to be regarded as the most elevated attribute" of God, but its precise meaning was much debated in early and classical Islam. Does it mean "the highest authority"? Or perhaps "made of solid beaten metal," as the earliest Greek translators had it? The great early tenth-century commentator al-Ṭabarī opted for "solid," which had many variants: "one that is not hollow," "without a body cavity," or "one who does not eat or drink."[41] A few centuries later Fakhr al-Dīn al-Rāzī (1149–1209) collated some of these possibilities and offered still others: "some later philologists," he explained "said that *al-ṣamad* is smooth stone upon which no dust settles and which nothing penetrates or comes forth from." One should not literally attribute a body to God, he commented, but understood "figuratively" (*majāz*) it was indeed valid "to say that the body that is like that smooth stone is not susceptible to reaction with, or influence by, any other, which is then an allusion to God and His being the necessary existent by virtue of His essence, unsusceptible to change either in His existence or in His permanence or in terms of all His attributes."

As so often in the long history of human thought, we find pebbles deployed as a model for the apathic. But as al-Rāzī suggests, even the most lithic and unchangeable of bodies are problematic as attributes of transcendent divinity. Less dangerous, at first blush, is number, and above all "the one." Here, too, however, there were questions. Why, for example, does the verse call God the one (*al-aḥad*) rather than the number one (*wāḥid*)? Abū ʿAlī al-Ṭabarsī (1073–1153) was a Twelver Shīʿī, but his response to this question is representative of Sunnī views as well: "It is said that He said *aḥad* and not *wāḥid*, because "one" is used in arithmetic to which another [numeral] may be added. But "the one" cannot be split into parts or subdivided, neither in terms of essence nor in the significance of His attributes . . . and so it is the more excellent."

It is easy enough to see why "the one" should be so attractive to a competitively monotheistic theology: competitive, in this case, among numerous Islamic sects as well as against various forms of Judaism, Christianity, Zoroastrianism, and multiple polytheisms. But number could be put to the work of addressing other theological problems in Islam as well, such as the thorny one we have already encountered of bridging the gap between unchangeable God and mutable divine creation. Here Muslim thinkers took up the work of the Hellenistic philosophers and went far beyond them, both in championing the claims of number and in criticizing them.

The very first chapter of the first article of the early "encyclopedia" known as *The Epistles of the Brethren of Purity*—provides a good starting point:

> It is the method of our noble Brethren, may they be aided by God, to study all the sciences of existent beings that are in this world . . . , and to inquire into their principles and the quantity of their species, kinds, and properties, and into their arrangement and order, as they are at present, as well as into the process of their originating and growing out of one cause and of one origin, by one Creator, may His loftiness be exalted! They rely in demonstrating these [things] on numerical analogies and geometric proofs, similar to what the Pythagoreans used to do.

This compendium of all sciences was composed (we are told) by a mysterious group of scholars working in the cities of Basra and Baghdad at some point in the fourth Islamic century (ca. 900 CE). The Brethren's quest was to find the "veracity in every religion" as well as in the pursuits of reason. They did not distinguish between the two as sharply as a modern researcher might. For them the pagans Pythagoras and Nicomachus of Gerasa (on whose *Introduction to Arithmetic* they heavily relied) were "wise monotheistic [*muwaḥḥidūn*] philosophers" involved in the same science as theirs, the search for "knowledge of the essence of the soul." They considered much of this knowledge accessible to reason, even before revelation: "When those philosophers who used to discuss the science of the soul before the descent of the Qur'ān, the New Testament [Gospels], and the Torah, inquired into the science of the soul with the natural talents of their pure minds, they deduced the knowledge of its essence by the conclusions of their reasoning." (Compare the brief history of knowledge provided by the apostle Paul in chapter 1 of his epistle to the Romans.)[42]

Number seemed to the Brethren a powerful path between the One and the multiplicity of creation. Like their Pythagorean, Platonist, and Christian predecessors, the Brethren demonstrated that the whole numbers can all be generated from the one: "the whole numbers are generated by increasing them one by one, ad infinitum." Therefore "plurality is an aggregate of ones." From here we can create a world of number. But the Brethren (like their predecessors) did not stop there: "Know, O brother, may God aid you and us with a spirit from Him, that the relation of the Creator, exalted be His praise, to all existent beings is like the relationship of unit one [to the other numbers]." For them, the logical moves for producing number provided a strong analogy for monotheistic creation: the one, "the Creator, exalted is His name," simply counted, and the first thing he "invented and innovated from the light of His unity was a simple substance called 'Active Intellect,' as He made two arise from one, by repetition. Then He made the Universal Soul arise from the light of the [Universal] Intellect, as He made three from the adding of one to two. Then He fashioned Prime Matter from the movement of the [Universal] Soul, as He generated four by adding one to three," and so forth.[43]

But what about the reverse? Could number help the human soul return to its creator? Is there a way to count backward toward eternity, so to speak? Here the Brethren are more ambivalent. On the one hand, the fact that they start their compendium with number suggests optimism. Like Augustine and so many others, they see in number a universal form of reason. Hence, in chapter 24 of epistle 1—temptingly titled "On the Science of Number and of the Soul"—they explain that they begin with number "because this science is potentially embedded in everyone," and from it "one takes examples for everything else that can be known." For this reason the science of number comes first in their system of knowledge (as in so many others), and then come the other sciences, all of which have the same ultimate goal: true knowledge of God. But though the Brethren do use number to conclude about the eternity of the soul, they do not go so far as to pave with mathematics a path toward immortality.[44]

The Brethren were not the best mathematicians or the most acute philosophers of their age. Their compendium contains little that is "cutting edge" in either department. Their attitude toward number is valuable not because it is particularly clear or well articulated but because it points to an ambivalence about the powers of "the one" and its sameness that became very sharp in these centuries of Islam, centuries in which terrific progress in various branches of mathematics, logic, and philosophy coexisted with radical articulations of both the possibilities and the limits of these sciences for the understanding and flourishing of the human soul.

Coherence or Incoherence?

One way of understanding this struggle: do the same rules of thought that apply to arithmetic, geometry, physics, and other rational sciences also apply to the soul and to God?[45] Can the study of the soul—and even of God—be put on the same stable footing as the study of the more predictable aspects of our physical world? Down that path lay a consistent science but also grave danger. For if these rules bound not only the world and the soul but also God, then how can we say that God is omnipotent? Could the Creator not have created the world other than it is? Was it impossible for God, for example, to create a world in which the Principle of Identity did not hold? If God cannot suddenly make two plus two equal five or prevent cotton from burning in proximity to flame, then what does "omnipotent" mean?[46]

Muslim scholars inherited these and many other questions about the relationship between God and reason from their predecessors and neighbors. Hence, the many debates over number and its role in leading the human soul toward the Divine that we saw in the writings of the Christian Church Fathers, for example, also claimed the attention of some of the greatest thinkers of the "classical age" of Islam.[47] Several early attempts to produce an Islamic-(neo-)Platonic-(neo-)Aristotelian science of the soul have reached us in a form approaching integrity, including those of Abū Yūsuf Ya'qūb ibn Ishāq al-Kindī (d. ca. 870), who came to be known in the Islamic tradition as "the philosopher of the Arabs"; Abū Bakr Muhammad ibn Zakariyya ibn Yahya al-Rāzī (d. ca. 925 or 932), or Rhazes, as he came to be known in Latin Europe; and Muhammad b. Muhammad Abū Nasr al-Fārābī (ca. 870–950), called "the Second Master" in the Islamic philosophical tradition (second, that is, after Aristotle).[48]

For the sake of brevity, we must let the Second Master's science of the soul stand for all. That psychology emerges in such works as

The Principles and Opinions of the People of the Virtuous City, the *Enumeration of the Sciences*, the *Treatise on the Intellect*, his *Book on the Conditions of Certitude*, and other like titles. In these we learn of that part (or faculty) of the soul that provides "the necessary, universal premises that are principles of the sciences." These principles are the usual Aristotelian and Euclidean suspects, "common notions" about sameness, such as "amounts equal to a single amount are mutually equal."[49] Like Aristotle, al-Fārābī seeks to demonstrate how, from these faculties and principles, certain knowledge of things that can be known is attained. This demonstration is in the service of an even higher goal: illuminating in both philosophical and Islamic terms Aristotle's obscure teaching about the immortality of the intellect in *On the Soul*. Across his oeuvre, in slightly varying ways, the Second Master explains how through knowledge gained by application of the sciences the individual intellect comes ever closer to—though never quite becomes exactly the same as—the active intellect, until at last it achieves human perfection and happiness (*saʿādah*) by becoming immaterial, incorporeal, existent in the afterlife (*ḥayāt al-ākhirah*).[50]

Al-Fārābī has served the recent history of ideas as a pioneer in the exploration of the tensions between divine revelation and the axioms of reason, a founding example of engagement with the dangers that that polities governed by God's law pose to the practice of philosophy. But from the perspective of our topic, it is less the tension between reason and revelation that is striking than the synergy. It is through an eternal sameness at the foundations of reason that al-Fārābī's human approaches the eternal sameness of the One God. Al-Fārābī combined a (neo-)Aristotelian model of the sciences and the soul with a (neo-)Platonic theory of hierarchical emanation (*fayḍ*, in al-Fārābī's Arabic)—First Existent, second intellects, active intellect, souls, form, matter—in order to preserve the apathic aspects of both God and the human. For the human, the foundations of reason contained within the soul provide a certain path to eternity. Meanwhile, al-Fārābī's God (or First Existent) remains one, unique, indivisible, perfect, without partner, opposite, or multiplicity, without purpose external to itself, uncaused and everlasting, suffering no addition or subtraction to its own perfection and oneness, even while serving as First Cause.[51]

Notice the balancing act. Through universal faculties of basic sameness and difference in the soul, the rational part of the soul comes to know eternal truths (recall Aristotle's "like is known by like"), and the more of these it learns, the more it becomes one and the same thing with them, becoming itself eternal.[52] What a happy progression! But the power of that sameness cannot progress forever without threatening blasphemy. Whatever eternal sameness the human soul can achieve cannot be or become the same as the unique sameness of the One God, the First Existent.

The challenge for a philosophically inclined Muslim looked something like this: how to maintain the powers of rational principles of sameness and oneness and the Identity Principle while at the same time maintaining the unique sameness and "oneness" of God? In other words, how to distinguish between different types of sameness and oneness? If we seek to discover in the human the truths that ancient philosophers had identified as eternal without encroaching on the divine truths contained in sūra 112 and many other passages of revelation, then this becomes an urgent question.

The greatest minds of Islam debated possible solutions, or whether solution was even possible. Just a generation after al-Fārābī (in 981 CE) Abū Sulaymān Al-Sijistānī put it pithily when asked about "the one." "One," he responded, "is an equivocal term having many senses."[53] He summarized a number of these senses, wondering which is most appropriate to the One God, the Primal Being. But no existing idea of the one can quite escape the dangerous idea of some common substrate capable of uniting us to God. Perhaps, he suggests, we should refer to the Primal Substance (*al-dhāt al-ūlā*) as abstract unity (*al-waḥda al-mujarrada*), a oneness not present in the soul nor in the substance of any existing entity.[54] We are not so far from Augustine's God as number without number.

Al-Sijistānī serves us by his conciseness, but far more influential was a philosopher of the subsequent generation, Ibn Sīnā (980–1037), or Avicenna, as he came to be known in Latin Europe.[55] Working in what is today Uzbekistan and central Iran, Ibn Sīnā built a comprehensive system of thought intended to illuminate all major philosophical subjects of his day—the natural sciences, epistemology, ontology, logic—and also all major religious ones—the nature of prophecy; of divination, miracles, and wondrous deeds; of life after death; and of God—all in terms of philosophical reason and its sciences.

Ibn Sīnā wrote hundreds of works of which many survive, but we need only peruse his commentary on sūra 112 in order to understand the approach.[56] Commenting on the first line—"Say, 'He is God, unique'"—he reminds us of the problem:

> Whenever one explains this being by the concomitant of divinity, a consequence of this is that He is unique [al-ahad], which is the utmost extreme of oneness [al-ghāyatu fi'l-waḥdāniyyati] There is resemblance in it, though it is the furthest of the utmost extremes of unity [fi'l-waḥdati].

From God's extreme unity, total sameness, and utter simplicity, plus one neo-Platonic, neo-Aristotelian (and very Avicennian) principle—ab uno simplici non est nisi unum (from one perfectly simple cause there can be only one effect)—Ibn Sīnā will generate a fully causal cosmos in which even the all-powerful God is bound by the rational laws of science.

Those of you eager to see how he does so can turn to other volumes in his vast oeuvre, especially his magnum opus, *The Metaphysics of "The Healing,"* with its magisterial review of the philosophical tradition on the nature of "identity" "number," and "mathematical objects," all in the service of articulating a systematic philosophy capable of teaching us necessary truths even about the One God. There you will find serried ranks of enticingly titled chapters: "On discussing the one"; "On ascertaining the one and the many and showing that number is an accident"; "On ascertaining the nature of number, defining its species, and showing its beginnings"; "On the opposition of the one and the many"; "On the characteristics of unity by way of identity and its divisions, and the characteristics of multiplicity by way of otherness, difference, and the well-known kinds of opposition;" "On relating the doctrine of the ancient philosophers regard-

ing the exemplars and principles of mathematics and the reason calling for this; revealing the origin of the ignorance that befell them, by
reason of which they deviated [from the truth]."⁵⁷ But the gist of it all
you will already have found in his brief commentary on the Qur'an's
briefest chapter. For according to Avicenna, sūra 112 contains within
itself revelation's own teaching about how God is conducive to the
sciences of reason and the sciences of reason to God:

> The furthest goal in the quest for the sciences in their entirety is
> the grasp of God's essence—may He be exalted—His attributes,
> and the modality of the emergence of acts from Him. This *sūra* in-
> dicates the manner of [their] exhibition and the sign concerning
> all that concerns God's essence—may He be exalted. Surely it is
> equal to a third of the Qur'ān.

How fitting that this commentary on the One concludes with a numerical statement about the proportional value of its revelation!

Avicenna famously represents those Islamic philosophers who
sought to build a system capable of consistently and rationally em-
bracing all that is, from God to the human and the physical world:
to bind all that is with the same laws of thought and principles of
sameness. But this movement was only one of the various powerful streams in Islam. There were other schools—just as powerful
and deeply rooted in philosophy and Islamic thought—that sought
to preserve the oneness and omnipotence of God by insisting on
the contingency of creation, including even the contingency of the
choice between sameness and difference.

According to the more radical thinkers of these "occasionalist"
schools—as they are often called by historians of ideas—there is no
absolutely necessary causality or continuity in creation except by
God's ongoing choice. God re-creates the cosmos in every instant
and could choose to do so differently with every re-creation. Should
God so choose, the properties (which they called "accidents") of
stone that make it sink could instantly change to properties that
make it float or fly. If there are any "laws" of physics or of thought,
any apparent causalities and continuities, it is because God generally makes that choice. But God could choose—and sometimes

does, for example in miracles—to deviate from or even shatter those habits. Since God determines the properties of every thing in every instant—"Not even the weight of an atom in the heavens or the earth escapes his knowledge" (Qur'an 10:61)—God could at any moment create any and every aspect of the universe differently.[58]

One of the many achievements of medieval Islamic theology and philosophy was to confront these two broad streams of thought with each other in a way that preserved, but also transformed and limited, the powers of each. The great Abū Ḥāmid al-Ghazālī, deeply learned in both the "occasionalist" and philosophical or "Avicennan" traditions, is the most famous impresario of this achievement. His *Incoherence of the Philosophers*, published in the year 1095, has often been treated as the definitive attack on philosophical thought in Islam, but it might just as well be understood as a defense of philosophy that proceeds by demarcating its limits. We could even argue that he intended his work as a nonaggression pact between reason and revelation, a treaty dividing up the world of physics and metaphysics, cosmology and ontology, into respective spheres of influence for divine omnipotence and for human logical knowledge of causality, delimiting the questions and forms of knowledge appropriate to each.[59]

A nonaggression pact was necessary because the tension between human and divine powers of "sameness" was potentially explosive in Islam as in Christianity. As an example of such explosions, consider the Almohad movement that sent Berber armies sweeping across North Africa and the Iberian Peninsula in the twelfth century. According to a later (and very improbable) tradition, the revolution's leader Ibn Tūmart (ca. 1080–1130) was moved to revolt by a rumor that authorities had burned one of al-Ghazālī books. Whether or not the tradition is true, the very name of that movement—Almohad derives from Arabic *al-Muwaḥḥidūn*, "those professing the unicity of God"—suggests that it was animated by the kinds of questions that have preoccupied us throughout this chapter.

The credo (*'Aqīda*) attributed to Ibn Tūmart reads from its opening line like a manual for radical sameness: "On the Merit of Unification and Its Necessity and the First Thing Needed for Its Attainment." The Almohads' military prowess and accompanying massacres—their bloody purges of other Islamic schools and their forced conversion of Jews and Christians in the regions they over-powered—earned them a historical reputation as fanatics. But the gap they were trying to bridge—between the one of divinity and the many of the human soul and world—was a gap common to mono-theism, and their sometimes murderous solutions were philosophi-cally quite up to date.

It was under the Almohad caliphs of the western Mediterranean that the last of the Muslim philosophers we will mention carried out their work on the relation between the human soul and the divine one. The most famous of these is Ibn Rushd of Cordoba (d. 1198), or Averroes, as he was called in the Latin West, where his fame was suf-ficient to earn him a distinguished place in hell (namely, in the high-est circle of Dante's *Inferno*). We know that we should at the very least discuss in some detail the assumptions about sameness, iden-tity, and number that Averroes deploys in his (no less than three) commentaries on Aristotle's *On the Soul* and describe his differences with al-Fārābī, Ibn Sīnā, and al-Ghazālī.[60] But many are the quar-rels of philosophers and time is short. So we will neglect Averroes and conclude this chapter with his predecessor, patron, and protec-tor, the polymathic physician Ibn Ṭufayl (ca. 1105–1185), whose ap-proach to monotheism's problem of sameness and difference seems to us more engaging and also more pointed and profound, at least when it comes to challenging what we think we know about same-ness and difference.[61]

Ibn Ṭufayl was not only the caliph's physician but also his con-fidant and his vizier or minister, deeply involved in the business of governance, war, diplomacy, and administration. "The Commander of the Faithful . . . loved him so well that he stayed with him in the palace night and day, not coming out for days at a time." Perhaps this intensity of engagement with the business of human society contributed to Ibn Ṭufayl's choice of setting for his philosophical tale: a desert island without human society at all. The tale came to be known in Latin as the *Philosophus autodidacticus*, "The self-taught philosopher," and in Arabic by the name of its protagonist, *Ḥayy ibn*

Yaqzān (roughly, "Life, son of Awareness"), a baby brought up not by humans and not by wolves (as in Virgil or Kipling) but by a gentle and philanthropic doe.[62]

The tale is as readable as the adventures of Robinson Crusoe (which it may have influenced) and as deep as you wish. We urge you to peruse its (short) pages. If you do, you will notice that by the time of our baby's twenty-first birthday, he has reached an admirably relative understanding of the importance of sameness and difference:

> Ḥayy considered all objects in the world of generation and decay ... [and] saw that while physical things differed in some respects they were alike in others, and after some study and thought he concluded that inasmuch as things differ they are many, but inasmuch as they correspond they are one. At times he would concentrate on the peculiarities that differentiate things from each other, and then things seemed to be manifold and beyond number. . . . But looking at it from the opposite point of view, he realized that . . . all could be said to be one.

So much for the physical world. But of course Ḥayy keeps learning. By the time he is thirty-five he has established for himself that the world must have a cause—the Necessary Existent—that transcends physical being in every respect and can therefore not be apprehended through the senses but only by something eternal within Ḥayy himself. If a mortal can achieve such knowledge before death, Ḥayy speculates, then surely his eternal part, "on leaving the body," "will live on in infinite joy, bliss, and delight, happiness unbroken."[63]

By his forty-ninth year, Ḥayy does indeed reach a vision of the divine world. We won't tell you how—for tips on eternity you must read the text yourself—but we will simply emphasize what he learns: that sameness and difference, one and many, cannot bind the created and the divine. This knowledge does not come easily. At first it was only by thinking in traditional terms of "identity" that Ḥayy could imagine access to God. But it was just as he was about to fall into that trap, whose power we have witnessed throughout this chapter, that "God in His mercy caught hold of him and guided him back to the truth." That truth is a shocking one:

> "Many," "few," and "one"; "singularity" and "plurality"; "union" and "discreteness," are all predicates applicable only to bodies. But those nonmaterial beings who know the Truth, glorified and exalted be He, precisely because they are free of matter, need not be said to be either one or many.[64]

Ibn Ṭufayl wants us to realize just how radical this lesson is, so he has a talking bat fly into the narrative, "its eyes blinded by the sun," to interrupt him with cries of outrage: "This time your hair-splitting [*tadqīq*] has gone too far. You have shed what the intelligent know by instinct and abandoned the rule of reason. It is an axiom of reason [*min aḥkām al-ʿaql*] that a thing must be either one or many!"[65]

The bat serves Ibn Ṭufayl (as it did al-Ghazālī before him) as a representation of the Aristotelian philosopher, whose eyesight is too weak to tolerate the brighter light of truth.[66] The "axiom of reason" it clings to is one we have found often enough throughout these chapters, a mishmash of the Principles of Identity, of Non-Contradiction, and of the Excluded Third into a "law of thought." Ibn Ṭufayl's response is stunning:

> [The bat] says I "left what every sound mind is born with and abandoned the rule of reason." I shall grant him that. I have left him and his reason and his "sound minds." What he means by reason—he and his ilk—is no more than the power to articulate, to abstract a general concept from a number of sensory particulars, and his "men of sound reason" are simply those whose minds work the same way.

Number, logic, the laws of thought: these are not absolute but relative, dependent on perspectives or habits and workings of mind that may not be universally shared among men, and certainly not between men and God. Divine beings can never be said to be one or many in any of the senses derived from our human sciences of sameness and difference. Those sciences can never bridge the gap between the divine world and the sensory one, "for the two cannot be joined in one state of being—like two wives: if you make one happy, you make the other miserable."[67]

Modern Faith in Number

From Ibn Ṭufayl's many teachings we would extract only two: (1) "same" or "different," "one" or "many," are relative to the question you are trying to answer, the perspective you take, the shape of your soul and experience; and (2) number is not a bridge between the oneness of things, of the soul, and of God. We highlight these lessons not because we take them to be true (we are apostles of the first, while on the second we remain agnostic) but because they fly so directly in the face of dogmas widespread in both faith and reason, in Christianity and Islam as well as in the philosophical and scientific cultures those faith traditions produced.

Of course Ibn Ṭufayl's critique was not the only or the last attempt to escape the difficulties we've been charting within Judaism, Christianity, or Islam. We have not even touched on some of the greatest of those efforts, such as Islamic Sufism, Jewish Kabbalah, and Christian negative theologies. The difficulties remain, and so do marriages of mysticism and mathematics.

Consider the still fashionable philosopher, political activist, and mystic Simone Weil (1909–1943). Weil was by all accounts an extraordinary figure. Her commitments could be ferocious and critical simultaneously. In 1933 she worked to save communists fleeing Nazi Germany while at the same time writing articles attacking the Soviet government. In 1936 she attempted repeatedly to enlist in desperate intelligence and combat missions on the Republican side of the Spanish Civil War, attempts frustrated by commanders reluctant to assume responsibility for an untrained and frail volunteer. Eventually she joined a French-speaking combat unit, but her service ended in October with an accident involving a cooking fire and her evacuation from Spain by her family.[68] It is at this point that Simone began the intensive dialogue with Catholic spirituality for which she is today best known. Her last years were spent in exile from Nazi-occupied France. Her writings from those years—on love, on God, and on the relation between human and divine—began to attract widespread attention after the war. She herself, however, died in an English sanatorium in 1943 at the age of thirty-four. The cause of her death is debated—tuberculosis? self-starvation?—but it quickly came to be seen by some as an act of love and an imitation of Christ.

Simone was not the only extraordinary figure in her Parisian Jewish family. She was one of two siblings. Her brother André (1906–1998) became a famous mathematician, a founder and guiding light of a group who, under the pseudonym "Nicolas Bourbaki," undertook collectively to write an encyclopedic account of math. Their goal was a sort of Euclid's *Elements* for the modern age, though unlike Euclid its logic would be impeccable, its standards of rigor unassailable (no legitimizing role for intuition, no pictures), and its level of generality as high as could be conceived. Simone appears in early photos of the group, the only woman.

Simone was not a mathematician, but she was very much a Pythagorean, not least in her insistence that mathematics, and specifically Greek mathematics, provides a form of relation to God. "There are texts that show with certainty that Greek geometry originated in religious thought: and it is quite apparent that it is a kind of thought almost identical to Christianity."[69] Consider, for example, how she approaches the important question of what might possibly *mediate* between mutable humanity and eternal God. With Christ in mind, Weil creates a numerical analogy. Suppose we have two different numbers and we want to define a "mediator" or "mean" between those two. There are many (indeed infinitely many) ways in which a number can be "in between" two other numbers. The ancient Greek mathematicians were particularly interested in three of these: the arithmetic, geometric, and harmonic means.[70] Weil focuses on the geometric mean. We read $h/a = b/h$ as "h (the geometric mean) is in the *same* proportion to a as b is to h," or, more briefly, "h is to a as b is to h." And she wants to say that the *same* is true when a is man, b is God, and h is Christ: we are to Christ as Christ is to God. Christ is the mediator, the geometric mean, between humankind and the Creator.

The general analogy has many historical precedents, including

Plato's characterization in the Symposium of the god Eros as the *metaxu*, the middle or in-between. Mathematically, however, it is arbitrary. Why the geometric mean instead of any of the many others? Even if we are willing to set the arbitrariness aside and embrace the choice, we notice a problem: when $a = 1$ and $b = 2$, the geometric mean h is the square root of 2, an irrational number. In fact, h is irrational for most choices, that is, unless ab is a perfect square. Should that mean that Christ is *alogos* (irrational)?

Weil dealt with that blasphemous thought with another analogy, this time between mathematical proofs and mysteries of the Christian faith. She interprets the scandal of the $\sqrt{2}$ and other surds as an opening of the human mind to its power to devise ideas that are perfectly and rigorously demonstrable but unapproachable by human imagination. "It is marvelous, it is inexpressibly intoxicating, to think that it was the love and desire for the Christ that caused the Greeks to invent math proofs. . . . That notion [of math proof] *forces the intelligence to seize with certainty* relations that it is incapable of representing or imagining. That is an admirable introduction to the mysteries of the faith."[71]

For all her attraction to Catholicism, Simone Weil considered herself a universalist, uncommitted to any tribal or national loyalties. She imagined her mathematical conception of a mediator as not *only* Christian, but as universal, expressed

> in the ancient mythologies; in the philosophy of Pherecydes, Thales, Anaximander, Heraclitus, Pythagoras, Plato, and the Greek Stoics; in Greek poetry of the Great Age; in the universal folklore; in the Upanishads and the Bhagavad-Gita; in the Taoist texts and in certain Buddhist schools; in the remains of Egyptian sacred scriptures; in the dogmas of the Christian faith and in the writings of the greatest Christian mystics, especially John of the Cross; in certain heresies, especially the Cathar and Manichaean traditions.[72]

She seems admirably willing to extend the reach of her arithmetical notions of transcendence even beyond the cultures of the axial age with which we began, but with two flagrant omissions. No Jews and no Hebrew Bible, for Weil declares that Judaism is tribal, exclusivist, the opposite of universal. And just as striking, but without explanation, no Muslims and no Islam.

How does Simone Weil's association of number and the divine compare to that of some of the thinkers in this chapter whom she might consider her predecessors? If anything, her identification of the two was more total. Earlier we heard Augustine expressing qualms about the dangers of identifying number, *arithmós*, with the (divine) word, *lógos*. Weil, for her part, fiercely *wills* the sameness of those two: "We must not forget that in Greek *arithmos* and *logos* are two terms exactly synonymous."[73] Exactly? Weil was a well-trained student of Greek, but a world of prejudice colors this claim to sameness, as she could have learned from one line of the poet Pindar, who uses both words in the same sentence: "For, in truth, / I could not have the skill to tell (*légein*) the number (*arithmón*) of the pebbles of the sea."[74]

The poet's logos is the action of *légein*, which in this case means *to tell*, and here what is *told* by the logos is the *number* of pebbles in the sea. If, in a religious key, we wish to push all this up to the divine, we may say that the divine logos can tell the number of pebbles in the sea, but also many other things besides. Or in terms drawn from Genesis and John, surely the Word includes Number, but it also includes infinitely else. All of that infinitely else is lost in Weil's "exactly synonymous."[75]

∗   ∗   ∗

We began this chapter claiming that science and religion have often shared a preoccupation with problems of sameness and difference and a predilection for finding relief from those problems in the enduring sameness of number. Simone Weil provides a nice example of that predilection from the modern mystic's point of view, so we should close with some examples from the modern sciences. There are many, for throughout much of modern Western history, nearly every advance in the sciences, from arithmetic and astrophysics to the zoology of mollusks, was often enough greeted as renewed evidence of "proof of the unity of the Deity" (in words of the British mathematical astronomer Mary Somerville celebrating the discovery of the planet Neptune in 1846).[76] The Harvard mathematician Benjamin Peirce put the view well in his presidential address to the American Association for the Advancement of Science of 1853: "Modern science has realized some of the most fanciful of the Pythagorean and Platonic doctrines, and thereby justified the divinity of their spiritual instincts."[77] The Roman Cicero observed before the birth of Christ that knowledge of the regularity of the heavens was "the most potent cause" of human belief in the existence of God, since it could not be thought of as "the mere effect of chance." That observation remains relevant twenty centuries later.[78]

A greater mathematician than Pierce made a related but much more restrained point in 1886, when the number theorist Leopold Kronecker told the Berlin Congress of Natural Scientists: "God made the natural numbers, all the rest is human handiwork."[79] Compare Kronecker—a Prussian Jew who converted to Christianity just before his death in 1891—and his imagination of eternity with that of one of our own contemporaries, the Oxford mathematical physicist Sir Roger Penrose, who coauthored (with Stephen Hawking) a famous 1970 paper that predicted the existence of singularities in space-time (a.k.a. black holes). Penrose is also among the relatively few physicists who have sought to contribute to the ongoing debate about the nature of the mind. In one of those contributions he asks by what right we call this realm of ideas, the natural numbers, an existing world. His reply:

> Its existence rests on the profound, timeless and universal nature of these concepts, and on the fact that their laws are independent of those who discover them. The ragbag—if indeed that is what it is—was not our creation. The natural numbers were there before there were human beings, or indeed any other creature here on earth, and they will remain after all life has perished.[80]

Penrose is a religious skeptic, and he does not attribute to God the *creation* of the natural numbers. Instead, he treats that whole infinite "ragbag" as existing, so to speak, *before* creation. His natural numbers are something akin to what medieval Muslims called "the necessary existent," always already before all else. With apologies to the Gospel of John, "in the beginning was the number."

In short, a common idea stretches from Plato to Penrose across all of the thinkers we've touched on in this chapter. The idea is this: for all the terrifying flux of the world and the mind, with their relentless generation of difference, there exists a grounding and eternal sameness that we can glimpse in number. In pointing to that commonality, we in no way intend to assimilate the truths on offer in, say, mathematics, physics, and theology to one another. We seek only to convince you that in the Western world at least, certain strategies and habits of thought are common to these great fields of human endeavor.

5

FROM DESCARTES TO KANT

An Outrageously Succinct History of Philosophy

Where [philosophy] was victorious, behold, it generally built its
throne on the ruins of mathematics and experiences from physics.

JOHANN GOTTFRIED VON HERDER[1]

Philosophers constantly see the method of science before their eyes,
and are irresistibly tempted to ask and answer in the way science
does. This tendency is the real source of metaphysics, and leads
the philosopher into complete darkness.

LUDWIG WITTGENSTEIN[2]

Enduring Revolutions

The previous chapter focused on Judaism, Christianity, and Islam in late antiquity and the Middle Ages. It stressed that these vast systems of thought shared a foundational concern with sameness (and difference) not only with each other but also with the philosophies and sciences, the mathematics, physics, and metaphysics, of their worlds. In this chapter we will shift our focus to Europe and to the periods historians of ideas have often called the scientific revolution and the Enlightenment (roughly, from the publication of the astronomer Copernicus's redescription of the universe in 1543 to Napoleon's conquest of Europe ca. 1800).

As their names proclaim, these periods encompass important transformations in the history of thought. Religious reformations, political revolutions, new worlds and new empires all contributed to these transformations. But sticking for now only to the history of science, it was in this period that the observational, experimental, and mathematical methods developed by storied figures like Galileo Galilei (1564–1642) and Sir Isaac Newton (1643–1727) destroyed some basic "truths" of previous ages (such as a geocentric universe) and erected new ones (such as the laws of gravity) in their place.

Poets and philosophers, physicists and theologians, mystics and mathematicians could all share a sense that these and other transformations represented an inflection point in the history of human knowledge. As Alexander Pope put it in his 1727 *Epitaph* for the most famous mathematician and physicist (and also a theologian) of his age,

> NATURE and Nature's Laws
> Lay hid in Night:
> God said, "Let Newton be!" and
> all was light.

Without in any way diminishing the importance of these transformations, our goal in this chapter is to suggest that the basic questions, problems, and habits of thought about sameness and difference persisted through all of them even as they were refashioned into novel forms that we today consider modern.[3]

That novelty is important. New mathematical methodologies and discoveries in physics and the natural sciences contributed to the emergence of new anthropologies, theologies, and philosophies (and perhaps vice versa as well).[4] The authors of these systems of thought did sometime imagine their relationship to the past as one of night and day. Immanuel Kant (1724–1804), for example, famously referred to his own contributions to philosophy as a "Copernican revolution."

But we want to convince you that there is also a strong continuity, both in terms of the fundamental questions these early modern thinkers were asking about the nature of human knowledge and in the type of answers they proposed. We will focus on a handful of thinkers who made prominent contributions to philosophy as well as to mathematics, physics, and theology. Though today we think of these disciplines as separate, when thinking turns to ultimate foundations we often find them huddled together, seeking warmth from common commitments.

In order to understand better those common commitments, let's return one last time (in this book) to a dialogue by Plato. In the *Phaedo* Socrates's skeptical interlocutor, Simmias, describes the search for fundamental truths in a heroic key: "it is either impossible or very difficult to acquire clear knowledge about these matters in this life. And yet he is a weakling who does not test in every way what is said about them and persevere until he is worn out by studying them on every side." In Simmias' view, all that study will yield three possibilities:

· We may discover how things really are.
· "If that's impossible, then adopt the best and least refutable of human doctrines, embarking on it as a kind of raft, and risking the dangers of the voyage through life."
· Or we may adhere to some divine doctrine.

We have seen, in all of the previous chapters, just how difficult it is to identify the boundaries between these possibilities. It is not easy to know the difference between a certainty, a dogma, and a faith. The *Phaedo* provides yet another example of that difficulty: which of the three strategies would Plato have thought he was following when he had Socrates convince Simmias that the soul is immortal by arguing that the idea of equality or sameness must be present in the mind independently of and before any experience in the world?[5]

Beneath Plato's search for an innate sameness lies the desire to fortify two certainties that are always in doubt: first, that there are eternal (unchangeable, noncontingent) truths accessible to the human intelligence, and second, that at least some part of the human is immortal. Those commitments remained common in the centuries of European history we are now entering, centuries in which (very) few thinkers emerged who were willing to challenge openly the second commitment. But even for them, the great quest remained that of Socrates and Simmias, Plato and Aristotle: to find kernels of eternal sameness, however small, out of which—through various maneuvers of unification and separation, synthesis and analysis—we can build knowledge.

You remember Plato's version of the story. All thought is based on the choice between two "opposite" operations that the mind performs on objects within it: association or unification into sameness and separation or division into difference. The thinker skilled in this dialectics will, like a good butcher, unify and separate "according to nature" or "in a natural way," careful not to splinter an essential underlying sameness.[6] Such models of thought can still be found in the most up-to-date schools of philosophy. When David Lewis, "one of the most important philosophers of the 20th century," writes of "natural properties" and "objective resemblances," he sounds much like a Plato who has read up on our latest physics and logic.[7]

Precisely because the vocabulary remains powerful today we repeat our warning: whenever a philosopher uses the expression "natural" or "according to nature" you should sniff for unexamined assumptions. There is no "natural" way to divide the body of an animal, for example, an ox, *independent of the question or purpose one has in mind*. For the purpose of studying the digestive system, for instance, the division will be different than for that of studying the circulatory system.

It is obvious, you might object, that what Socrates had in mind was cutting up an ox for consumption. But even then, context matters. It is not only that every ox is different, as the great Chinese philosopher Zhuang Zhou (ca. 369–286 BCE) is said to have stressed a century or so after Socrates, in a story about a master butcher who feels with his entire being the particularities of every beast he carves. It is also that different cultures, different countries, divide animals differently. The barbarian nomads in Kafka's "Pages from an Old Text" (*Ein altes Blatt*) do not even bother to carve, instead taking bites out of the living animal. On the other hand Tristan, the hero of Gottfried von Strassburg's medieval romance, rises to glory because he teaches the hunters of King Mark's court a new (to them) way of dressing deer. Without fashions in butchery, no Tristan and Isolde. Nor, despite Socrates's prejudices, need our incisions be made at the "natural joints." Osso buco, for one, is cut across the shank bone.[8]

Descartes: "One Thing Certain and Unshakeable"

The thinkers in this chapter are remarkable because—challenged in part by the astounding victories of mathematics and physics over the world in which they lived—they searched so vigorously for new kernels of certainty and sameness on which these new powers were built, and in the process questioned so relentlessly the "joints" that might relate human thought and the cosmos. René Descartes (1596–1650), to pick a famous example, wrote that his goal was like that of the Greek mathematician and physicist Archimedes, who needed only one firm and immoveable point in order to shift the earth. So, too, wrote Descartes, he could hope for great things if he could "discover one thing only, however slight, that is certain and unshakeable."[9]

Everyone remembers that the first indubitable thing he finds is his own existence—"I think therefore I am" (*cogito ergo sum*)—thereby founding what is sometimes called "subjective" philosophy, that is, grounded in the individual person.[10] Fewer people recall that a little further, in the Third Meditation, Descartes discovers a second thing that he finds indubitable and which is (he maintains) independent of, and before, experience. This second thing is God. For if, Descartes argues, he has an idea of an infinite being, it can only be an infinite being who can be the cause of his having that idea. Since Descartes has in his mind the innate idea of a being absolutely perfect, infinite, omniscient, and so forth, that in itself "provides a very clear proof that God indeed exists."[11]

It is because the argument begs the question that it has come to be called the "Cartesian circle."[12] Like the monotheistic mathematics we encountered in the previous chapter, that circularity is symptomatic of the desire for an eternal truth, a stable sameness, on which certainty can be grounded. But note that although he was himself a distinguished mathematician—you will recall from chapter 1 that Husserl and Heidegger blamed his contributions to geometry for the crisis of modernity—Descartes himself deliberately refuses to build certainty directly on mathematics. He is not a Pythagorean, identifying being with having number.[13] For Descartes, God is the "necessary existent" (as al-Ghazālī would have put it), not number. Indeed, for God (Descartes writes) it is not necessary that $2 + 2 = 4$. Commentators debate whether Descartes also held that if God had so willed it, he could have made $2 + 2 = 5$, as the "occasionalists" we met in the previous chapter might have done. But what is clear is that Descartes's divinity, unlike Plato's, is not bound by the necessity of number.

It is true that Descartes can then derive all the truths of number from the innate certainty of God's existence and goodness. (Those who wish to can watch him doing so in works such as his *Rules for the Direction of the Mind*, as well as in the *Meditations*.[14]) Still, we think the difference is meaningful insofar as it reveals a preoccupation that will be shared by many philosophers of the scientific revolution. Like their Christian, Jewish, and Muslim predecessors, many of these philosophers want their systems of thought to preserve the omnipotence of God, the freedom of the human, and the powers of mathematics simultaneously, and this at a time when those last were growing dramatically.

Descartes certainly believed that the study of mathematics habituated the mind to reason. To that end he even developed a new method for grasping geometry. But he did not make mathematics the cornerstone of certainty or of virtue.[15] As we have seen, he preferred two other ideas to do this kind of work. Those two ideas, however, are of quite different types. Descartes's first basis for certainty, his *cogito ergo sum*, can be thought of as the product of our particular language rather than as a universal necessity that preexists particular forms of human culture and experience. Because our languages have developed in particular ways across history, many of them require any use of the verbal form "I think" to be attached to an existing subject. We could call this a historically contingent necessity but not an absolute one.[16]

But Descartes's God is a certainty and a necessity of a different order. Like Plato's (or Penrose's) natural numbers, Descartes's God is not the product of human language or experience. It preexists the human altogether. By the time we reach the end of this chapter we'll have learned from Immanuel Kant to apply the term *a priori* to such ideas and truths believed to preexist our experience and the term *a posteriori* to ideas and truths that follow from, or are the product of, our experiences. *A priori* will become yet another term — like *natural, necessary, innate* — to characterize those apathic and indivisible kernels of certainty out of which thinkers have hoped to build a universe of knowledge. Sometimes those kernels have been drawn from mathematics and logic, sometimes physics or psychology, sometimes theology or metaphysics, often a blend of some or all of these. And of course all the philosophers we'll meet disagree as to which truths are in fact a priori and which are not. Without taking sides in those debates, we propose the following:

1. All these a prioris are attempts to find some eternal sameness from which we can draw confidence in the certainty of some aspect of our knowledge.

2. Conviction about the necessity of any given a priori — any necessary truth or idea innate in us but before our experience — is generally a dogmatic position about the nature of human nature and human knowledge.[17]

"Human Understanding": Locke vs. Leibniz

One way to make visible the role of dogmas about sameness within those debates is to compare John Locke's *An Essay Concerning Human Understanding* (1690) with Gottfried Wilhelm Leibniz's response, aptly entitled *New Essays on Human Understanding* (finished in 1704, first published in 1765).[18] Locke held, famously and boldly, that there are no innate ideas, no "primary notions ... characters as it were stamped upon the mind of man, which the soul receives in its very first being; and brings into the world with it" (1.1.1).[19] There are no "imprinted," innate speculative principles (1.2); no innate moral principles (1.3); no innate ideas of God; of sameness (identity); or of the impossible or inconceivable (1.4). The mind is ab initio a blank slate.[20]

According to Locke it is only through sensorial experience and reflection that the mind acquires simple ideas, which it can then combine, compare, and generalize into complex ones. Locke did believe that mathematics was an excellent way to train the mind into "a habit of reasoning closely" so that it could make sound combinations and generalizations.[21] But he didn't believe that any of its truths were innate. For Locke, all of our knowledge—including that of space, time, number, morality, and identity—is produced by combining these simple "atoms" of ideas. Some of the ideas we produce have the "Certainty of true knowledge," but this is only because they are built on sound reflections on and combinations of "simple ideas," not because they are innate to the mind.[22] It is for this reason that Locke is often honored among the founding gods in the pantheon of British empiricists.[23]

Leibniz combatted these views in a book length commentary on Locke's work titled the *New Essays on Human Understanding*, presented as a debate between Philalethes ("Lover of truth," i.e., Locke) and Theophilus ("Lover of God," i.e., Leibniz). Leibniz had many reasons for opposing Locke's views, among them—as the names of his characters make clear—what he feared they implied for the human soul. According to him, Locke's denial of innate ideas was the leading edge of a materialist's argument against the immateriality of the soul, a potential challenge to belief in its eternity and therefore a mortal danger to religion, decency, and peace.[24]

Leibniz challenged Locke precisely at the point that we've come to expect: through "necessary" truths.[25] Empiricism might account for particular or contingent truths, but it could not account for necessary ones, truths that are universal. These could not be arrived at by individual sensual experience but only by a preexisting disposition that "determines our soul and brings it about that [necessary truths] are derivable from it."[26] What truths might these be? Leibniz's essay "First Truths" (*Primae veritates*) of 1689 rounds up the usual suspects, the "sameness" truths of identity and noncontradiction:

> First truths are those which predicate something of itself or deny the opposite of its opposite. For example, A is A, or A is not non-A; if it is true that A is B, it is false that A is not B or that A is non-B. Likewise, everything is what it is; everything is similar or equal to itself; nothing is greater or less than itself. These and

other truths of this kind, though they may have various degrees of priority, can nevertheless all be grouped under the one name of identities.[27]

These "sameness" truths or identities were, according to Leibniz, written on our soul by God, but he realized that they were not themselves sufficient for the proof of God's authorship.[28] For that, one or two further truths were "necessary." As he put it much later in a marvelous epistolary exchange in which he sought to warn Princess Caroline of Wales against dangers that the materialists posed to Christian souls:

> The great foundation of mathematics is the *principle of contradiction or identity*.... This single principle is sufficient to demonstrate every part of arithmetic and geometry, that is, all mathematical principles. But in order to proceed from mathematics to natural philosophy, another principle is required . . . ; the *principle of sufficient reason*, namely, that nothing happens without a reason why it should be so rather than otherwise. . . . Now, by that single principle . . . one may demonstrate the being of God and all other parts of metaphysics or natural theology and even, in some measure, those principles of natural philosophy that are independent of mathematics; I mean the dynamic principles or principles of force.[29]

Identity, noncontradiction, and sufficient reason: from these basic principles of sameness and difference, Leibniz sought to prove that our souls are not corporeal, that they were created by a God who is perfect and does not deceive, who fashioned a world governed by necessary and universal laws, a world that is the best of all possible worlds. In other words, through these principles of sameness and difference, Leibniz sought a Christian unification of physics, metaphysics, and psychology (in the sense of a science of the soul).[30]

The "laws of thought" that Leibniz developed in order to achieve this unification—such as the Principle of the Identity of Indiscernibles and the Principle of Continuity—became (and remain) very influential in logic, philosophy, physics, and other fields. Consider the Principle of the Identity of Indiscernibles, sometimes called "Leibniz's Law," which still looms large in our discussions of modern physics and quantum mechanics. Leibniz states it charmingly in his correspondence on behalf of the Princess Caroline by describing an earlier debate before another philosophically inclined princess:

> There is no such thing as two individuals indiscernible from each other. An ingenious gentleman of my acquaintance, discoursing with me in the presence of Her Electoral Highness, the Princess Sophia, in the garden of Herrenhausen, thought he could find two leaves perfectly alike. The princess defied him to do it, and he ran all over the garden a long time to look for some; but it was of no purpose.

If two things resemble each other completely and differ *only* in number (*solo numero*, or as Leibniz sometimes suggests, only in time or space), then they are in fact the same: one thing, not two.[31]

Among other things, Leibniz believed that this principle dealt the death blow to the "atomism" of his materialist rivals. In this he turned out to be wrong, although the principle he articulated continues to pose basic challenges to our understanding of particle physics today (see chapter 7). But this "law" was as much a consequence of his theological commitments as of his logical ones. If a creator God adheres to "first truths," such as the Principle of Sufficient Reason, then that God must have a reason for creating things as they are.

If indiscernible individuals existed, then God could just as easily have created a world in which the two were switched. There would be no sufficient reason for God having made the world as it is. Therefore, such sameness in created things cannot exist: there cannot be indiscernible individuals. With this "law" in hand, Leibniz could now prove to his satisfaction that a God who abides by the Principle of Non-Contradiction made this world by necessary and universal laws, made it the best of all possible worlds, and that in it every soul, every individual, is in some necessary way unique.[32]

Leibniz's achievements were monumental. Perhaps more than any other early modern philosopher, he managed to hold together the demands of a perfect God, the diversity and constant change of creation, and the mathematical and metaphysical discoveries of his own day (a number of which were his own).[33] Rejecting the mind-body dualism of Descartes, the materialism of Hobbes and Locke, and the pantheism of Spinoza (with whom he met shortly before Spinoza's death), he insisted instead on a divinely preestablished harmony between matter, mind, soul, and Creator.

The task required engagement with many of the deepest questions we've encountered, and Leibniz developed new approaches to many of them. His mathematical work on problems of the continuum and on infinity was groundbreaking.[34] The "laws of thought" that Leibniz developed in order to achieve this unification became (and remain) very influential in logic, philosophy, physics, and other fields. We are in awe of the mind capable of such systematic thought across so many fields of inquiry.

But even as we admire, we may also quail before the laws Leibniz legislated. "Everything in the world," he wrote in 1695, "can be explained in two ways":

> through the *kingdom of power*, that is, through *efficient causes*, and through the *kingdom of wisdom*, that is, through *final causes*, through God, governing bodies for his glory, like an architect, governing them as machines that follow the *laws of size* or *mathematics*, governing them, indeed, for the use of souls, and through God governing for his glory souls capable of wisdom, governing them as his fellow citizens, members with him of a certain society, governing them like a prince, indeed like a father, through *laws of goodness* or *moral laws*.[35]

Two laws and two kingdoms, but one common and underlying commitment to identity and noncontradiction at the foundations.

Logicians and analytic philosophers today have long forgotten the commitments to divinity that shaped Leibniz's thought, but they have kept its formulations and even radicalized them. A logical principle such as the Identity Principle is, they like to say, true "in all possible worlds." Leibniz did not himself use this now common expression, but it bears his mark: *ex ungue leonem*. To our mind, the phrase exhibits its hubris in sequins and faux fur. It is difficult to imagine anything more subject to historical change than that which is humanly conceivable, hence considered possible, hence a feature of a possible world.[36] But the hubris has a familiar ring. Of the gigantomachia we have just staged we might say, returning to our pre-Socratic terms, that Locke and the empiricists sought to push as much knowledge as possible toward the vulgar side of Parmenides's tracks, that is, toward *dóxa*, experience, and opinion, whereas Leibniz strove to drive what he could toward *alḗtheia*, eternal truth.[37]

Hume's "Simple Ideas"

These differences between so-called rationalists and empiricists were strongly felt and strenuously fought. But we would like to stress as well a basic similarity. It is true that the rationalists' mania for necessity pushed them into theology: the systems of a Descartes or a Leibniz all depend on a God for their functioning (in this sense as in so many others, Spinoza was exceptional). But the empiricists also have their dogmas, which amount to almost the same thing. For if the a priori commitment of the rationalists was to the constancy and perfection of God-given principles, the chief dogma of those like Locke and Hume, who maintained that sensory impressions are the only sources of knowledge, was to the constancy and perfection of our representations. Constancy and perfection—that is, forms of sameness and identity—are at the bottom of both.

This sameness is the starting point of Hume's *A Treatise of Human Nature* (1738–1739): "When I shut my eyes and think of my chamber, the ideas I form are exact representations of the impressions I felt [when my eyes were open]; nor is there any circumstance of the one which is not to be found in the other."[38] He concedes that complex impressions, such as that of the city of Paris with all its streets and buildings, cannot be reproduced exactly in one's imagination. But he takes it as an indubitable axiom that with *simple* impressions the shut-eyed idea is exactly and exhaustively *the same* as the sensory impression or idea: "All our simple ideas in their first appearance are deriv'd from simple impressions, which are correspondent to them, and which they exactly represent."[39] Hume pronounced this the "first principle" in the "science of human nature." Historians of philosophy today call it the "Copy Principle."[40]

And what are these simple or atomic impressions and ideas that copy so perfectly? They are, for Hume, "those that admit of no distinction nor separation." In other words, the "simple" or "atomic" ideas of a Locke or a Hume are like Plato's "natural joints," an irreducible, stable, and certain foundation of sameness on which all else can be built. "The same idea," writes Locke, "will eternally have the same habitudes and relations" (4.1.9). In this empiricist psychology our perceptions and our concepts are at their foundations clear and distinct, without ambiguity, unchanging, always identical to themselves. Our ideas obey the Identity Principle for these psychologists just as strictly as they do for the rationalists. The empiricist preaches the primacy of *experience*, but it is experience of a very special kind, the ancient one of playing with real or imaginary pebbles.

This "atomism" is closely related to that of the physicist, and not only in its vocabulary. The original, dominating intention in both cases, physical and psychological, was to reduce the observable phenomena to the operations of putting together and separating *apathic* elements—indivisible and unchangeable—so as to explain the world and the mind mechanically.[41] Here Hume the student of the mind agrees with Newton the student of the universe. As the latter had written in the preface to the first edition of his *Principia* (1687),

> For the whole difficulty of philosophy seems to be to discover the forces of nature from the phenomena of motions and then to demonstrate the other phenomena from these forces. . . . If only we could derive the other phenomena of nature from mechanical principles by the same kind of reasoning![42]

"If only . . . !" Hume took to heart Newton's emphasis on the discovery of forces (in Newton's case, gravity, the attractive force between point masses) so foundational that their origin cannot be explored any further and applied it to the workings of the mind.[43] He thought he had found the psychic equivalent of Newton's physical "gravitation" in what he called "association" (hence his teaching is often called "associationism"): a basic force, intrinsic to human nature, whose laws governed the mind's bringing together of "simple" perceptions into "complex" ones and separating them back again into "simples" in the process of thought.[44]

Hume's laws of thought were designed to solve what was to him—and to us as well—a pressing problem, a vulgar version of which today underlies many an attack against the humanities. Why is it that the strides we have made in our understanding of psychology and human nature seem so much smaller than those we have made in our understanding of the physical world? Hume's treatise began with that question. Attacking the "airy sciences" of the philosophers, metaphysicians, and theologians, he claimed to have discovered "a compleat system of the sciences, built on a foundation entirely new." The foundation he proposed was a Newtonian science of human nature based only on observation, unlike the "airy sciences" in being entirely without hypotheses.[45]

But from our point of view, Hume's whole edifice is built on two very large "hypotheses," or rather assumptions. The first, scarcely acknowledged, is the untroubled rule of the Identity Principle over the "simples" and our perception of them, putting all of human nature under the universal sign of apatheia. The second, almost explicitly stated, is the equivalence of logical and imaginative possibility: we are able to imagine a situation if and only if it is logically consistent. Hume does not consider the possibility that what is or is not imaginable is very much subject to change. Hume himself, for example, could not imagine infinite sets as logically consistent, and so he rejected their possibility, a rejection that is itself difficult for a philosopher today to imagine as logical.[46]

Like every idea in every chapter of this book, Hume's strategy also has a long history. Already in 1817 the poet Samuel Taylor Coleridge suggested with only some tongue in cheek that Hume had plagiarized his "universal law of the passive fancy and mechanical memory" (as Coleridge called it) from the ancients and medievals:

In consulting the excellent commentary of St. Thomas Aquinas on the *Parva Naturalia* of Aristotle, I was struck at once with its close resemblance to Hume's Essay on Association. The main thoughts were the same in both, the order of the thoughts was the same, and even the illustrations differed only by Hume's occasional substitution of more modern examples.[47]

According to Coleridge, Hume's personal copy of Aquinas's commentary had since been found, "swathed and swaddled" in the Scottish philosopher's marginalia. Coleridge is making an important point. Philosophers in the age of scientific revolution drew on a long medieval Christian tradition of thinking about the relationship between the one and the many, sameness and difference.[48] But for once we will resist the temptation to follow the footprints of ideas into the past in favor of reinforcing the general point.

Whether rationalist or empiricist, skeptic or divine, all of these thinkers share the need for an irreducible, indivisible, unchangeable, ever certain kernel. Call it the "atomic," the "simple," the "monad," the "singleton," the "unit," or the "one," what is sought is some basic particle at the bottom of it all that remains always and eternally *the same as itself* and can thereby serve as a foundation for our knowledge of ourselves and the universe. The words of criticism that Leibniz addressed against mathematicians in 1716 could be used of all these thinkers (including Leibniz) as well: "the Mind, not contented with an Agreement, looks for an Identity, for something that should be truly the same."[49]

Kant: Out of the Frying Pan and into the Fire

This desire was shared by physics and psychology, by metaphysics, mathematics, and theology, a common yearning that explains in part why these disciplines have been so codependent for so long. Whenever what seems like a firm base has been found for any one of them, it has immediately attracted the attention of the others. And vice versa, as we saw in chapter 1: a crisis of confidence in the foundations of any one of them has often proven contagious, capable of infecting the others.

There were certainly many in the eighteenth century who thought that mathematics should be the model for philosophy. In Denis Diderot and Jean D'Alembert's *Encyclopédie*, a work whose volumes (published between 1751 and 1752) became emblematic of the European Enlightenment, the article on "Méthode" proclaimed that "transporting the mathematical method into philosophy, one finds that truth and certainty will manifest themselves." We've deliberately avoided the partisans of such transportation (such as the German Christian Wolff) and focused instead on philosophers who thought deeply about mathematics and physics but insisted (as did Bishop Berkeley and Immanuel Kant alike) that "by means of his method the mathematician can build nothing in philosophy except houses of cards, while by means of his method the philosopher can produce nothing in mathematics but idle chatter."[50]

Why have we spent our time with these thinkers? Their abiding influence on the future of philosophy is only part of the reason. More important for us is the intellectual experience they exemplify of avoiding one danger by falling into a similar one, albeit with a different name. For at the same time that they attempted to limit the authority of the new mathematics and physics over philosophy, they seized on certain kinds of sameness and laws of thought that lent those disciplines their authority, imported them into the human (and divine) mind, and thereby reinforced the power of those identities over the realms of philosophy and psychology (as well as theology) for centuries to come.[51]

These thinkers might have offered a shared handful of reasons for their attraction to the subject of mathematics even as they have varied sharply in the strength of that attraction:

- Mathematics is in some sense a priori. At a minimum (Hume) this means that some aspects of it are dependent only on "the mere operation of thought" within our minds and are independent of our experience of the world. (Even if, for example, one finds no perfect Euclidean triangles in the "real" world, it nevertheless remains true that the interior angles of a Euclidean triangle add up to 180°.) At a maximum (Leibniz), this may be

taken to mean that certain mathematical truths are engraved
a priori on the human soul.[52]

· Mathematics is exact. Within it we can establish strict equality
and identity, a perfect type of sameness. Hume the minimalist
again: "Algebra and arithmetic [are] *the only sciences*, in which
we can carry on a chain of reasoning to any degree of intricacy,
and yet preserve a perfect exactness and certainty. We are pos-
sest of a precise standard, by which we can judge of the equality
and proportion of numbers; and according as they correspond
or not to that standard, we determine their relations, without
any possibility of error. When two numbers are so combin'd,
as that the one has always an unite answering to every unite of
the other, we pronounce them equal; and 'tis for want of such a
standard of equality in extension, that geometry can scarcely be
esteem'd a perfect and infallible science."[53]

· Mathematics is necessary, meaning that its truths seem to force
themselves on us independently of culture, contingency, or
convention. No amount of "experience" in the world would
convince us that those truths can be denied without a fall into
contradiction.[54] If we find, for example, a triangle in the real
world whose angles do not add up to 180°, we would doubt the
geometer, builder, or surveyor who designed or built it, but we
would not question the Euclidean truth.

· As Newton had made manifest to so many, mathematics is mysteriously applicable to the physical world. How is it that these
"mere operations of thought," internal to the human mind,
can, when combined with observation and experience, provide
powerful and even predictive explanations of the workings of
the world external to that mind? At the very least, we can say
with Bertrand Russell that mathematics' "apparent power of
anticipating facts about things of which we have no experience
is certainly surprising."[55] It is this surprising power that has
tempted so many to draw on the attributes of number in order
to build stable bridges between our minds and the world.

Consider as our last example the case of the great German philosopher Immanuel Kant. On the one hand, Kant was deeply im-

pressed by Hume's emphasis on the mind's construction of the known. In his modestly titled *Prolegomena to Any Future Metaphysics*, he wrote that Hume's work had woken him from the "dogmatic slumber" that had enveloped all previous metaphysics. But Kant also felt that the Scot had left the mind in an unbearable position.[56] If all of our basic principles and truths originate in experience gained through our mortal and imperfect senses, what warrants their absolute validity?

Take the case of Euclidean geometry. As we saw above, Hume taught that geometry failed the test of a perfect and infallible science. Where, Kant wondered, did this leave the rational mechanics of a Galileo or a Newton?[57] How could such powerful systems of knowledge rest on anything but "certainty" and "clarity," that is, on "pure reason," independent of experience?

Unlike Hume, Kant took Euclid's axioms to be as unassailable as "two plus two equals four." He saw in the Greek Thales (or whoever it was who "demonstrated the isosceles triangle") the agent of a scientific revolution, the dawning of "a new light" in the mind.[58] For him, geometric propositions like "between two points there can be only one straight line" were so "immediately certain" that "no further mark of proof" could be given for them beyond "what they themselves express."[59] (It did not occur to Kant that between two diametrically opposite points on a sphere there are infinitely many geodesics, or great circles.) Euclidean "truths" were, for him, infallible and immediate.[60]

We can see the influence of these mathematical and physical truths on Kant's thinking in the questions he posed against Hume at the beginning of his *Critique of Pure Reason*:

"How is pure mathematics possible?"

"How is pure science of nature possible?"[61]

Answering these questions required, according to Kant, establishing a priori powers of pure reason, independent of experience. With those powers in hand, one could then even ask Kant's third and (for him) key question in a new way: "How is metaphysics as a natural predisposition [of human reason] possible?"[62]

The answer that Kant advanced to these three questions in his "Transcendental Aesthetics," the beginning part of his *Critique of Pure Reason*, is perhaps his most famous teaching: it is the mind that spatially and temporally orders the world and not the other way around. Kant repeatedly maintained, not only in the First Critique but already in his *Inaugural Dissertation* of 1770, that we cannot know anything about a thing "in itself" independently of its perception by a mind, yet he also maintained (perhaps contradictorily) that among those things "in themselves" there are no spatial or temporal relations—neither are they at distances from one another nor do they succeed one another in time. Everything that impinges on our minds from the mind-independent world arrives as a "manifold of raw intuition" that is not embedded in any spatial-temporal matrix. It is not the world but the mind that organizes this manifold, using two pure *forms* of intuition, space and time, that are innate within it. (*Pure* here means "not derived from experience.")

Kant modestly called this his "Copernican revolution," and we can see why. Much as Copernicus had reversed the relationship between earth and the universe, so Kant's proposal entailed the reversal of some very ancient and influential models of the human mind. Ever since the invention of devices for impressing, writing, or recording, people have tended to take these as models of the mind, especially of memory. The camera and the computer of the present are only the latest avatars; for a long time, we made do with clay seals, wax tablets, marble, slate, parchment, or paper as model. In his treatise *On the Soul*, for example, Aristotle compares thoughts to letters on a tablet. The ancient Stoics, too, favored wax tablets; the medieval, seals and sealing wax. The idea was that the soul or mind receives "impressions" from the world outside, some clear, some indistinct. This is how Diogenes Laertius explained the Stoic version of this doctrine:

> There are two types of impression . . . : the cataleptic, which they hold to be the criterion of matters, is that which comes from something existent and is in accordance with the existing thing itself, and has been stamped and imprinted; the non-cataleptic either comes from something non-existent, or, if from something

existent, then not in accordance with the existent thing; and it is neither clear, nor distinct.[63]

There were discussions among the Stoics about some questions of detail, like how could the mind retain several impressions simultaneously when a wax tablet had to be smoothed out, erased, before a new impression could be stamped on it. But you get the point: the mind, in these models, is the passive recipient of impressions from the world. And regardless of the details, the common, often unexpressed assumption is that the mind's impression is, ideally, a copy of the thing in the world.

But Kant could not agree with the Stoic vision of a passive material mind catching and recording outside impressions. The Stoic vision was not a secure foundation for the new sciences. As we have seen, Descartes had argued that deceitful gods can fool us in respect to "outer" impressions but that the "inner" consciousness is a trustworthy guide, and he had persuaded many rationalists (we do not here include Spinoza). And so Kant turned things around: it is the mind that acts on the world's passive raw material. Instead of being like a wax tablet, the mind is like a ravioli *form* that presses down on ravioli dough. A Copernican revolution indeed. But note what did not change in this revolution: the demand for an unchanging sameness beneath it all, what we are calling apatheia.

Just how insistent Kant was in that demand is perhaps most evident in the importance he ascribed to the apriority of space and time, that is, to making all space-time properties and relations independent of experience and original to the nature of the mind. This step, which moves in the same direction as making logic normative for thought, gives us a comforting sense of certainty and safety. Certainty and safety about what? Not about peace and solidarity in our community or about our physical or psychic well-being or moral well-doing or our good standing with the gods. It gives us the certainty that mathematics is perfect and that the foundations of the new science of nature are safe. You may rest assured that Euclid's fifth postulate is so because our mind is so. (And vice versa, which is for some equally calming.)[64]

You won't be surprised that we think the calm is deceiving or at least contingent on assumptions and definitions. Contrast, for example, the conclusions of another philosopher (and great mathematician), Blaise Pascal, about space. Pascal famously looked at the "infinite number of representations" of space, saw the endless depths that surround us—infinite vastness on the one hand, infinite divisibility and smallness on the other—and concluded that in the face of such terrifying abysses our only sane choice is to sacrifice reason and put our trust in God.[65] Kant chose instead to drain the terror through a startling act of renaming. "No concept, as such, can be thought as if it contained an infinite set of representations *within itself*. Nevertheless space is so thought (for all the parts of space, even to infinity, are simultaneous). Therefore the original representation of space is an ***a priori* intuition**, not **a concept**." In the difference between concept and intuition, we find the miraculous conversion of abyss into foundation.[66]

Today (meaning after 1854, the year of Bernhard Riemann's Göttingen lecture "On the Hypotheses Which Lie at the Bases of Geometry"), no knowledgeable person believes in the necessary and universal truth of Euclid's axioms or of their corollary, the proposition that used to be always quoted as a paradigm of such truths, namely, that "the sum of the angles of any triangle equals two right angles." What would Kant have said had an advance copy of Riemann's lecture somehow fallen into his hands? Presumably he would have simply chosen a different mathematical example. To him mathematics was primarily useful as the confirming example of a deeper truth that

> Time and space are accordingly two sources of cognition from which different synthetic cognitions can be drawn *a priori*, of which especially pure mathematics in regard to the cognitions of space and its relations provides a splendid example.[67]

Time, space, and mathematics: these are for Kant independent of sensory experience. They are a priori (albeit in different ways), and their truths are not merely dependent on our language (as when we say that I think therefore I am, or that a widower is a man whose wife has died) or our logic. They are, in Kant's vocabulary, synthetic rather than analytic. We can argue, as later philosophers will do, with the clarity of Kant's influential distinction between synthetic and analytic.[68] But his intention in establishing the distinction is clear enough: to make the truth of mathematical propositions necessary and universal.[69] And his motive is equally clear: to find within the human a foundation of eternal sameness, of absolute identity and strict *apatheia*.

What that foundation looks like appears most clearly in the *Critique*'s appendix—the sting is always in the tail—titled "About the Amphiboly of the Concepts of Reflexion Arising Out of a Confusion of the Empirical and the Transcendental Utilizations of the Understanding." Here Kant presents his view of sameness at its starkest:

> 1. **Identity and Difference**. If an object is presented to us several times, but always with the same inner determinations (quality and quantity), then it is always exactly the same if it counts as an object of the pure understanding, not many but only one thing (numerical identity); but if it is appearance, then the issue is not the comparison of concepts, but rather, however identical everything may be in regard to that, the difference of the places of these appearances at the same time is still an adequate ground for the **numerical difference** of the object (of the senses) itself.[70]

Perhaps Kant should have placed this statement at the beginning of his *Critique*, instead of relegating it to an obscure corner 263 pages later. For it makes absolutely clear his view of the distinction between sameness and difference, which amounts to a primer for counting pebbles or an (incomplete) set of axioms for psephology.[71]

Some further commentary is necessary in order to clarify why this view seems to us so problematic. Let's start with what Kant calls a phenomenon or an appearance (*Erscheinung*). He gives the example of two drops of water that may be identical in all their "inner determinations," but the fact of appearing in different parts of space *at the same time* makes them numerically different; that is, they should be counted as two, not one. What that means is that for Kant any unit, any physical object that is to be counted as one, must be topologically connected (or at least appear so). This, for him, seems to be an axiom just as obvious, in his view, as "there's only one straight line between two points."[72] Kant's stated intention here was to correct Leibniz's Principle of Identity of Indiscernibles, in the case of phenomena, by adding location as a criterion. Ironically, he does that by tacitly assuming another Leibnizian principle that would prove untenable in modern physics: the Principle of Continuity, or that nature never jumps—*natura non facit saltum*—whence his unexamined belief that unity implies connectedness.

That Kant was unaware of the developments in physics occurring in the two centuries plus after his death is not the point we want to make. Our point, rather, is that his engagements with physics and mathematics (and indeed much else) eliminate human freedom from his account of sameness and difference. An object is one or is several independently of our judgment or feeling, and the verdict is fundamentally never in doubt. But if we go back to our metaphor of the ravioli form for the Kantian model of the mind and imagine that we press it onto a perfectly homogenous sheet of dough, do we have just one sheet of dough or twelve (say) ravioli? Or do we have to wait until we have finished rolling our wheel cutter to have the ravioli? Kant doesn't say.

A less culinary example: assume Kant had listened to J. S. Bach's *Goldberg Variations*, composed some forty years before the *Critique of Pure Reason*. Did he think that the theme, the Aria that is played at the beginning, is the same as the Aria that is played at the end? The notes, the tempi are exactly the same; in fact one could switch them, and no difference would result. But even if the two are reproduced exactly (as with recording technology they can be), the listener's experience remains very different because the final aria comes after thirty variations on the beginning one. The total effect is like starting from home, going through wondrous adventures and changes and then, like Odysseus or like Hegel's *Geist*, returning home *and knowing it for the first time*. The different effects on the listener are not due (only) to any internal difference in the two executions of the aria but to the context and opportunity in which they are played. To claim, like Kant, that sameness or difference can be determined from repetition independently of context is to claim that the mind can make such determinations apathically, that is, mechanically, like a machine.[73]

We know that Kant's "model of the mind is quite implausible," as Walter Kaufmann put it. But the point here is that, as Kaufmann also pointed out, "while few of his successors or admirers have accepted it, many have accepted the need for an essentially unpsychological, unempirical theory of the mind."[74] And we know that "many" does not mean "all." How different a Bergson, for whom flux is part of the fathomless logos of our psyche; how different a Heidegger, Patočka, or Sartre, for whom time and historicity are inseparable from our being![75]

But when the subject is arithmetic (the natural numbers and their properties), the ontological consensus remains strong. Henri Poincaré, who maintained that geometry is conventional, was not willing to go so far with arithmetic. Indeed, he claimed that if there is any such thing as an a priori synthetic principle, that is, a feature of the mind, it is the Principle of Mathematical Induction, the axiomatic tool to prove arithmetical propositions, due to Pascal: if $p(n)$ is any proposition about the natural number n, and we can prove two things—(a) $p(0)$ is true and (b) $p(n)$ implies $p(n + 1)$—then we can conclude that $p(n)$ is true for all n. "It is a feature of the mind/spirit (*l'esprit*) itself."[76] Kurt Gödel thought that *all* the axioms of set theory "force themselves upon us as true."[77] Roger Penrose, quoted at the end of the previous chapter, goes a good deal further than Kant and Hume: the natural numbers and their properties, that infinite labyrinth, are eternal and independent of the existence of any mind and of the existence of any being other than themselves. They are the ultimate, all-underlying *substantia* out of which cosmos and the mind may—or must—be built, step by irrefragable step. If this is true, then we should all agree with the recent and blunt assessment of Penrose's sometime coauthor, the physicist Stephen Hawking: "Philosophy is dead."[78]

So is this true? Our answer is *no*. Numbers have needs. We must study those needs if we wish to set boundaries to number. Kant again provides an example, both positive and negative. He certainly felt the imperative to set boundaries, and he set out to study the needs. Kant asks the question, what does it take for $7 + 5 = 12$? But—here comes the negative—he did not answer it. Instead, he proclaimed that the sum, and all pure mathematics, was an example of a synthetic a priori judgment. That is, for Kant, $7 + 5 = 12$ is a truth that is absolutely necessary, independent of experience. It is not analytic in that it is not "contained" as a concept in other concepts (in this case, concepts of 7, of 5, of sum, and the Principle of Non-Contradiction). In order to accede to 12, "one must go beyond these concepts, seeking assistance in the intuition that corresponds to one of the two [7 or 5], one's five fingers, say . . . and one after another add the units of the five given in the intuition to the concept of seven."

From a logical point of view such a demonstration does not *prove* anything. At most it *defines* the concept 12 by adding, one after the other, five units to the concept 7. But it certainly does not prove that adding 7 to 5 *in two separate lumps*, so to speak, will yield 12. Nor does it tell us anything about the conditions necessary for that to happen. So in the next chapter, we'll set out to explain the needs of number, and our question will be even simpler than Kant's. When does $2 + 2 = 4$?[79]

6

WHAT NUMBERS NEED

Or, When Does 2 + 2 = 4?

*A metaphysician is one who, when you remark that two times two
makes four, demands to know what you mean by twice, what by
two, what by makes, and what by four.*

H. L. MENCKEN[1]

Over and over again in the previous chapters we have been watching age after age creating knowledge through assertions of sameness or difference, putting to work those basic operations of human thought we call association and separation, and finding in number the best guarantee of the truths such operations produce.[2] Over and over again we have suggested, sometimes coyly, sometimes crassly, that all these ages erred and that we are all the inbred heirs of that error. But what exactly is the error? The foregoing pages have been pointing at dark corners rather than filling them with light, partly in order to convince you that there really is a fright of some importance lurking there and partly to screw up our own courage for the fight. (If the monster were an easy one to combat, it could not hold humanity so powerfully in its thrall.) Now, liquored up by so much philosophy, we are ready to confront the question head-on. What do numbers need in order to produce their truths? Only with an answer to this question in hand can we begin to ask the next, which will govern the chapters that remain: what should we do when those needs are not met?

Counting Leaves

Mathematics is a vast realm, much of it beyond our (and anyone's) ken, so we will do no more than enter at its humblest postern gate, the one through which we all have passed at some time or another, and learn once again to count, holding in mind Plato's two basic operations of association and separation. Let us undertake together a (barely) imaginable task, that of counting the leaves on one of the two venerable maples growing outside our window, while asking what does it take to do so? Assuming for the moment that we know *what* a leaf is and what counts as *one leaf*, we'd have to go through the leaves one by one, and first of all we would have to decide whether a given leaf "belonged" to the tree we had in mind (or rather, whose trunk we had in mind) or to the other maple tree (for the trunks are about twenty feet from each other, and many branches intertwine) or to one of the several vines whose tendrils so invasively abuse arboreal patience.

This corresponds to Socrates's first operation, association, unification, or bringing together, and it might not be easy, for it involves, for each leaf, tracing its petiole to a twig, that twig to a branch, that to another branch, until we reach a trunk. If that is the trunk we had in mind, we would "count that leaf in," that is, we would "bring it together" into the collection that we want to count: we might mark it with a special mark. Then there is the problem of encountering cases among the marked leaves where we would not be sure whether it is a leaf or two or more or no leaf. For example, more than one blade might be attached to a single petiole, or a leaf might be not yet fully developed, or it might be decayed, or split.

This involves Socrates's second operation, separation or division: distinguishing one leaf from several, maple leaves from vine leaves, and leaves from not leaves. Notice that we could have started with the separation or division and then proceeded to the unification or association, thus inverting the order of the two operations. The whole task of counting the leaves of one of the two maple trees involving the two Socratic operations, association and separation, may be long and cumbersome. It may involve some difficult or even arbitrary reckoning choices. But we are not saying that it is impossible, as we did say that counting our thoughts is impossible.

The operations of separation and association remain the same, but the difficulties disappear if we abandon trees (and indeed every other thing in the world) and stick entirely to the realm of numbers, which unlike leaves are not subject to development or decay (or indeed any change at all). Ask a mathematician "what is counting," and she will likely say something along the following lines: "to count a finite set is to assign to its elements, in a one-to-one manner, the numbers 1, 2, 3, . . . , *n*, without missing any one of the latter." The task of counting a finite set is a typical math problem in that everything is *given* except for what we are supposed to find on the basis of the given, in this case the number of elements in set *A*. We say, "given a set *A*," or "given the elements of the set *A*," and "given the natural numbers 1, 2, 3, 4, and so on," and we are supposed to find the one-to-one assignment or correspondence and the number *n*. Having done which, we conclude, "The set *A* has *n* elements."

We have counted. Instead of having to establish through our two Socratic operations the elements belonging to the set of maple leaves we wanted to count, as mathematicians we simply assumed that the set is given and that there is no question as to what elements belong to it. All we needed were the most basic logical requirements for talking about numbers (we will only deal with natural numbers here), namely, there should be a first number, say 1 (if you want to start from 0, be our guest), and it should always be possible to speak of "the following number," or "follower," or "this number plus 1." So the follower of 1 is a number, and the follower of the follower of 1 is a number, and so on. Since those chains of followers become unwieldy and hard to follow, we give them names: the follower of 1 is named 2, which can also be written as 1 + 1; the follower of 2 is named 3, which can be also be written as 2 + 1, and so on. But all we really needed was the 1 and the possibility of speaking of the 1 + 1.

We have not said much thus far that the Muslim Brethren of Purity did not say more than a millennium ago when they extended this logic in order to describe the One God's acts of creation:

The Creator . . . invented and innovated from the light of His unity . . . a simple substance called "Active Intellect," as He made two arise from one, by repetition. Then He made the Universal Soul arise from the light of the [Universal] Intellect, as He made three from the adding of one to two. Then He fashioned Prime

Matter from the movement of the [Universal] Soul, as He generated four by adding one to three, and so forth.[3]

Many today might smile at this passage, and yet insist "that twyce
twoo are foure, a man may not lawfully make a doubt of it," or (in
the more modern locution of Alfred North Whitehead) that "of all
things it is true that two and two make four."[4] So let us see what
sort of truth is "two plus two is four" and whether or not it deserves
the halo of absoluteness. (Our conclusions will hold as well for St.
Augustine's "seven plus three is ten" or any such assertion.) We
promise to play nice: were we to bring up the flow of our consciousness, or some other flow in which it does not make sense to count,
and present it as a counterexample to "two plus two is four," you
would object with all justice that we are cheating.[5] We must present counterexamples, if at all possible, of collections of things that
can be counted but for which "two plus two is four" is false. Through
such examples, we can increase our awareness of just what it is that
number needs.[6]

When Does 2 + 2 = 4? A Non-Platonic Dialogue

First of all, we should briefly notice that it is "two plus two" and not
"three plus one" that has become shorthand for absolute truth. Why
is that? Remember our basic axioms of counting: we started with 1,
and the infinite possibility of saying "the following number," or "this
number plus 1." Since those chains of followers become unwieldy,
we gave them conventional names: the follower of 1 is named 2,
which can also be written as $1 + 1$; the follower of 2 is named 3,
which can be also be written as $2 + 1$. Thus, the assertion that "two
plus one is three" is true by virtue of naming: *two plus one* and *three*
are synonyms. In other words, when we say "two plus one is three,"
we are uttering a tautology that is true by linguistic convention and
brings us no new knowledge, no new comfort of solid ground. (It
is what Kant would have called an analytic rather than a synthetic
a priori.) But the situation with "two plus two is four" is entirely
different. We know from our rules for counting that adding 1 to a
number means just taking the following number, but what is the

meaning of adding 2 to a number, for example, of adding 2 to 2, and saying it is 4?

The question is important enough to have deserved its own dialogue in the ancient academy. Perhaps it is a symptom of that academy's dogmatism about number that it never received one.[7] So let us approach it here through our own not-very-Platonic dialogue between an anonymous teacher (T) and a pesky student (PS).

T: Let me write it down for you, step-by-step. First of all, what is 2? It is 1 + 1, so we have

$$2 = 1 + 1. \tag{A}$$

Next, what is adding 2 + 2? Let's write that down:

$$2 + 2 = (1 + 1) + (1 + 1). \tag{B}$$

These parentheses mean that you are supposed to add 1 + 1 in the first parenthesis, keep this first result, then add the 1 + 1 in the second parentheses, keep that second result, and, finally, add the first and the second results. On the other hand, let us write what 4 is. We saw that we call 4 the result of adding 1 to 3, so here it is:

$$4 = 3 + 1. \tag{C}$$

Now, what was 3? We call 3 the result of adding 1 to 2:

$$3 = 2 + 1. \tag{D}$$

Let us substitute this result for 3 in the previous equality; we get

$$4 = (2 + 1) + 1. \tag{E}$$

Then, finally, substituting 2 = 1 + 1 in this last equality we get (mind the parentheses!)

$$4 = ((1 + 1) + 1) + 1. \tag{F}$$

And now let us compare the right-hand sides of (B) and (F) above: $(1 + 1) + (1 + 1)$ and $((1 + 1) + 1) + 1$. They both contain four 1s, but they differ in the location of the parentheses. We have a law about these, called the Associative Law of the Sum, and this law tells us that the location of the parentheses does not make any difference when you are adding numbers.[8] So we conclude that expressions (B) and (F) are the same, and since the left-hand side of (B) was $2 + 2$ and that of (F) was 4, we have proved that $2 + 2 = 4$, do you see?

PS: Kind of. First you say, "mind the parentheses!," then you say that the location of the parentheses doesn't make any difference. This law you mentioned, that's what I don't quite understand.

T: The Associative Law. What is it that you don't understand?

PS: What kind of law is it: a federal law?

T: No, of course not. It is just part of the definition of the sum. Better call it the Associative Property of the Sum.

PS: I don't remember having ever seen a definition of the sum including this Associative Property.

T: There's the Peano definition of the sum, but I don't want to get into that because it involves mathematical induction, which is hard to explain, and takes too much time.[9] Easier just to show you how it works with sets. Here we have some objects on the table, in the event, poker chips. I take a chip, and put another on top of it, so here I have two chips, representing the number two. Now, separately, I take this chip and put another on top of it, so now I have two piles of two chips each, each representing the number two. If I lift this pile with two chips, and put it right on top of the other pile, that's like doing two plus two. On the other hand, if instead I pick this single chip on top and put it on the other pile of two, I'll get a pile of three, and then I place this remaining chip on top of the three, and I get four. In both cases I get the same pile with four chips, so doing two plus two is the same as doing two plus one and then plus one again. Got it?

PS: I'm not sure. Maybe it just works with poker chips.

T: Ha! That's a good one. You understand, of course, that the poker chips are only to help you visualize how it works, like when you were studying geometry and you drew the lines, or triangles, or circles, with chalk, or red pencil, or blue, and it didn't make any difference. Well, I did this demonstration here with poker chips, but I could have done it with pebbles, or buttons, or beans, or whatnot.

PS: I don't see that. If the maximum weight you can lift is one hundred pounds, and instead of these poker chips you are doing this demonstration with sixty-pound stones, you could put four of them on top of each other when you do it one by one, you could put a thousand or more stones on top of each other one by one given enough time, but you would not be able to add two stones plus two. Does that mean two plus two would be impossible?

T: Wait, no. It would just mean that—look, this is just like in high-school geometry—remember? One should not totally trust the diagrams; one should logically deduce everything from the axioms. Diagrams or poker chips are heuristic aids whose purpose is only to help you prove theorems from the axioms, logical step by logical step. In geometry you worked with five Euclidean axioms; now here, too, there are axioms, you see. There are the axioms of set theory, eight of them, and they are called the Zermelo-Fraenkel axioms.[10] And one of those axioms, called the Axiom of Unions, says that when you have any collection of sets, any sets, you can always take their union. The "union of those sets" means a set that contains all the elements of the sets in that collection, and only those elements. So, to return to your sixty-pound stones, this axiom of set theory ensures that even if you give me two piles of stones weighing a ton each, I will be able to put them together, or take their union or unification. Not with my hands, of course, not bodily, but in my mind. Got it?

PS: So you're saying that in your mind you can put together all sorts of things?

T: Right.

PS: And nothing happens to those things in your mind when you put them together?

T: That's a good question, and the answer is no, nothing happens. They don't change at all. They are always identical to themselves. You can take that as part of the Axiom of Unions, although one generally does not bother to mention it explicitly because it's true by virtue of the Identity Principle, which is even more basic: a logical principle that's before all these other axioms of set theory. It is the most basic of all logical principles. A law of thought, which states that every thing is identical to itself. We studied it last semester, in the introductory logic course. "For all x: $x = x$"—remember?

PS: Yes, and that's kind of funny, because I also remember you talking about another law of thought, and you also said that it was the most basic.

T: You are thinking of the Principle of Non-Contradiction, which states that we cannot hold a meaningful proposition to be both true and false at the same time and in the same respects.

PS: But this holding you're talking about, is it with your hands?

T: Of course not. It is with your mind.

PS: But you said just now that in your mind you can put together all sorts of things and that nothing will happen to the mental objects as a result. Right?

T: Yes.

PS: That is what doesn't make sense to me. I've got an example. Can you tell me of an important mathematical theorem that has not yet been proven?

T: Of many. For instance, there's the famous twin primes conjecture, that there are infinitely many pairs of prime numbers that differ by 2, like 3 and 5, 11 and 13, and so on.

PS: Well, imagine you come up with a proof that twin primes are infinite but then you drink so much in celebration that you wake up having forgotten all about it. You return to the problem and a month later you come up with a proof that there are only finitely many. In your euphoria you suddenly recall your previous happiness and the first proof, and so now you hold in your mind two proofs that are in contradiction. Wouldn't that change those two statements in your mind? It would put one or both proofs in doubt, it seems to me. Otherwise, if nothing happens and both proofs stay unchanged in your mind, all the proofs of all theorems in the world become nonsense. Didn't you tell us in Logic 100 that just one contradiction will make the whole edifice of math collapse?

T: Yes, that's true. But you see, I was speaking loosely when I said that in the mind we can put together or take the union of all sorts of things and nothing will happen to them. To be precise I should have said that the things math is concerned with, the objects we should be talking about, are of a very specific kind.

PS: So the axiom about the union of things works only for those things of a very specific kind?

T: That's right.

PS: And so the Associative Property, and two plus two equals four, also work only for those things of a very specific kind?

T: That is also right.

PS: Well, can you tell me more about that very specific kind?

T: I was trying to keep things simple, not counting on your clinging so stubbornly to details. But let's try a more logical route. Assume that there is a set called the empty set or null set, ø, which contains no elements. That's its definition: it contains no elements. And there's no arguing whether it exists or not, like people used to argue at least since Aristotle about the void—a space with nothing in it. There's no arguing because its existence is axiomatic, it is our starting assumption, directly guaranteed by one of Zermelo-Fraenkel axioms. This empty set ø is our initial object, our starting point that you must keep in mind. Now, unless you believe that All is One and the multiple is only an illusion, we will need more objects. Where will we get them? We cannot just pick the first thing that comes under our noses. For first of all, such things may not last, and we need eternal stuff. Pebbles, stones, even the stars in the universe have a finite shelf life. Furthermore, as you rather crudely pointed out, these objects must not ever change, not when you put a bunch of them together, nor when they are already together and you separate them. Unfortunately those two conditions are not easily met.

PS: Funny that you say that. Just before coming here I was reading that putting together and separating are the two basic operations of the Socratic and Platonic dialectics.

T: Well, that's ancient stuff better left to historians. The important thing is that the invariance of sets under separation is assured by another Zermelo-Fraenkel axiom, called the Axiom of the Power Set. It states that if you're given any set A, you can take all the parts, in other words all the subsets, of A and form a set with them. So, as a particular consequence of that, whenever you put together a bunch of objects, you can always still separate and recover any single object without change.

PS: Okay, but you still didn't tell me about what other objects to pick for the specific kind besides the empty set.

T: I'm coming to it. But first, do you agree that the empty set never changes?

PS: I guess so. What's there to change in it? Nothing will always be nothing.

T: Very good. And do you agree that if you hold in your mind the empty set and no matter what else at the same time, this will not affect your idea of the empty set?

PS: Yes. I don't know if I can form any idea of this empty set, but if I could, I think it would not be changed by other ideas, because, again, there's nothing to change in it.

T: Good. Now comes the masterstroke, so to speak. In set theory, the most basic operation is "to take the set of," "to collect into a set," or "to gather together," and it is usually symbolized by curly brackets: {}. Perhaps this is what your ancients were trying to get at by their putting together and separating, I don't know. What I do know is that if you have a bunch of things, say a, b, c, and you put brackets around them, $\{a, b, c\}$, this means "the set whose elements are a, b, and c."

PS: I remember this from Logic. By the way, your saying "to collect into a set" just reminded me that I'm taking a humanities elective on Greek and Latin etymology for premeds, and in it we learned that the words "collect," "lecture," "elect," "select," all come from the Latin verb *legere*, which means to gather or to pick. There is a related Greek verb, *legein*, which also means to gather or to pick, but it means many other things as well, like to speak, to read, to count or tell or reckon. Apparently the name of what we're having here, dia*log*ue, comes from that word and means "a talking through." Inte*llig*ence also comes from this word, and so does Plato's dia*lec*tics, which, as I just mentioned, is supposed to consist of two activities: col*lec*tion and se*lec*tion, putting together and separating. The corresponding noun, *lógos*, is also very interesting and was apparently even equated by the early Christians with God. Apparently all those words come from the same root!

T: I would advise you to stay clear of theology and the humanities. Stick to logic and mathematics and you'll get much farther. Now, to continue with our masterstroke. We are going to take a set of some things. I mean we'll look around for elements to collect in a set. So far we only have at our disposal the empty set ø, which, we agreed, never ever changes, right?

PS: Yes.

T: Well, since the empty set never changes, what happens if we take a set whose only element is the empty set, I mean, if we take {ø}: will it ever change?

PS: The way I figure it, for a set to change there must be a change in one or more of its elements, but since ø never changes, {ø} will never change either.

T: Very well! I want you, though, to appreciate the enormous jump we have effected: from the empty set without any elements, to a set with *one* element. From ø to {ø}. From zero to one. It's like the first step in the creation of a world, like a big bang! With this difference: the world we are creating here, this math world, is eternal and unchangeable. It is an independent world, created out of pure intelligence.[11]

PS: Excuse me, but is there such a thing as an impure intelligence?

T: You must stop focusing so much on the meaning of words. I'm using "pure" in the same way as when one says, "pure mathematics," meaning it's not applied to anything at all in the outside world. Anyway, so far we've got two eternal and unchanging math objects: ø and {ø}. If we want more objects—and we do—now, for the first time in eternity, we have two options: either choose as our third object the set {{ø}}, "the set whose only element is the set whose only element is the empty set," or {ø, {ø}}, "the set with two elements, namely, the empty set and the set whose only element is the empty set." As you agreed before, both of those choices have the virtue of being eternal, and both can be taken as definition of the number 2. For technical reasons, however, it is better to follow the second option and define 2 as {ø, {ø}}. This option was the idea of John von Neumann, the awesomely beautiful mind who engendered the Hilbert space formulation of quantum mechanics, the Theory of Games, and the hydrogen bomb. From now on, we will follow the same method, and define adding one to any number *n* as taking the set of all numbers from zero up to and including *n*. Thus, to list the first few:

0 is defined as ø;
1 is defined as {0}, that is, {ø};
2 is defined as {0, 1}, that is, {ø, {ø}};
3 is defined as {0, 1, 2}, that is, {ø, {ø}, {ø, {ø}}};
4 is defined as {0, 1, 2, 3}, that is, {ø, {ø}, {ø, {ø}}, {ø, {ø},
   {ø, {ø}}}};
and so on.

To define each number, throw all the previously defined ones into a new set, that is, put them, separated by commas, between brackets. Notice that it is enough to count the uninterrupted extreme-right-hand-side brackets to know which number is denoted by each set. Got it?

PS: I think so. So those are the natural numbers?

T: Right, plus the fact that there is another Zermelo-Fraenkel axiom that guarantees you can take a set, of course infinite, that contains all of them, so no smart aleck may object that there's not enough time in our lives to go on like this forever. Forever is already here.

PS: And so these are the objects of the very specific kind that you were saying math applies to?

T: These, plus all other objects that can be formed from the infinitely many I've just mentioned, by applying the operations guaranteed by the Zermelo-Fraenkel axioms: pairing, taking parts (subsets) of any set, taking the union of any collection of sets, and so on. In that way one gets *all* mathematical objects. Unless one postulates further ones, of course.[12]

PS: And if we work with only those objects we can always be sure that two plus two is four?

T: Exactly, because the Associative Property holds true.

PS: And all of that is based on the empty set ø.

T: On the empty set and on the postulation of the set consisting only of the empty set and so on, plus on all the other Zermelo-Fraenkel axioms. You could hardly ask for a more parsimonious, elegant, and environmentally friendly foundation for the most elaborate and immense construction ever produced by the human mind. Don't you think?

PS: It is amazing! But . . . doesn't that also mean that for all the other objects we can think of that are *not* built out of the empty set and these interesting axioms, 2 + 2 = 4 doesn't necessarily hold?

Our student could have asked the teacher yet another difficult question about sameness and difference here. The teacher presented two of what are in fact infinitely many ways of constructing the set of natural numbers 0, 1, 2, 3, . . . from the empty set. She offered

a. ø, {ø}, {ø, {ø}}, {ø, {ø, {ø}}}, . . . (von Neumann);
b. ø, {ø}, {{ø}}, {{{ø}}}, . . . (Zermelo).

Do these two *different* definitions define the *same* infinite set of natural numbers? And what is the meaning of the word *same* in that context? For instance, if m and n are *any* two natural numbers as defined by the sequence of sets in (a), then either $m \subseteq n$ or $n \subseteq m$, where the sign $\subseteq$ means "contained in." But this is not true when m and n are defined by the sequence of sets in (b). We will not pursue this question, but others have.

See Paul Benacerraf, "What Numbers Could Not Be," *Philosophical Review* 74 (1965), 47–73; collected in P. Benacerraf and H. Putnam, eds., *Philosophy of Mathematics: Selected Readings*, 2nd ed. (Cambridge: Cambridge University Press, 1983), 272–94; Barry Mazur, "When Is One Thing Equal to Some Other Thing?" in *Proof and Other Dilemmas: Mathematics and Philosophy*, ed. B. Gold and R. Simons (Washington, DC: Mathematical Association of America, 2008), 221–42; Philip Kitcher, "The Plight of the Platonist," *Noûs* 12 (1978), 119–36.

Wherein We Comment on Our Own Dialogue

The student's point seems sharp enough, but we will nevertheless expound on it. (For what would a pseudo-Platonic dialogue be without professorial commentary?)

Regarding the right-hand sides of equalities (B) and (F) above, $(1 + 1) + (1 + 1)$ and $((1 + 1) + 1) + 1$, the pesky student had remarked that they would not be the same—that is, we could not rearrange the parentheses—if we were dealing with sixty-pound stones (a far cry from the empty set!) but could lift no more than one hundred pounds. The example might have seemed far fetched, but there are in fact countless situations in our physical world in which the objects we are trying to add are not built out of the Zermelo-Fraenkel axioms and do not meet the conditions for the Associative Property. It is worth putting ourselves in place of PS and trying to imagine some of them—again, we will not bring up uncountable examples like "our own thoughts" but focus on countable things in the world—in order to develop a sense of just how often two plus two might not equal four.

Suppose we have a supply of items lying on the ground, "items" purposely unspecified, with the only proviso that they can be placed on top of, or next to, or together with each other. Imagine we have a mechanism, for example a crane or a forklift, with which we can move items and stack them on top of one another. By lifting *one item at a time* we can thus form stacks of height one, two, three, four, and more. The instruction "Do 2 + 2, that is, (1 + 1) + (1 + 1)," however, tells us to form two stacks of height two, then use the crane or forklift to lift one of those stacks and place it on top of the other. Now, suppose that you try lift two items: what might happen? Here are just a few possibilities:

a. Our forklift does not have the power to lift more than one item, in which case 2 + 2 is impossible. This was our Pesky Student's original idea.

b. When two items are lifted together, an interaction occurs, and they annihilate each other: then 2 + 2 will be just 2.

c. Two items lifted together fuse into one, so that 2 + 2 = 3.

d. If the two items are unchanged when lifted together, we do have 2 + 2 = 4.

e. Two items lifted together produce another, so that 2 + 2 = 5.

We have by no means exhausted the possibilities. 2 + 2 = 4 is in fact only one of untold possible situations for the interaction of objects in our physical world, and there is no reason to believe that reason must prescribe it. You might object that the only situation you have ever encountered in the everyday world is situation (d) until you reflect about all sorts of chemical reactions: for example, two molecules of hydrogen peroxide will split, if exposed to light, into two molecules of water plus one molecule of oxygen, an example of situation (e), if one takes "items" to mean "molecules" and "lifting" to involve "exposing to light." Or you may think of physical interactions such as the interference of coherent light beams, where two beams annihilate each other, an instance of situation (b). At very low temperatures you might suddenly encounter the fusion of two bosons into one, an example of situation (c).

We find situation (d) commonsensical because we tend to think of beings as solid bodies for which any interactions or changes (other than locomotion) are too slow for our senses to perceive. "Pebbles," for example. Or cement blocks. But even in the physical world, there are many things for which those conditions do not hold (we've just seen examples including subatomic particles, molecules, and waves, and we could have gone on and on), and the same is true in many other "worlds," including the ones we create in all of the languages in which we think and speak every day. (We mean our "natural languages," as they are sometimes called: the nonformalized, nonmathematical ones.)

Many a thought experiment could quickly confirm what may already be obvious to you: that the associativity for the composition of functions of the sort necessary for the truth of our $2 + 2 = 4$ often fails when we are communicating in ordinary speech. Perhaps even more obvious is the fact that our "laws of thought" do not apply to linguistic objects in the same way that they do to mathematical ones. For example, a word can mean one thing and its opposite simultaneously, thereby violating the Principle of Non-Contradiction.[13] Nor is a word's meaning always the same as itself (for a given word x, x need not equal x), not even at a given time within a specific context. Polysemy, ambivalence, the simultaneity of literal and metaphorical meanings, all these and many more features of our language are violations of the strict Principle of Identity.

So the student's point, at its simplest, is that objects, items, things, concepts, categories, beings, do not necessarily conform to the Zermelo-Fraenkel axioms, or even necessarily behave like pebbles or cement blocks. And when they do not—as they so often do not—many of our "certainties," even 2 + 2 = 4, need not hold.[14]

We insist on that point, ringing it with as many changes as possible, because it is forgotten so often that alternatives have become unimaginable, even (or especially) to those philosophers whose profession it is to imagine them. "Consider the following statements," writes Martin Stone:

> 1. 2 + 2 = 4. That is a fact.
> 2. 2 + 2 = 4. That is a "fact," given our shared interpretive framework. . . .

> (2) implies that we can make sense of a counter-factual possibility: If our interpretive assumptions were different, then 2 + 2 might equal 5. . . . When we start to explore these things, however, we find that we can't make sense of them as genuine possibilities. . . . We remove the quotation marks from "fact." . . . We realize that the very notion of "interpretive framework" is an illusion too. For if that notion means anything, it should be possible to make sense of our having a different framework.[15]

Those who cannot imagine a "different framework" have fallen victim to the situation so well described in Borges's "Blue Tigers," whose narrator could sooner imagine unicorns on the moon. They have again forgotten Nietzsche's warning: "without these articles of faith nobody now could endure life. But that does not prove them."[16]

The Choice between Pathic and Apathic

Mathematicians and logicians design axioms (e.g., the Zermelo-Fraenkel axioms) and rules (e.g., the Associative Law of the Sum) to guarantee that the mathematical objects they are working with suffer no change when we put any collection of them together in our mind (take the union of) and then separate them. Mathematical objects are always strictly *the same*, not out of necessity or because an alternative is unimaginable but simply as a consequence of the choice of mathematicians and logicians to play only by these axioms and rules and to deal only with objects conforming to them. Within these axioms and rules, mathematical objects can indeed become "something proved and demonstrated." But what about everything else?

We should pause for a definition here, one we have been anticipating in all previous chapters. We will call objects, items, things, categories, concepts, beings *apathic* if and only if whenever collected together or separated they remain the *same*. Everything else we will call *pathic*, meaning that it admits *difference* or suffers change. Our basic point throughout this book is that as we set out to ask any particular question of ourselves and the world, we risk falling into fundamental errors if we do not also inquire into our assumptions about the kinds objects of thought we are dealing with and about the approaches appropriate to those objects.

But please note: our definition does not imply that these two "opposite" conditions, *apatheia* and *patheia*, and their related sameness and difference, are *absolute*. On the contrary, we maintain that these conditions are absolute only within the conventions we have created in order to think about purely mathematical objects. In all other cases they are always *relative* to a situation, position, or question. "Asleep in your bed you are, from an earthy point of view, relatively apathic with respect to position: for eight hours (if you are lucky) more or less in the same spot. But seen from the sun, you keep appearing at different places." Items may be usefully thought of as apathic or same for the purpose of some questions and as pathic or changing for the purpose of others. "From the chemist's standpoint, putting together KOH (potassium hydroxide) and H_2SO_4 (sulfuric acid) results in thorough change: we get molecules of K_2SO_4 (potassium sulfate) and H_2O. It is violently pathic. But from the nuclear physicist's standpoint, the same atoms are still there: it is apathic." Indeed shifts in the emphasis of our attention, from sameness to difference, or vice versa, can themselves bring vast new fields of science into view. "For centuries, biology and natural history focused its attention on identifying the essential sameness that defined organic species and other "natural kinds." Darwin's attention to difference, that is, to the minute variability within conspecifics, discovered the preconditions for the operation of natural selection."[17] Items may be apathic in one circumstance but pathic in a different circumstance; they may be apathic when collected and separated by a certain mindset or in a certain time frame, but pathic in another mood or time frame.

In the remaining chapters we will provide numerous nonmathematical examples, ranging from physics to psychology to poetry, in order to stress that the pathic and the apathic are always relative. And the reason that they are relative is because—we return to the basic point of this book—sameness and difference are not absolute. Or to put it more dramatically, the choice between sameness and difference, between the apathic and the pathic, between emphasizing abiding sameness or the inescapable potential for difference, is not forced on us by nature or necessity. We are always free to make that choice differently or even not to make it at all. Often the customs and common sense of our cultures, the demands of our disciplines, our habits and predilections, the shapes of our souls and our relations, conspire to make us forget this liberty of choice and thereby to surrender a basic attribute of what we mean by human freedom.[18]

As we've seen in each of the previous chapters, many ancient and early modern philosophers were willing to give up much of that freedom, preferring to posit some kernel of unchanging sameness at the foundation of their questions.[19] In our own terms we might say that they preferred to extend *apatheia* and the Identity Principle— any being whatsoever remains the same as itself when undergoing union or separation—to as many types of being as possible, even those where this is not necessarily (or even plausibly) the case.

This has been the preference of important strands of modern philosophy as well. We have already encountered Gottlob Frege, whose *Basic Laws of Arithmetic* (1884) articulated the project of proving— as Zermelo-Fraenkel and von Neumann did above with the empty set—"that all pure mathematics follows from purely logical premises and uses only concepts definable in logical terms." But you will also recall that at much the same time Frege was musing about how the apathic conditions of logical and mathematical objects could be extended to natural language. His example: "the morning star" and "the evening star" refer to the *same* thing (the planet Venus) but have *different* "senses" in that they do their job of reference by pointing to *different aspects* of that *same thing*.[20] With such distinction between substance and aspects, Frege claimed to preserve in sameness the stable essence of *things*.

Since Frege there have been numerous attempts to bridge the broad ditch between the pathic and the apathic. Some, like Frege, continue the effort to build foundations for philosophy out of sameness and its derivative, similarity. Consider again David Lewis, whose writings are among the more influential and systematic of recent attempts to bind together all different registers of being, from the strictly logical and mathematical to the physical, the linguistic, the psychological. Throughout those writings, sympathy for the power of sameness limits the ability to imagine difference. In *Parts of Classes*, for example, he finds that the first operation of the Platonic dialectics, putting things together, is not problematic at all: it is "the innocent business of making many into one."[21]

"Cats" is the fetchingly feline category that serves Lewis as example for this innocent business of making classes. He creates the category by taking the "fusion" or "sum" of all cats, consisting in putting together—into the same sack, as it were—all the cats there are in such way that all the parts of cats—whiskers, tails, front halves, back halves, right sides, left sides, even each molecule in any cat—will be in this fusion or sack. We might want to ask further questions (are we talking of all cats imaginable—blue tigers, say, or poetic ones of fearful symmetry—or of biologically defined exemplars exclusively? Of live cats only, or of dead cats too? And what of Carroll's Cheshire Cat, or Schrödinger's cat, undecidably living and dead?[22]) But for Lewis this "cat fusion" presents no problems: "take them together or take them separately, the cats are the same portion of Reality either way."

Reality is capitalized in the original, maybe to signal its independence of the mind that's gathering the cats, although in Reality, if those cats were thrown into the same sack, there would be not just fusion but violent confusion from which no cat would emerge unchanged and in which it would even become difficult to decide whether a given molecule belonged to one cat or other. (The potential for unpleasant change in such situations is what prompted our premodern predecessors to drown parricides in sacks accompanied by animals: the medieval Saxons favored cats.)[23]

Curiously, what Lewis finds less innocent is the set-theoretical possibility of taking a thing x, then taking the "singleton" $\{x\}$ (the set containing only x), and then iterating the operation: $\{\{x\}\}$, etc. This, he says, is making many things out of one, which he finds much more problematic that making one out of many. He also finds distasteful the very idea of the empty set, with what he calls its whiff of nothingness. But since the empty set is the price we have to pay for the foundations of mathematics, he concedes we have no choice. There is no disputing about taste. But just as distasteful, from our point of view, is the suggestion that cats, humans, indeed everything that is should be considered apathic, unchanging under the basic operations of the mind.

As a last example, consider yet another effort to resolve the sharp division between apathic and pathic, this one by a mathematician, Gian-Carlo Rota (1932–1999). With this one we are more sympathetic, perhaps because one of us knew him, or perhaps because Rota understood the divide as a split in the psyche, one that has to be dealt with.[24] But how to deal with it? Like Husserl, Rota neither wished to curtail the certainty of mathematical truth nor to diminish the power and architectural beauty of mathematical constructions. Rather, he hoped to build bridges between apathic reason and everyday experience, to connect more firmly the fundamental concepts of logic and mathematics with the *Lebenswelt,* and so conjure the dangers of dehumanization, which lurked, as he (and Husserl) thought, under "our excessively abstract modernity." The question, of course, was how that could be achieved. How could the axioms of logic and math be so modified as to start building those bridges and, dare we hope, conciliate rigorous thought with human freedom?

Rota's idea was not to replace the axioms, since they are the fountainhead of mathematical certainty, but to add to them further axioms related to our human experience. Rota reasoned more or less as follows. Set theory is built on two basic formalized relations, $x \in y$ (x belongs to y) and $x \subseteq y$ (x is contained in y); why not formalize other relations and perhaps add them to the axioms? Some possible relations, now "concealed" (so Rota and Husserl thought) by the exclusivity of the above two, are "*x* lacks *y*," "*x* anticipates *y*," "*x* haunts *y*." In a paper published in 1989, Rota held one of Husserl's relations as especially promising, namely, the relation of *Fundierung*, or "backing up."[25]

Fundierung here is conceived as a binary relation, that is, a relation between two elements, of which one is called the *facticity* and the other the *function*. The facticity is a self-standing support for the function. For example, the queen of hearts in the French deck, the *factual* card, can serve as support or backup for different functions, for example, in canasta, poker, or bridge. This written text, this paragraph, can serve as *Fundierung* or support for its content, viz an explanation of Husserl's *Fundierung* (other texts can do that too). Incidentally, Frege's distinction between *Sinn* and *Bedeutung* is a strategy related to *Fundierung*, one that can be put in its terms as well: the planet Venus is the one facticity that supports the two different functions, morning and evening star. Rota concluded his paper by asking "In the formalization of logic (or of set theory) what would result by incorporating the notion of *Fundierung*? Would the notion of set have to be radically altered or even entirely done away with?"

The answer to that question must depend on *how* one incorporates the notion of *Fundierung* to, say, the Zermelo-Fraenkel axioms. But we can already eliminate the hope for conciliation. If, in order to try to get the conceptual edifice of logic and math "closer" to our everyday world (as was Husserl's and Rota's ambition), one imposes the restriction that all axiomatic concepts of set theory be *fundiert*, that is, be supported by something factual, the first thing one would have to delete from the Zermelo-Fraenkel axioms would be the empty set, which in its utter lack of content is the least factual "thing." In other words, in order to bring logic and mathematics closer to our everyday world in this way, we would have to give up the object that guarantees the apathic foundations of mathematics and thereby assures us that there are any apathic objects in the world at all.

What We Are Saying and What We Are Not

In this chapter we have stated what we take to be two truths. First, number (by which we mean here axiomatized mathematics more generally) requires *apatheia*. Second, with the exception of number, nothing is absolutely pathic or apathic, but only contingently so. Contingent, that is, on our perspective, on the questions we are asking and the shape of our attention, on time and time frame, and indeed on an infinity of other things.

What does this mean for what we know and how we know it? Here we have to be very careful, for it is easy to become dogmatic. Honoré de Balzac, for example, began in much the same fashion as this chapter did, but he reached quite a different conclusion: "you will never find in all nature two identical objects; in the natural order, therefore, two and two can never make four, for, to attain that result, we must combine units that are exactly alike, and you know that it is impossible to find two leaves alike on the same tree, or two identical individuals in the same species of tree." Balzac may be right about the limits of strict identity among leaves (he is repeating Leibniz), but we do not find it helpful to ignore, as he so willfully does here, the choices through which we can fruitfully apply number to the natural world.

Balzac's next sentence is equally problematic: "That axiom of your numeration, false in visible nature, is false likewise in the invisible universe of your abstractions, where the same variety is found in your ideas."[26] That depends on the ideas. Number, for example, is an idea, and it is one that, as you've just read at some length, can be axiomatically rendered apathic. The power of that axiomatization is immense, and we gain nothing—indeed we lose an enormous amount—by denying it. It is an error to consider the truth of 2 + 2 = 4 to be absolute, but it is just as much an error to belittle the power of that truth.

If most of the people we have encountered thus far cleave to the first error, there are also plenty who, like Balzac, incline toward the second. In 1826, a few years before the publication of Balzac's musings, Goethe is reported to have said that

Mathematics . . . is totally wrong in claiming to provide infallible conclusions. Their whole certainty is nothing else than identity. Two times two is not four, but it is just two times two, and that is what we call four for short. But four is in no way anything new.[27]

We can feel in this passage Goethe's antagonism toward the colonizing power of Newtonian science—an antagonism he shared with other contemporary poets (such as Blake and Coleridge) and philosophers (such as Schelling and Hegel)—spilling over into his views on math. Against Kant's granting to math the status of synthetic a priori truth, Goethe claims instead that mathematical truths such as 2 + 2 = 4 are analytic, that is, mere tautologies.

In this he was wrong, for as you know, it is by virtue of the Associative Property of the Sum that 2 + 2 = 4 holds true, and this property is not a tautology, as Goethe maintained, or a synthetic a priori truth, as Kant wrote. It is not even a universally verified empirical fact (like, say, gravitation). It is rather a convention we enter into, a postulate we accept in order to do math, just as we accept certain rules in common when we sit down to play poker.[28] Still, Goethe is right in one sense. These common rules require our commitment to apatheia and the Principle of Identity, that is, to our thinking of objects as if they remained strictly the same as themselves whenever separated or combined. In this sense, the whole certainty of math indeed "is nothing but identity" (weiter nichts als Identität).

We interpret this "nothing but" much more generously than Goethe did. Nor can we agree with Nietzsche's critique in *Human, All Too Human*: "Number.—The invention of the laws of numbers was made on the basis of the error, dominant even from the earliest times, that there are identical things (but in fact nothing is identical with anything else); at least that there are things (but there is no 'thing')."[29] Unlike Goethe or Balzac, Nietzsche's aphorism, if rightly interpreted, cannot be said to be wrong. We disagree with it in only one respect: its *categorical* rejection of the Identity Principle and corresponding affirmation of flux: "but there is no 'thing.'" Our position is rather that it is up to us whether to wager on sameness or difference, identity or flux. The decision about whether there is or is not a "thing" in our world is our own terrifying choice to make.

There is an immense amount of utility in opting to treat certain objects of thought as apathic, and unlike these last writers, we should not deny that utility. But neither should we forget that the choice between the apathic and the pathic, sameness and difference, is just that, a choice we must make as we ask questions of ourselves and of the cosmos. For many questions it is extremely helpful to extend the power of apathic sameness into realms where it does not strictly hold—as we do, for example, when we go to bed unconcerned about the velocity of our movement relative to the sun. For other questions—including many basic questions concerning our fears and loves, our conflicts and desires—it may be less helpful, indeed dangerous.

The task before us, then, is above all to become more conscious, more critically self-aware, about how and when we make that choice, for what purpose, and with what attendant consequences for our perspective onto the questions we are asking. In strictly apathic circumstances the axioms of set theory may, as Gödel proclaimed, "force themselves upon us as true."[30] In all other circumstances it is we who choose to force them on a pathic world. What is gained by such force, and what is lost? The answer to the first question is much easier to see than the answer to the second. Human knowledge, indeed the possibilities of life (and not only human life) on this planet, have been dramatically transformed by our mathematical extension of the laws of sameness and apatheia to an overwhelmingly complex and ever-changing cosmos where those laws do not strictly apply. Only a god or an ideologue would dare offer "cost-benefit analyses" of such vast phenomena. But in the chapters that follow, we will focus on areas where we think it particularly important not to forget the nature of our choices: namely, in the study of our physical world, of our social world, and of our psyches, or inner world.

7

PHYSICS (AND POETRY)

Willing Sameness and Difference

Theory first decides upon what one can then observe.

ALBERT EINSTEIN to WERNER HEISENBERG[1]

We must bridge this gap of poetry from science.
We must heal this unnatural wound.

JOHN DEWEY[2]

The previous chapter concluded with two truths that are not self-evident but that are vital for human flourishing:

- Number (by which we mean here axiomatized mathematics more generally) requires *apatheia*.
- Nothing is absolutely pathic or apathic but only contingently so.

By contingent we meant dependent on our perspective, on the questions we are asking and the shape of our attention, on time and time frame, and on many (indeed an infinity of) other things. Outside of the realm of purely mathematical objects of thought, in which we have agreed by convention to sacrifice contingency to axiom, statements such as 2 + 2 = 4 do not express absolute and necessary truths. They are not the product of the way things are and ever were or of divine decrees or of the inescapable nature of our mind. They are rather the product of human habit and choice, of a preference for the apathic or the pathic. In this sense we can say that all of our knowledge is conditioned by our will to sameness or difference.[3]

Given how fundamental the choice is, and how rarely we are conscious of it, it is tempting to stage the choice as some eternal conflict between conflicting drives: a will to sameness locked in struggle with a will to difference. But we must not forget that the two wills are in inseparable relation. Sameness and difference are always simultaneously expressed.

This is true even in pure mathematics. When, in set theory, we take the set of all objects that have a certain property p, we are exercising our will to sameness, since we have put all those objects together under the same umbrella, as it were. But by the same action, we have established a difference between *those* objects and all the others. "Wait!" you will object. Had we taken the empty set ø, we would have established no difference at all between *some* objects and *the other* objects, since in the empty set there are no *somes* or *others*. You are right that the empty set is the exception: the only noetic object that is pure sameness with no admixture of difference. But the "taking" of the empty set was itself a choice between "that which has no some and other" (the empty set) and all *others* that do have some as opposed to other. The sameness of the empty set is produced by a willing of difference.

For every object of knowledge we have a choice of focus and attention, a choice between willing sameness and willing difference (or perhaps not willing at all). That choice is, however, often prejudiced in the literal sense of the word: prejudged, the product of prior habits and commitments, theory laden. Among the more notable of those prejudices is the yearning for an enduring sameness at the foundations of knowledge. "Human intelligence is given to abstraction by its very nature, and it pretends to find constant those things which are in flux." Thus Francis Bacon in his *Novum organum scientiarum* (The new instrument of the sciences) of 1620, where he presented this yearning as one of the "idols" whose worship misled science.[4]

If those we nowadays call scientists have often stood proudly among the idol worshippers, it is because their faith has been amply rewarded. The discovery of "samenesses"—laws, equations, constants, regularities, repetitions, identities—has produced momentous advances in physics, chemistry, and many other sciences. Our goal in this chapter is not at all to question those advances. It is only to point out that each of these "samenesses" is contingent. The truths of the physical sciences, unlike those of the empty set and the mathematical logic built on it, are the product of the shape of our attention to sameness or difference. Since this may seem ridiculous—can we possibly mean, for example, that the laws of physics do not hold everywhere and always the same?—we will dedicate much of this chapter to describing in what senses it is true.

Let's begin by invoking authority. Émile Meyerson's *Identity and Reality* appeared (in French) to great acclaim in 1908. Albert Einstein read the work with appreciation, and made sure to visit the chemist-philosopher on his trips to Paris. Like Francis Bacon centuries before, Meyerson (1859–1933) noted the human thirst for constants—"identity is the eternal framework of our mind"—and suggested that "science is penetrated" by this framework. He organized his chapters around important manifestations of this will to sameness in the various sciences: "Law and Cause," "Mechanism," "The Conservation of Matter," "The Conservation of Energy," "The Unity of Matter." He demonstrated how each of these powerful principles of sameness in the world turns out, with a shift in question or perspective, to dissolve into the discovery of new differences as well. Meyerson sought to show that for all the productivity of "identity" and "equality" in every science, every science that wishes to gain purchase on reality must resist the urge to sameness on which it simultaneously depends:

> I break a plate, I put together its pieces, none is lacking, so I shall not hesitate to express this fact by an equation: calling the plate A and the pieces B, C, and D, I shall write A = B + C + D. This equation seems to affirm that there is equality between the two conditions of the plate; yet I know very well that if I were to try to use it now I would have a mishap. These things do not shock us; we have not for a moment forgotten that at bottom it is here a question of a diversity, not of an equality. Science, however, often goes far in this line of thought.[5]

Where Meyerson distinguished between different types of commitments to "sameness," different sciences, and different objects of knowledge, other philosophers of science continued to dream of the unity of all knowledge and searched for a "sameness" capable of serving as a foundation for everything that can be known. We have already met some of these fanatics of sameness circling round Vienna after World War I, among them Rudolf Carnap, who maintained in his modestly titled *The Logical Structure of the World* (1928) that "there is only one domain of objects and therefore only one science."[6]

Like Meyerson, we have repeatedly insisted that this yearning for "the one" is blind to some basic truths about the nature of our knowledge of the world. In this chapter we will insist on the point once again by focusing on the physical world. Can our knowledge of space, of energy, mass, and momentum, of electrons and other elementary particles, or of any other physical aspect of the cosmos be fully grounded on an apathic, unchanging sameness, as John von Neumann and other mathematicians succeeded in doing for their knowledge of mathematical objects with the empty set?

We've already told you that our answer will be no. The application of mathematics to physics has produced great monuments to the power of human knowledge and given us deep insight into the universe, from its smallest to its largest scale. But mathematics has not eliminated the choice of sameness and difference in human investigations of the universe. In this sense, and despite all efforts to identify physics (and other sciences) with mathematics, physics (like the other sciences) retains a kinship with other domains of knowledge, such as poetry and myth, that take more pathic routes in their explorations of our relations to our own being, to other people, to other forms of life, and to the world. That is the point we seek to make in this chapter. You will notice that in it, as we talk about science, we will sometimes do so in terms of poetry. We do so in order to perform our point: even the most mathematical of physical sciences retain some relation to more pathic ways of knowing, such as poetry, and can learn from them (and vice versa) how to reflect on sameness and difference.

The Celestial Abode of Sameness

Across the millennial history of knowledge we have skimmed thus far, we've already found numerous candidates presented as foundations of sameness. All of them have turned out to be false, the

product of ignorance, habit, prejudice, or theoretical commitments. Recall, for example, the Pythagorean, Platonist, Aristotelian, and Scholastic belief that the heavens were the abode of pure
sameness: that the sun and the other stars moved in periodic paths
as immutable as the laws of number (and music). The contrast between those eternal and mathematically regular heavens and this
lowly earth of ours, this world of pain and anguish where beauty
and lust are usually followed by decay and oblivion, has been the
theme of much religion and poetry as well as philosophy and science. Equally familiar is the ontology, the hierarchy of truths about
being, based on this distinction between the mutable and the immutable, between "the great concert of eternal lights wheeling
about the heavens" and the baseness of the earth. Above our world
of vain appearances, the heavens and their astronomical laws promise eternal truths.[7]

In November 1572 something hitherto unnoticed appeared in the
sky. A bright new star, today known as SN 1572 (a.k.a. Supernova
B Cas or Tycho's Supernova), was observed in the constellation
Cassiopeia. The Dane Tycho Brahe was one of several sky watch-
ers who discovered the phenomenon more or less simultaneously.
Another was Jerónimo Muñoz, a Valencian mathematician, astronomer, and scholar of Hebrew who published his account of
the discovery in 1573 under the title *Libro del nuevo cometa* (Book
of the New Comet).[8] Comets were a familiar phenomenon, long
ago (albeit incorrectly) assimilated by Aristotelian theorists into
the "sublunary" realms of flux rather than the eternal heavens. But
despite the title of his account, Muñoz's own parallax calculations
showed this to be a star much farther out than the moon. Muñoz did
not shrink from the momentous implications of his findings: stars
change. The heavens are not the abode of sameness. Aristotelian
(and Platonic) cosmology should be rejected.[9]

This is an example of the collision between one foundational
fantasy of sameness—in this case that of the pure sameness of the
extralunar heavens—and one mathematically assisted discovery in
astronomy and astrophysics. Galileo Galilei's condemnation by the
papal Inquisition in 1633 represents an even more famous example
of such a collision, one often repeated in the more heroic narra-

tives of science's struggle against religion.[10] But scientists are just as prone to the yearning for foundations of sameness as theologians are. The scientific revolution canceled the Platonic-Renaissance polarity, the charged difference between sky (always the same) and earth (ever different), and replaced it with an even more *universal* claim to sameness.

The laws of motion worked out by Galileo and Newton, as well as the pull of gravitational forces, held *equally* for heaven and earth. A new synthesis became possible wherein heaven and earth are ruled by the *same* laws. In his *Rules for the Direction of the Mind* René Descartes, Galileo's great (but more discreet) contemporary, gave as his first rule the intrinsic unity and unchanging sameness of all true knowledge:

> For the sciences as a whole are nothing other than human wisdom, which always remains one and the same, however different the subjects to which it is applied, borrowing no more distinctions from them than sunlight does from the variety of things it illuminates.[11]

Or in the words of the greatest rationalist of the next generation, Gottfried Wilhelm Leibniz, "Whoever understands any part of matter, understands the whole universe on account of the mutual union and communion of which I spoke. My principles are such that they can hardly be separated from one another. Who understands well one thing, understands them all."[12]

What is wrong with this fantasy of the sameness of all knowledge? First, and perhaps least important, it founders in the expanding ocean of our knowledge about the physical world. Every candidate for foundational unity disappears under those ever-rising tides. Descartes's invocation of sunlight as evidence of the sameness of our knowledge is in this sense typical. Unlike Descartes, we now know that light does not remain the same regardless of the things it illuminates. If it falls on a mirror, the reflected light remains very nearly the same. But if it falls on a leaf, a good portion of the light is put to work in the process of photosynthesis that sustains the plant.

Could it be that human wisdom, too, in spite of Descartes, Leibniz, and many others, differs according to the subjects to which it is applied? If so—and this is the second problem—then succumbing to fantasies of *sameness* undermines our ability to cultivate our capacity for such *difference*.[13]

Space: Indifference at the Frontiers of Geometry

Now that the stars have failed us, what else in the universe could be thought of as an abode of pure sameness? Another venerable candidate has been the idea (or intuition, as Kant called it) of space. Not the space of the universe crowded with mutable stars and galaxies but space as we might imagine it used to be before creation: starless, dark, empty, a container never corroded or in any way affected by whatever is put in it, perfectly apathic, always the same.[14] The science of such space the Greeks called geometry, and its elements have been passed on to us by Euclid. So let us return for a moment to the first pages of Euclid's *Elements*, a book whose influence on Western thought could be compared to that of *Genesis*.

In the beginning we have points and we have lines. That those two kinds are deficiently defined by Euclid need not trouble us, for when it comes to basic, elementary stuff, we cannot expect it to be reducible to things even more basic. What really matters is how points and lines are related. We would like to state, "If P and Q are two *different* points, then there is one and only one straight line passing through P and Q," and "If P and Q are the *same* point, then there are any number of straight lines passing through P and Q."

Those are fundamental propositions; indeed, they are taken by Euclid as axiomatic. But in fact he was not, nor are we, logically allowed to utter them because we have not yet defined the meaning of *same* and *different* in reference to points. Should we then try to define those notions in a manner that accords with our geometric intuition? We direct our eyes to a region in space and imagine a point there, P; then we move our eyes to a different region of space (the sensible motion of our eyes tells us that the second region is *different* from the first) and imagine there a point Q. We define P as *different* from Q since they belong to different regions. If the points P and Q are too close for us or other humans to perceive that we have to move our eyes to go from one to the other, the situation is more delicate. But if we can imagine some instrument that can register the motion required to go from P to Q or vice versa, we will still say that P and Q are *different*. In the unreal, ideal, limiting case, if no difference can be perceived, we will say that P and Q are the *same* point.[15]

Here, however, our troubles begin. Let P and Q be different points in the above sense. Now let us imagine that we interchange P and Q. We cannot tell the difference. Interchanging the actual entities P and Q has no more effect than interchanging their names given by us. In conclusion, P and Q are indistinguishable. Points in Euclidean space carry no marks or qualities whereby we could tell them apart. As Euclid put it, "Point is a sign that has no part," or perhaps, "point is a sign of that which has no part" (σημεῖόν ἐστιν, οὗ μέρος οὐθέν).[16] The problem then is how can *indistinguishable* entities be *different*? That would violate Leibniz's principle about the identity of indiscernibles.[17] When David Hilbert axiomatized synthetic geometry for the modern age, he did not provide a strict logical definition of points, nor did he define when two of them are the same. Instead he finessed the problem and simply wrote, "For every two points A and B there exists no more than one line that contains each of the points A and B." Perhaps he was assuming that since we name the two points with different letters, and since we say that there are two of them, that *by itself* implies that A and B are *different*?[18] That might please dame Intuition, but it offends dame Logic.

These difficulties with points have been discussed for more than two millennia. Is a point a real entity or a mere name, a *flatus vocis* ("air of the voice," as medieval nominalists called words without corresponding realities)? Or perhaps, it is just a *flatus*? Jonathan Swift characterized the debate with his usual wit in 1722: "The Mathematicians steer'd a middle Course between the Naturalists and the MetaPhyz-icians; they own'd a fart to be a Quantity yet Indivisible, and gave it the name of Mathematical Point, as having neither Length, Breadth, nor Thickness." Two centuries later, the

French poet Paul Valéry looked back across those millennia and summed them up in two sentences: "Greece founded geometry. It was a senseless venture: we are still *discussing* the *possibility* of that folly." But since this is a matter of math, perhaps we should give the last word to a great mathematician rather than to a poet he inspired. Henri Poincaré pointed straight at the extrapolation of human habits and familiar experiences of sameness into a mathematics of space: "Geometry would not exist if there were in nature no solid bodies that move without being modified."[19]

Hilbert and Poincaré, Husserl and Valéry: these mathematicians, philosophers, and poets of modernity were struggling with a question very similar to that of the ancient Greeks. Recall the *ápeiron* (the unbounded), offered by Anaximander as the foundation of all things. We can view the *ápeiron* as the collection of points in Euclidean space, before the introduction of lines or planes. In such a collection there are no limits or boundaries, and any distinction or individuation must arise from the arbitrary introduction of lines or planes as boundaries. But as candidates for the title of abode of sameness, both the *ápeiron* and our classical, Euclidean idea or intuition of space fail the tests of logic (specifically the Principle of Sufficient Reason and the Identity of Indiscernibles). They also fail the test of our expanding knowledge: nowadays the physicists' conception of space-time is not apathic. General Relativity, for example, conceives of space-time *not* as a huge, empty stock pot but as a Riemannian manifold whose shape and curvature depends on the masses contained and distributed therein.

Sameness and Difference in Mechanics

The logical difficulties involving the sameness or difference of points are not confined to geometry. They afflict many aspects of the physical sciences, including the mechanics so famously proposed by Isaac Newton in 1687 in his *Philosophiae naturalis principia mathematica* (*Mathematical Principles of Natural Philosophy*), where he stated his three "Axioms, or Laws of Motion." Let's look only at the Third Law, which holds that when one body exerts force on another body, the second body simultaneously exerts a force on the first body equal in magnitude and opposite in direction. Newton offers some examples to explain this law of action and reaction:

> If a horse draws a stone tied to a rope, the horse will (so to speak) also be drawn back equally towards the stone, for the rope, stretched out at both ends, will urge the horse toward the stone and the stone toward the horse by one and the same endeavor to go slack and will impede the forward motion of the one as much as it promotes the forward motion of the other.[20]

Many students of physics have wondered how Newton's Third Law could be true, since if any force applied to a body elicits an equal force in the opposite direction, it would seem to follow that the body would not move. Yet bodies in this world *do* move. If we have been fortunate in our teachers, someone explained to us that, in the example above, Newton did not bother to mention another pair of forces, equal but opposite: the force the horse exerts on the ground, directed backward, and the opposite force that the ground exerts on the horse, toward its head, hence opposite to the stone and tending to move it. When we add the two forces acting on the stone, if the sum is greater than the force that the stone exerts on the horse, the stone will move. Everywhere, though, we must bear in mind that the Third Law applies *only* when we have two *different* bodies, like the horse and the stone, or the horse and the earth. If it were mistakenly applied to the *same* body, it would indeed follow that the body in question cannot move.

The problem may seem trivial when we are dealing with horses and stones, but it becomes serious when we deal with objects whose sameness and difference we feel less confident in judging. The great nineteenth-century Viennese physicist Ludwig Boltzmann well understood that the ghosts of indistinguishability haunting geometry "are transmitted equally to all views concerning the principles of mechanics since all must start from space and time."[21] Boltzmann realized that if we imagine the tiniest physical objects (e.g., molecules and atoms) as mass points, then points endowed with the same mass, imagined at rest, are *not distinguishable*, violating Leibniz's

Principle of the Identity of Indiscernibles. Hence, Boltzmann held that the *distinguishability* of each mass point had to be a fundamental assumption of statistical mechanics, which deals with molecules and their motions. In other words, Boltzmann exercised a priori his will to difference, providing distinguishability as an axiom.

By now you won't be surprised if we point out that the choice was not the only possible one. Like Boltzmann, James Clerk Maxwell (1831–1879) had a keen interest in the kinetic behavior of gases (many contributions to thermodynamics bear his name, such as Maxwell-Boltzmann statistics and Maxwell's demon). He was also among the first to recognize fundamental relationships between electricity, magnetism, and light, and his mathematical treatment of electromagnetism ("Maxwell's equations") provided a powerful model for describing the propagation of waves in electrical and magnetic fields. But the reason we mention him here is that, unlike Boltzmann, Maxwell was willing to embrace the possibility that there is "no variation, or rather no difference, between the individuals of each species" at the molecular or the atomic level of a thermodynamic system. He was even willing to extend this elementary "sameness" throughout the universe, pointing to the astounding fact, only just then becoming established through spectroscopy, that light from the most distant stars displayed exactly the same spectral patterns as light on earth, implying that the atoms that produced that light were exactly the same.[22] Perhaps even more astoundingly, Maxwell was willing to conclude that if the physical sciences had reached the limits of distinguishability, then humanity had reached the limits of the laws of thought and reason.[23]

In short, Maxwell was willing to entertain the insufficiency of the Principle of Identity and other laws of logic at the physical foundations of the universe. (We should point out that this commendable lack of scientific dogmatism was in part facilitated by a different kind of dogmatism: Maxwell's desire to preserve the place of a creator God.) Others, less willing to give up on the distinguishability of points, tried to solve the difficulty by appealing to motion. Geometric points per se do not move, they maintained, but elementary particles, atoms, and molecules do. We could try to individualize such particles by their movement, creating devices to follow a particular trajectory through space in time. But this solution, too, runs into difficult problems of sameness and difference. We follow our particle (let us call it *P*) up to a point where it vanishes for a split second then reappears nearby (let us rename it *Q*) and keeps going. We may conjecture that *P* is the *same* as *Q*, but how do we know? It is perfectly possible that *P* vanished from sight for good and *Q* is a *different* particle that happened to be nearby.

It was in order to address this kind of problem that Boltzmann made a further assumption for his mechanics: the trajectories of his particles had to be continuous. This assumption was not new. Its general formulation, *Natura non facit saltum*—Nature does not jump, but moves continuously from one state to another—is attributed (yet again) to Leibniz. But this principle, too, would soon prove untenable with the discovery of the world of quantum physics.[24]

But wait (you might ask): *in reality*, are those atoms and particles different or not? To which we can only reply with Meyerson: "considerations of conservation, of identity, intervene at every step in empirical science, which is, in spite of appearances, saturated with these *a priori* elements."[25] Or to quote Einstein (as reported by Heisenberg), "Theory first decides upon what one can then observe." And among the first decisions theory makes for us is the choice between sameness and difference.[26]

Does Entropy Depend on the Observer?

The previous example focused on the contingency of knowledge on the sameness or difference of the object observed. We can also focus on the contingency that arises from the sameness or difference of the observer. Entropy provides a good example. One way of defining *entropy* is as the measure of how much mechanical work can be extracted from a system or apparatus. Ever since the Industrial Revolution, humans have become familiar with how differences in temperatures can be used to produce mechanical work: a hotter gas exerts net pressure on one side of a piston (pressure being roughly proportional to temperature), the piston moves, transmits its motion to a wheel or an axis, then moves back, enter more hot gas, and the cycle repeats.

Many basic concepts of thermodynamics have been derived from thinking about engines of this sort, among them the Second Law of Thermodynamics: the total entropy of an isolated system cannot decrease over time, a law first formulated in 1824 by the French engineer Sadi Carnot in his *Reflections on the Motive Power of Fire*.

Another way of conceiving of entropy is by interpreting temperature as a measure of the average kinetic energy of the molecules. When the hot gas is separated from the cold gas by a diaphragm, we have most of the high-energy molecules in one container, and most of the low-energy molecules in the other—there is a measure of discrimination among the molecules according to their kinetic energies. The greater the discrimination, the lower the entropy and the higher the amount of work we can extract from the system. Accordingly, entropy is often called a measure of the disorder or lack of discrimination in a system. If we let the gases intermix until a state of equilibrium is reached, entropy is at a maximum and no work can be extracted from the system. Conversely, this increase in entropy is a measure of how much work from the outside would be necessary to *return* the system to its original state, before the diaphragm was opened.

Circa 1875 the Yale professor Josiah Willard Gibbs discovered the following remarkable fact. Put gases at the *same* temperature and pressure into two separate containers connected by a closed valve. Now open the valve, allowing the gases to diffuse through each other. If the gases are *different*, after equilibrium is reached there is an increase in entropy that does not depend on the nature of the gases but only on the fact that they are different. When the two gases are the same, there is no increase. The results were later considered paradoxical because they suggest that as gases become indistinguishable, entropy becomes indefinable. This question of indistinguishability is clearly important for anyone concerned with sameness and difference.[27] But for the moment we will stress a different implication: here, in this measure of entropy, we seem to have a purely *objective* distinction between the sameness or difference of the gases![28]

But like so many things that seem to be purely objective, this one turns out to be only apparently so. Consider the scenario imagined by another American physicist, Edwin T. Jaynes, more than a century after Gibbs's thought experiment.[29] Today, so far as we know, there is only one kind of argon. But we may imagine that in the future, say fifty years hence, two different kinds will be discovered, "A_1 and A_2, identical in all respects except that A_2 is soluble in Whifnium, whereas A_1 is not (Whifnium is one of the rare superkalic elements; in fact, it is so rare that it has not yet been discovered)." Physicists of the future will be able to use the apparatus with A_1 in one container and A_2 in the other to identify an increase in entropy from the difference in the gases—from which they may also be able to extract a degree of work—which we, today, are unable to do, since as far as we know all argon is the *same*.

"In this scenario," Jaynes writes,

> our greater knowledge . . . leads us to assign a different entropy change to what may be in fact the identical physical process down to the exact path of each atom. But there is nothing 'unphysical' about this since that greater knowledge corresponds exactly to — because it is due to — our greater capabilities of control over the physical process. . . . It would be astonishing if this new technical capability did not enable us to extract more useful work from the system.

To put it in terms of our discussion, far from being objective, entropy turns out to be subjective in the sense that it depends not only on our own size, nature, and senses but *also* on how much we know or what it is we want to know about the objects we are studying.[30]

"Entropy is an anthropomorphic concept," Jaynes concluded, attributing the remark to his teacher Eugene P. Wigner (1902–1995).[31] Wigner was, like John von Neumann—with whom he had shared a math teacher during their time as schoolboys in Hungary—a giant in the application of mathematics to physics. (He was awarded the Nobel Prize in Physics in 1963 for his "contributions to the theory of the atomic nucleus and the elementary particles, particularly through the discovery and application of fundamental symmetry principles.") His use of the word "anthropomorphic" is more significant than it might seem at first blush.[32] Max Planck, "founding father" of quantum mechanics, had proclaimed in a 1909 lecture at Columbia University that modern science is characterized precisely by the fact that it is moving ever farther *away* from "anthropomorphic" considerations.[33] (In other words, if there is intelligent life elsewhere in the universe, or if divine beings exist, those beings ought increasingly to agree with us on our math and our physics.)

Wigner certainly hoped that the powers of math would extend to more and more domains of learning. As he put it in one of his more famous essays,

> The miracle of the appropriateness of the language of mathematics for the formulation of the laws of physics is a wonderful gift which we neither understand nor deserve. We should be grateful for it and hope that it will remain valid in future research and that it will extend, for better or for worse, to our pleasure, even though perhaps also to our bafflement, to wide branches of learning.

And yet he did not shrink from consigning entropy to the subjective realm of humanity. Indeed in his final writings he extended this judgment to the foundations of physics more broadly: "It was not possible to formulate the laws of quantum mechanics in a fully consistent way without reference to consciousness."[34]

Does Logic Govern the Foundations of the Universe?

Thanks to the so-called two-slit experiments, physicists have long known that elementary particles are not like pebbles. Here is the basic setup for a two-slit experiment, conducted in this case with electrons (but photons, protons, and other elementary particles would work as well). A source of electrons, like an electron gun, is placed symmetrically in front of an impenetrable wall provided with two slits, A and B, through which the electrons can go through and finally be stopped by a final wall, where they are detected, localized, and counted.[35]

First, we cover slit B, so that electrons can only go through slit A, and we count how many electrons are detected at the final wall. Then we do the same, covering slit A, and counting. We might expect that when we leave both slits uncovered, the total count would be just the sum of the two previous counts, but it is not so. It would be so if we were dealing with pebbles or the like, but it is not so with electrons: when both slits are open, electrons behave like waves, that is, they interfere.[36] The results of these experiments, performed in the first third of the twentieth century, were shocking. At the detector level, electrons do not behave as if they had only two possibilities: passing through either A or B. They violate, so to speak, the law of thought we've been calling the Principle of the Excluded Third.

What if we redesign the experiment so that we can know, for each electron, which was the chosen slit? Imagine that you add a light source near the back of the first wall and at equal distances from both slits: when an electron goes through a slit it scatters a flash of light to the observer's eye. If for each electron reaching the detector you keep a record of its origin in either A or B, you will find that the interference phenomenon has disappeared. The electrons are back to behaving like pellets or pebbles. We've answered our question about which slit, but our achievement of that knowledge itself alters the behavior of the electron, making it behave like a particle rather than a wave.

The question is disturbing enough to bear repeating. Could it be that the behavior of an electron—which path it takes, whether it acts as a particle or as a wave, etc.—depends on what the observer is trying to know about it (as Niels Bohr, the great pioneer of quantum theory and founder of the "Copenhagen" interpretation, first insisted)? That, as Carl von Weizsäcker put it in 1941, which "quantity is determined" depends on the observer's "act of noticing"?[37] Or that quantum objects somehow know what we have decided to measure and change their behavior accordingly? Beginning with Einstein and Bohr, physicists have proposed experiments, such as "delayed choice" experiments, to test this seeming observer-object codependency at the smallest scales of the universe.

Imagine, asked Einstein, that we do not "decide" what to measure until the quantum object has presumably (under the classical view) already passed through a slit. According to Bohr, this would make no difference. The outcome would still be determined by what we ultimately chose to measure, as if our choices have "an unavoidable effect on what we have a right to say about the *already* past history" of a quantum object. John Wheeler, whose words we've just quoted, proposed one such "delayed decision" thought experiment with photons in the late 1970s. That experiment became possible in the 1980s, with advances in laser and fiber-optic technology. (We won't provide a description, but there are excellent ones readily available.)[38] And indeed, the realization of these experiments repeatedly demonstrates Bohr's conjecture: it makes no difference when we intervene before measurement takes place at the detector. It is as if quantum objects intuit what information we are going to ask of them and alter their behavior accordingly.[39]

This "wave-particle duality" and its associated phenomena turn out to be true not only of electrons but of many other quantum objects, from the very smallest, such as photons, to molecules and nano structures consisting of hundreds of atoms. Are the most elementary objects in the universe pathic waves or apathic particles? Or are these both aspects of the same thing whose manifestation is disturbingly dependent on our own interaction with them?

Those are some of the challenges that the emerging field of quantum physics posed early to previous intuitions of physical causality. Perhaps the most famous of these challenges is called entanglement: the correlation of states between quantum objects. Einstein found this phenomenon so contrary to the logic of causality that he dismissed it as "spooky action at a distance." As we write, it is the object of billions of dollars of competitive strategic investment by China, the United States, and the European Union.[40] And there are other challenges as well. Consider just these implications about quantum objects that flow from our double-slit experiments:

· *Superposition.* Quantum objects are in more than one state simultaneously. In the words of Paul Dirac, "whenever the system is definitely in one state we can consider it as being partly in each of two or more other states."[41] But what sense does it make to speak of a thing as being both heads and tails at the same time or traveling along two paths at once?

· *Uncertainty.* There is a limit to the precision with which any two complementary properties of a quantum object, such as position and momentum, can be simultaneously known. (This is commonly referred to as "Heisenberg's uncertainty principle.")

· *Disturbance by observation or measurement.* Our observation of a quantum object affects its state. For example, when we observe an electron's position we seem to make its wave function disappear. ("Wave-function collapse" was one early and influential way of thinking about this problem of disturbance. For a more recent and increasingly influential approach, interested readers may investigate "quantum decoherence.") But if the world changes in response to our choice to observe it, how is "objective" observation possible?

· *Individuality versus identicality.* Except in some probabilistic sense, can we say that "this" quantum object detected here is "the same" as "that" quantum object detected at another point on a trajectory? Quantum objects do not seem to be reidentifiable. Conversely, can we say, as Erwin Schrödinger did in 1952, that all electrons, protons, or other quantum objects are identical? But then, by the Principle of Identity of Indiscernibles, must we say that there is only one proton or electron in the universe? How do we decide if electrons are one or many?[42]

We hope you can already see from these aspects of quantum physics how nature at its tiniest and most fundamental casts into crisis some of the logical axioms, laws of thought, and causal explanations that humans have developed from our experience of larger structures of the world. The Principle of Identity (a thing x is identical to itself), the Principle of Non-Contradiction (a thing cannot be both x and not x at one and the same time), the Principles of the Excluded Third, of Sufficient Reason, of the Identity of Indiscernibles: all of these rules that work well for pebble-like objects in an apathic world can sometimes fail us at the quantum level.

Given our focus on the limits of number and logic as foundations of sameness and certainty, we should highlight that one result of these failures is a crisis of counting. Why a *crisis*? Recall the assumptions we made in order to count the leaves of our twin maples, A and B, in the previous chapter. We assumed that no question arises as to whether a leaf is one leaf or none or more. We assumed of each leaf that it must belong either to *A* or to *B*, but not to both (it seems to be a botanical fact that there are no maple leaves ultimately attached to two different trunks). We assumed that leaves retain their identity throughout the counting, so that the operation of adding numbers of leaves is associative. We are, or think we are, so familiar with leaves and piles of them that it is hard not to take such properties of identity for granted. But these are precisely the assumptions that do not hold at the quantum level, where the cognitive habits necessary for counting become deeply misleading. A recent thought experiment of Yakir Aharonov and his collaborators illustrates the crisis well: in our macro world, if you have two pigeonholes and three pigeons, at least two pigeons must share a pigeonhole. But if you shoot a trio of electrons with parallel trajectories at a double-slit apparatus, this is not necessarily true.[43]

The failure of identity and other "laws of thought" at the quantum level means that, thus far, no one has been able to develop compelling logical and set-theoretical axioms—like those von Neumann developed for mathematical objects in the previous chapter—for quantum mechanics and its mathematics. (That challenge does not seem to have troubled von Neumann himself, who in 1932 produced an early and important textbook of quantum mechanics.[44]) This is not for want of trying: readers interested in some of the exotic logics—such as qua and quasi sets—may turn to several excellent expositions.[45] Perhaps someday some physicist or philosopher will discover a set of axioms capable of providing logical foundations for the quantum world, and (who knows) even for the quantum and classical worlds simultaneously. But for our purposes, it is enough to insist that for now the case remains as it has always been: no one has yet succeeded in making identity and apatheia (or any of that coupling's cognitive children) entirely sovereign over the foundations of the universe.

From Physics to Psyche

Discoveries in physics often have consequences for thought about humanity and vice versa. From the pre-Socratics, Pythagoreans, and Platonists to the medieval Christian Aristotelians and the Islamic atomists to the physicists, philosophers, poets, and polemicists we encountered in the environs of World War I: over and over we have seen humanity tempted by the urge to understand the nature of its possibilities in terms of its understanding of the cosmos just as we have seen physics shaped by human habits of thought and experience. A longer book with more learned authors could offer examples from many other cultures, from the Aztecs to the Zulus.

Often enough, the temptation to move from physics to psyche has been animated by a debate over the nature of the human subject. Are we determined by laws of nature? Do those laws govern even the gods? Or does our will (or that of the gods) have the power to break the causal principles by which we come to explain the physical world? In this sense (among many others) each moment in the history of physics can be said to be potentially related to each moment in the history of philosophy. The writings of the ancient "atomist" philosopher Lucretius in the first century BCE provide a good example both of the temptation and of its historical potentials. In *On the Nature of Things*, Lucretius argued for an element of chance at the atomic foundations of the universe, which he called a "swerve." "If there is no atomic swerve to initiate movement that can annul the decrees of destiny and prevent the existence of an endless chain of causation, what is the source of this free will possessed by living creatures all over the earth?"[46]

Histories have been written about the influence of Lucretius's atomist theories on ancient thought and—after they were rediscovered in the early fifteenth century—on the natural philosophy and theology of the European "Renaissance." In our own day, the philosopher of science Michel Serres has drawn on Lucretius in order to champion a "liquid" thought opposed to what he takes to be the epistemological tyranny of solid bodies. One more example (as if another were needed) of the enduring but not always self-conscious interaction between the history of our physics and that of our philosophy.[47]

We have already seen some of the early debates between Planck, Einstein, Born, Heisenberg, Pauli, Schrödinger, and many others over the implications of their mathematical models for how we should think about the nature of human knowledge. Now, having spent more time with the "paradoxes" of identity, causality, and objectivity raised by quantum mechanics, we will be even less surprised to see that the challenges that quantum mechanics raised to questions of identity and difference at the level of elementary particles were translated to questions of identity and difference at the level of the human subject. Does I = I? What is the relationship of my mind to other minds? What is the relationship of mind to matter, of the human subject to the world?

Some of the pioneers of quantum mechanics were remarkably self-conscious about this interaction between how we think about physics and how we think about the psyche, taking up such questions throughout their careers, often in explicit conversation with the long history of philosophy. Werner Heisenberg—he of the "uncertainty principle"—collected his own engagement with these questions into a set of (as yet untranslated) physico-philosophical essays titled *Der Teil und das Ganze* (The part and the whole). Wolfgang "exclusion principle" Pauli's efforts to reconcile *physis* and *psychē* included an analysis and a collaboration with the psychologist Carl G. Jung. They coedited a book on the subject— *The Interpretation of Nature and the Psyche*—in which Jung set forth his ideas on "Synchronicity: An Acausal Connecting Principle" (an essay that became popular among those seeking a scientific basis for psychic and paranormal phenomena), and Pauli discussed "The Influence of Archetypal Ideas on the Scientific Theories of Kepler." Their last exchange of letters continued to explore the theme, with Pauli relating two of his dreams, which he associated with the nonconservation of parity experiments conducted in 1956 by Chien-Shiung Wu, who had apparently deeply impressed Pauli, both as a physicist and as a woman.[48]

Schrödinger, too, took up these questions. He demanded a reconsideration of key "Western" assumptions about the nature of subjectivity and knowledge, urging attention to the Vedanta traditions of India as models for the relation between mind and matter, self and world. He, too, was fascinated by the question of the synchronicity of ideas and the connectedness of minds, even going so far as to argue that "the over-all number of minds is just one."[49]

Over and over he stressed the codependence between models of the world and of the human and insisted that the implications of the new sciences might require a reconstruction of both.

Schrödinger wanted to alert humanity to its "unawareness of the fact that a moderately satisfying picture of the world has only been reached at the high price of taking ourselves out of the picture." According to him it was the Greeks who had removed the subject and the psyche from scientific discourse in search of "the song seraphically free / Of taint of personality" (to quote a later poet). Schrödinger felt that this erasure was no longer scientifically or psychologically tenable but also not easily mended. "A rapid withdrawal from the position held for over 2000 years is dangerous. We may lose everything."[50]

How to confront the problem of this split between the thinker of things and the things thought? Mathematicians and physicists with philosophical inclinations tend to choose between two approaches, one focused on sameness, the other on difference, both insufficient to the problem. As an example of a focus on sameness, we offer a great quantum physicist of the second generation, the American David Bohm (1917–1992). Like Schrödinger, Bohm stressed that in light of the new physics, "one can no longer maintain the division between the observer and the observed." "What is needed in a relativistic theory is to give up altogether the notion that the world is constituted of basic objects or 'building blocks.' . . . One has to view the world in terms of universal flux of events and processes."[51] How to achieve this? The most general way of perceiving order and structure within reality, according to Bohm, is "to give attention to similar differences and different similarities."[52]

Though the approach may seem admirably suspended between sameness and difference, it is everywhere evident that Bohm is emphasizing oneness, beginning with the name he assigned to his approach, Undivided Wholeness in Flowing Movement. He opposed his approach to the fragmentation he thought dominant not only in science but "in every aspect of human life." That "fragmentation is in essence a confusion around the question of difference and sameness (or oneness). . . . *To be confused about what is different and what is not, is to be confused about everything.*" The consequences of such confusion were apocalyptic, "leading to such a widespread range of crises, social, political, economic, ecological, psychological, etc., in the individual and in society as a whole."[53]

Like some quantum-mechanical version of an ancient Greek sage, Bohm exhorts us to mind the all, care for the all. As an antidote to our fragmentation, he even proposed a new language he calls the Rheomode (from the Greek ῥέω, I flow). In it, rather than saying "An observer looks at an object," one would say "Observation is going on, in an undivided movement involving those abstractions customarily called 'the human being' and 'the object he is looking at.'"[54] The name is new, but the strategy is old (Aristotle already remarked in his *Physics* that out of fear of the Eleatic dogma of the unity of being, Lycophron and others banned the use of the copula "is").[55] Past failures are not sufficient proof of future ones, but we doubt that such linguistic strictures can go to the heart of logical or metaphysical problems any more than new set theories such as *Fundierung*, qua sets, or quasi sets will successfully supplement the apathic, standard set theory with pathic objects.

On the other side, as standard bearer for advocates of fragmentation, we have chosen Hermann Weyl, an intellectual giant whom we encountered in Zurich in chapter 1, as colleague and intimate friend of Schrödinger. A great innovator in the mathematics of quantum mechanics (among numerous other areas) and author of one of the first and most famous "textbooks" in the field,[56] he succeeded his teacher David Hilbert as professor of mathematics at Göttingen and then moved to the Institute for Advanced Study in Princeton—together with Einstein, Gödel, and von Neumann—in 1933. But we will skip over all these years and Weyl's many scientific achievements and jump instead to the address titled "The Unity of Knowledge" that he offered at the bicentennial celebration of Columbia University in 1954.[57]

From the word *unity* in the title you might think that Weyl was in the same camp as Schrödinger or Bohm. Quite the contrary. You may recall that Max Planck, in his own lecture at Columbia in 1909, had already characterized modern science as knowledge that moves ever away from anthropomorphism. Two world wars after Planck's discourse, Weyl is making the same point, warning the world that not all human languages are equally truth-able. Poetry, myth, religion, history, philosophy: unlike mathematics, all of these are tainted by "man's infinite capacity for self-deception."[58]

Blessed with an astonishingly wide-ranging mind and equipped with an excellent education in the German fashion, Weyl had assimilated a great deal of literature and philosophy at an early age and received a large dowry of phenomenology through his marriage to the philosopher Friederike Bertha Helene Joseph (1893–1948; student of Husserl, translator of José Ortega y Gasset). Now in New York, he took up the work of Ernst Cassirer, whom we met debating Heidegger at Davos and who had himself died on the steps of Columbia's Library just a few years before Weyl's speech.

Cassirer's great work, the *Philosophy of Symbolic Forms*, had argued that the natural and the human sciences, myth, religion, and the arts all fit under the same ample umbrella of human creativity with symbols. Weyl agreed, but only to a point. All *natural sciences* operate, says Weyl, on the same principle: they are symbolic constructions, and the resulting science is a "formal scaffolding of mere symbols." The emphasis here is on *formal* rather than *mere*, and the difference is important. Weyl starts his explanation at a place we should by now be used to, with "pure" numbers. Unlike Husserl, Heidegger, and others, he thanks Descartes for the invention of analytic geometry, which replaced "intuitive" space and time by sets of sets of numbers. "Intuitive space and intuitive time are . . . hardly the adequate medium in which physics is to construct the external world. . . . They must be replaced by a four-dimensional continuum in the abstract arithmetical sense."[59]

Let us translate. As we saw in chapter 6, numbers can be built on just two symbols, the empty set ø and the symbol for gathering together {}. To summarize Weyl's doctrine in those terms: all natural science is a formal scaffolding that can be built on just two symbols, ø and {}, and this logos is, precisely, what makes it objective, indisputable, always progressing. Its foundations and its scaffolding are apathic, transmittable without noise from mind to mind, from generation to generation. "The whole edifice rests on a foundation which makes it binding for all reasonable thinking: of our complete experience it uses only that which is unmistakably demonstrable [*aufweisbar*]."[60]

In short, through mathematics human knowledge becomes unified and universal, or rather, some forms of human knowledge and not others. Toward the beginning of his lecture Weyl had announced that confronted with the dilemma of sameness or difference, the unity or the multiplicity of Knowledge, he would "take cover behind the shield of that special knowledge in which I have experience through my own research: the natural sciences including mathematics."[61] That shield he proceeds to describe as strong and intact, built out of symbols clear and true.

But when Weyl turns to the human rather than the natural sciences, he notes that something else is required, something *other* than the capacity to construct indefinitely tall scaffoldings of symbols: "Here its place is taken by *hermeneutic interpretation*, which ultimately springs from the inner awareness and knowledge of myself." The work of a great historian, for example, "depends on the richness and depth of his own inner experience."[62]

With this difference, according to Weyl, we have begun to leave the world of "Knowing" and entered the world of "Being." And here comes the fragmentation:

> Being and Knowing, where should we look for unity? I tried to make clear that the shield of Being is broken beyond repair. We need not shed too many tears about it. Even the world of our daily life is not *one*, to the extent people are inclined to assume; it would not be difficult to show up some of its cracks. Only on the side of Knowledge there may be unity.

The shield of knowledge is intact, the shield of being fragmented beyond repair. It is a fool's errand to try to bring them together. It is in this emphasis on demonstrable knowledge, this refusal to include the symbolic creations of Being in the armory of unity and truth, that Weyl sees the superiority of his teaching:

I am closer to the unity of the luminous center than where Cassirer hoped to catch it: in the complex symbolic creations which this lumen built up in the history of mankind. For these, and in particular myth, religion, and alas!, also philosophy, are rather turbid filters for the light of truth, by virtue, or should I say, by vice of man's infinite capacity for self-deception.[63]

Fair enough: there are, after all, important differences between physics and poetry or philosophy, math, and myth. The problem comes in the claim that one leads to our unity and the other to our undoing. Here, unity comes from cultivating demonstrable knowledge, whereas religion, philosophy, and the other symbolic systems of "the Shield of Being" are deceiving and fragmenting, as is our daily life, our *Lebenswelt* and experience, Being itself. Between the two, "Knowledge" and "Being," an unbridgeable gap.

Once again we find a deep ditch dug within the human, running along by now familiar lines. We should worry about that ditch not least because the pursuit of "Knowledge" in the absence of attention to humanity may itself produce horrifying results, however unintentional. Perhaps that is what W. H. Auden had in mind in a poem written in 1952, a couple of years before Weyl's speech. In "The Shield of Achilles," Thetis inspects the armor that Hephaistos is fashioning for her son. The shield is richly depicted in the eighteenth book of the *Iliad*, but what Auden's goddess sees is not in Homer:

> She looked over his shoulder
>   For vines and olive trees,
>   Marble well-governed cities
>   And ships upon untamed seas,
>   But there on the shining metal
>   His hands had put instead
>   An artificial wilderness
>   And a sky like lead.
>
> A plain without a feature, bare and brown,
>   No blade of grass, no sign of neighborhood,

Nothing to eat and nowhere to sit down,
   Yet, congregated on its blankness, stood
   An unintelligible multitude,
A million eyes, a million boots in line,
Without expression, waiting for a sign.[64]

Bridging the Gap between Poetry and Science

By putting the poet's shield side by side with the scientist's, we do not mean to imply that one is safer or truer than the other or that the two can be beaten together into unity. ("Unity" is not something we prioritize.) But we do want to suggest that if the two reflect on each other, then the ditch between the two, and between the pathic and the apathic aspects of humanity, might be more often crossed. In this sense we agree with John Dewey's impassioned manifesto of 1891/1929: "this present separation of science and art, this division of life into prose and poetry, is an unnatural divorce of the spirit.... It exists because in the last few centuries the onward movement of life, of experience, has been so rapid, its diversification of regions and methods so wide, that it has outrun the slower step of reflective thought." Dewey concluded in the imperative: "We must bridge this gap of poetry from science. We must heal this unnatural wound."[65]

The task is not easy, not least because it requires us to interrogate the conviction that reason can give us unmediated access to the foundations of reality. Richard Rorty has described the challenge: "the difficulty faced by a philosopher who . . . thinks of himself as auxiliary to the poet rather than to the physicist . . . is to avoid hinting that this suggestion gets something right, that my sort of philosophy corresponds to the way things really are."[66] According to Rorty, many philosophers are really on the side of physics insofar as they assume that at the bottom of things there must be some necessary kernel of "sameness," something that "is" itself in an irreducible rather than a contingent sense. "At the heart of philosophy's quarrel with poetry is the fear that the imagination goes all the way down—that there is nothing we can talk about that we might not have talked of differently." It is in order to protect themselves from this fearful possibility that so many have opted to divide the mind, as we just saw Weyl do. "Before we can rid ourselves of ontology," says Rorty, "we shall have to give up the picture of the human mind as divided into a good part that puts us in touch with the really real and a bad part that engages in self-stimulation and autosuggestion."[67]

Well put. But this division of the poets on the side of imagination and physicists on the side of the real is a bit exaggerated. We have already met a few of both professions who don't fit so easily into such schematizations. In this chapter, for example, we heard from the physicist Schrödinger, more willing than many a poet to abandon these divisions of the mind; and in chapter 1 we met the poet Paul Valéry, every bit as attuned to the "ontological" challenges faced by physics and mathematics as he was to the possibilities of verse. Perhaps more important, Rorty also omits the depressing fact that plenty who stand on the side of poetry are just as determined to divide the human mind as those who stand on the side of physics even as they reverse the values assigned to each part.

Consider the critical fate of Paul Valéry's poem *The Cemetery by the Sea*, which we began to read in chapter 1. Writing in 1964 the French poet Yves Bonnefoy (1923–2016) took both poet and poem to task for reaching into the realm of reason. Poetry can only be "disappointed, deluded by a certain terrain" of knowledge. According to Bonnefoy, conceptual knowledge of the real "carries a great risk for poetry, that the word will no longer be a scandal." "The distance between the word and *this* real thing," "the confrontation of intellectual knowledge and this invention of the object which may surely be called love," these are the prerequisites for poetry. If you attempt to diminish that distance or that confrontation you destroy "the mystery of presence," "the living water" with which poetry heals the human. Poetry must maintain its distance from reason. The mysteries of presence must be coddled and cloaked from clarity. (Or as the poet Czesław Miłosz put it, poetry should "approach the mystery of existence more directly than through *mere* concepts.")[68]

This view of poetry also digs a ditch within the human and along the same familiar lines, albeit from the opposite side. It, too, has powerful philosophical allies, among them Martin Heidegger, for whom the poet's task is to call our attention away from the conceptual thought and technologies of reason that block our orientation toward our own finitude and prevent us from realizing our authenticity as humans. The poet—Heidegger tells us in "What Are Poets For?"—directs us away from the "reckoning time" we create by conceptually dissecting "what is" and directs us instead toward a relation to language, world, and temporality that makes us capable of our own mortality. This is what Bonnefoy means when, in his outrage at *The Cemetery*'s attempt to bridge the abyss rather than plunge into it, he claims that Valéry must be "unaware that death had been invented."[69]

Valéry did find more sympathetic readers, including T. S. Eliot, W. H. Auden, Wallace Stevens, and Jorge Guillén, to name only poets. Among the most inspired was the German poet Rainer Maria Rilke (1875–1926), a poet concerned, like Valéry, with the basic conflict confronting us throughout this book, the rift separating a Weyl from a Schrödinger or Bohm, that is, the rift and conflict between being a creature and being a knower. It is difficult to exaggerate Rilke's place in the German poetic pantheon, but it is as bard of that epic conflict that he merits our attention here.

In a eulogy given in January of 1927, shortly after the poet's death, the novelist Robert Musil honored "the greatest lyric poet that the Germans have possessed since the Middle Ages." It was for the twentieth anniversary of that death that Heidegger wrote "What Are Poets For?," with Rilke as protagonist.[70] Rilke's legacy might have been lesser had he not come across, in the spring of 1921, a copy of the *Nouvelle Revue Française* containing Valéry's *Cemetery by the Sea*. At that point neither of what would become Rilke's most famous works had yet appeared: the *Duino Elegies* (begun in 1912), and the *Sonnets to Orpheus*. Both would be completed in the same month and year, February 1922, the latter in a creative storm lasting just a few days. But when Rilke first stumbled on Valéry's *Cemetery*, he had been unable for almost a decade to complete a significant work of poetry, consumed instead by depression and inner struggle, a struggle about, among other things, the truth, the worth, and the future of poetry.

"I was alone. I was waiting, my entire work was waiting. One day I read Valéry—I knew that my waiting was at an end." These are Rilke's words as reported by his friend Monique Saint-Hélier. Struck and enchanted by this kindred spirit with whom he was in "unequivocal" agreement, he experienced a transformation, "a blood transfusion," as Saint-Hélier put it. Rilke translated *Cemetery* and many other of Valéry's works into German, and he took up again the completion of his own *Duino Elegies*. And then came an enormous burst of creativity, "a boundless storm, a hurricane of the spirit," as he put it to a former lover. In this boundless storm new poems and new poetic truths all came to him torrentially, including, in the space of just a few days, the *Sonnets to Orpheus*.[71]

Rilke's Orpheus reveals his poetic teaching right at the start of the *Sonnets*. For the poet-god, song is a link between earth and heaven, a bridge, and it is on that bridge that something *new* approaches or goes forth—a new beginning, a sign, a transformation. The human poet is exalted by a new possibility for being but daunted by the enormity of the task:

> Song, the way you [Orpheus] teach it, is not covetousness,
> nor publicity for something to be finally attained;
> song is existence. For the god that's easy enough.
> But when *are* we? And when turns he
> toward our being both the earth and the stars?
> (1.3, lines 5–9)

The focus here is on the split (*Zwiespalt*), the gap between earth and heaven, between our mutable earthly existence and the eternal motions of the stars. This split and its many siblings—contingent and necessary, love and knowledge, subject(ive) and object(ive)—have rent the hearts of those humans who, like Rilke, cannot truly *be* unless earth and heaven, seeming and essence, being and knowledge, both shields unbroken, are somehow reconciled in mind. But how?—"And when turns he / toward our being both the earth and the stars?"

We have already noted that there are many on both sides of the ditch who do not care to cross this gap.[72] Rilke was not among them.

Consider his version of our human relation to the stars in Sonnet 1.11:

> Look at the sky. Is no constellation called "Horseman"?
> For it is strangely stamped in us,
> this earthy pride. And a second,
> who drives and reins it and whom it carries.
>
> Isn't it thus, hunted and then tamed,
> this sinewy nature of Being?
> Road and turning. Yet a pressure establishes accord.
> New expanses. And the two are one.
>
> But *are* they? Or does neither mean
> the road, which they do together?
> Already table and pasture namelessly divide them.
>
> Even the starry connection is deceiving.
> Yet we enjoy it for a while,
> believing in the figure. That's enough.

Here the relationship between the starry heavens and our human "being" is repeatedly transformed from union to abyss and back again, the ditch crossed, redug and crossed again. Path and pivot. Rilke does not choose a side or revel in the abyss. As Musil put it in his eulogy, never before Rilke or after has any poet achieved such a bridging tension (*Spannung*) through "never ceasing movement" across these divides.

But what kind of truth is on offer in this poetry? Can it really be that the enjoyment of our deceit about connection is "enough"? *Trügt* rhymes with *genügt*, deceit with sufficiency. Was Hermann Weyl then right to say that this is what poets do, advise people to indulge their "infinite capacity for self-deception"? Rilke provides something of an answer in the next sonnet (1.12):

> Glory to the spirit, the power that's able to connect us;
> for it is truly in figures that we live.

And with tiny steps the clocks advance
alongside the day that is properly ours.

"It is truly in figures that we live." Just as the human spirit (*Geist*) connects the stars in constellations, it connects *us* through figures (*Figuren*). This is an important word, one whose history encompasses the action of fashioning or putting together.[73] It comes from the Latin root *fig-* or *fing-*, which gives *fingere*, "to shape" or "fashion," originally "to model in clay," "to knead bread." Related Latin words are *fictile*, "made of clay" or "able to be shaped," *figulus*, a potter, and *figura*, "shape," "body," "form," the source of English *figure* and German *Figur*. The past participle of *fingere*, *fictus*, shaped into *fictio*, a feigning or invention, is the direct source of English *fiction*. When Rilke says that even the figure deceives, it is equivalent to saying that the figure — any figure — is made up: it is a fiction.

We live in figures, these four lines tell us, and we live in time, in the tiny steps of the clock hand. And just as our "starry connection" with a figure like the Southern Cross is not the same as the astronomers' view, our lived time is not encompassed by the measure of clocks. In the poem, as in life, the two — the "figural" and the "rational" or "real" — are not easily separable. The mechanical hand of the clock can be thought of figurally as taking human steps. And conversely the constellations are not only figural projections onto the sky. They also serve to orient us in space, as when sailors, for example, navigate the pathless seas. They organize our world.[74]

Figures and stars, nature, knowledge, time, death and eternity, these permeate the sonnets. They come together in the work's penultimate poem:

Oh come and go, you, almost a child still, complete
for the blink of an eye the dance figure
to the pure constellation, one of those dances,
in which we fleetingly overtake
dull ordering nature.
(2.28)

The reference is to a young dancer Rilke had known whose death at age nineteen (in 1919) animates the *Sonnets* from their very subtitle: "Written as a Funeral Monument for Wera Ouckama Knoop." When the dancer "completes" her *dance figure*, she bridges by that act, however fleetingly, the space between us and the stars, between death and life, and transcends for a moment the laws of nature. She becomes a constellation, a *stellar figure*, and by the same act (the poet goes on to say) she becomes a figure of Orpheus, and of poetry, of the lyre that emerges from "the unheard-of center."

Can we speak, within this poetics, of a "theory of knowledge" of the sort Valéry undertook, an exploration of the relation between the pathic and the apathic, between poetry and logic? Should we hear hints of geometry in the "unheard of center," and of arithmetic in "completes"?[75] Look at the last three lines of Sonnet 2.13:

> To the used-up as well as to the dull, mute
> stock of nature's fullness, to the unsayable sums,
> add yourself jubilantly and abolish number.

What are these unsayable sums? They contain the used-up stuff as well as all that's natural. The unsayable sums include the jingles of our childhood, now disused but still resounding with nostalgia, and Yeats's foul rag-and-bone shop of the heart, where all the ladders start. They include, as well, stars, trees, crocuses, worms, and tigers: dull and mute because they haven't yet been enchanted by the Orphic song. And why does the poet say that those sums are unsayable? It cannot be because they are enormous: already Archimedes knew how to tell numbers as large as the number of grains of sand in the world. Perhaps it is because all that stuff thrown together does not constitute a collection of apathic elements, a set. We cannot count them, just as we cannot count our thoughts. In the final line we are urged to add ourselves jubilantly to those unsayable sums.

The German *vernichte die Zahl* is usually translated as "cancel the sum," "cancel the cost," "cancel out the count," "nullify the sum," "nullify the score," even "nullify the account." Idiomatically, those translations are unimpeachable, but they don't make sense in the context. To cancel or nullify a sum or a count, to get a total of zero, we must add to it another sum or count of equal absolute value but opposite sign: can we, readers of Rilke, be enjoined to add ourselves, jubilantly, as negatives of the original sum? Hard to make sense of that. And if the German words are translated as "abolish the sum," it would mean "abolish, i.e., kill yourself, and abolish the world." This cannot be the fulfillment of Rilke's previous imperative, "Be, and know at the same time the condition of Not-Being." Our "unsayable sums" suggests something different: in the jubilant joining we step out of the Pythagorean pebble world and for that instant "abolish number."[76]

That abolition is not permanent, but only for an instant, in the song. We want to claim that if Rilke has a theory of human knowledge, it is a contingent one with choice at the foundations: choice between sameness and difference, the pathic and the apathic. He offers his poetry as a bridge between becoming and being. All of this is audible in his parting words to us, the last lines of the last sonnet:

> And should the earthly forget you,
> say to the ground: I flow.
> speak to the swift water course: I am.
> (2.29)

What advice is on offer in these words? It is not (as Heidegger would have it in "What Are Poets For?") that Rilke's poetry rejects "calculation" and invites us to dwell in the abyss of being that number, physics, and other forms of reason blind us to. Nor does it quite offer what Musil praised in his eulogy, "A clear stillness in a never-ending movement." Rather, Rilke's poetry shuttles us back and forth across the abyss between sameness and difference, an abyss it cannot itself bridge.

Imagine this exhortation to the physicist: "when it looks like a particle, ask after the wave, and when it looks like a wave, do not forget the particle." To the philosopher: in your moment of despair, speak to Parmenides about Heraclitus, and to Heraclitus talk of Parmenides." And to the poet: do not (pace Bonnefoy and Miłosz) alienate concept from experience, the eternal from the ephemeral. Rilke's lines ask us to remember that every separation we make between sameness and difference estranges us from something, and that simply in the remembering there is new knowledge. Paul Valéry would later (ca. 1940) offer similar advice to a world again consumed by war:

I will summarize the doctrine for you. It consists of two precepts:

All things different are identical.
All things identical are different.[77]

8

AXIOMS OF DESIRE

Economics and the Social Sciences

*Whatever may be the practical value of a true philosophy
of these matters, it is hardly possible to exaggerate the mischiefs
of a false one. . . . And the chief strength of this false philosophy
in morals, politics, and religion, lies in the appeal which it is
accustomed to make to the evidence of mathematics and
of the cognate branches of physical science.*

JOHN STUART MILL[1]

*It seems to me that many of the current disputes with regard to
both economic theory and economic policy have their common
origin in a misconception about the nature of the economic problem
of society. This misconception in turn is due to an erroneous
transfer to social phenomena of the habits of thought we have
developed in dealing with the phenomena of nature.*

F. A. HAYEK[2]

An Avalanche of Numbers

Across previous chapters we've witnessed various attempts to fit the human psyche and human society into procrustean beds of reason, necessity, axiom, or law. There have been countless more. We will focus here on one whose influence is still felt everywhere in our modern faculties, namely, the rise of what in English we call—without anyone of us really knowing quite what it means—"the social sciences." And among these we will single out economics, the *nomos* (the Greek word means law or order) of the *oikos* (family, household, home), in order to demonstrate the powers and the limits of these sciences.

Like any massive phenomenon of thought, neither economics nor the social sciences have an "origin," a singular birthday or beginning point. But the European eighteenth century is as good a narrative starting point as any, with its attempts to build theories of social and political life out of the postulates of Newton's mechanics and to develop a calculus of probabilities capable of guiding decision making by rational individuals and states (whence the term *statistics*).

The Enlightenment produced an exuberant faith that number could bring new rule to society, providing accurate measure of the good. Even some of its more subtle and revolutionary psychologists, such as Jean-Jacques Rousseau (1712–1778), did not hesitate to claim that the welfare of the polity could simply be measured by number: "Calculators, it is up to you: count, measure, compare." A generation later and on the other side of the English Channel, the proclamation of the so-called founder of utilitarianism, Jeremy Bentham (1748–1832), sounds surprisingly similar: "It is the greatest happiness of the greatest number that is the measure of right and wrong."[3]

By 1789 and the French Revolution we can begin to speak, albeit still with anachronism, of "social scientists." The Revolution provided an opportunity for enlightened thinkers like Nicolas de Condorcet (1743–1794) to test their dreams of calculations binding on the shared rationality of humanity in society. Almost simultaneously the Revolution's murderous companion, the Terror, called into question the assumptions of a common rationality that underpinned those dreams (not least by killing many of the dreamers, including Condorcet).[4] Yet Revolution, Terror, and the Napoleonic empire and its aftermath only reinforced the usefulness of number as a ruler for society. Not only in France but across Europe the first half of the nineteenth century published so many surveys, censuses, and statistics that one historian speaks of an "avalanche of printed numbers." And atop the avalanche rode new advocates of positivism—such as Auguste Comte (1798–1857)—offering new sciences with new promises to help govern the newly empowered masses.[5]

It would be difficult to exaggerate the importance of the development of these social sciences. They provided powerful new tools suitable for an endless variety of human problems. Where to build a bridge. How to determine the effects of a disease or evaluate a cure.

What price to charge for insurance, or how to design an electoral system. Just as significantly, the application of these tools in social, political, and economic life transformed the relation between *omnes et singulatim* (as Michel Foucault put it), between the many and the one, between society and polity, family and the individual. Over the course of the nineteenth century, the social sciences contributed to the creation of new forms of power, new political and economic technologies, new state formations and forms of subjectivity. We could even say that they dictated new "laws" to humanity, ranging from the newly democratic "Principle of Identity" encapsulated in the slogan "one person, one vote" (eventually encoded in the jurisprudence of the United States by the Supreme Court) to the deterministic materialism of Karl Marx's "economic laws of motion of modern society" (the subject of his magnum opus, *Capital*).[6]

These and many other "laws" of social life were themselves animated by new forms of knowledge, new ways of thinking about the power and purchase that our own thought has on the world, on society, and on the human. And again, number and its principles occupied a large space within that thinking about the power and purchase of thought. As an example, let's return for a moment to the "fundamental axiom" of Bentham's utilitarianism cited above. If we rephrase it slightly as "maximizing happiness is the measure of right and wrong," the axiom starts to look like a projection of the most up-to-date principles of mathematics and physics onto the moral ground.[7] Consider the following problem: given two points P and Q that do not lie on a vertical line, find the trajectory on which a mass point will travel by the force of gravity from the higher to the lower point *in the least time*.

Mathematicians discovered a solution to this problem just as the eighteenth century was born: this trajectory (dubbed brachistochrone, Greek for minimum time) turns out to be a piece of a known curve called the cycloid. Soon the best mathematical minds were occupied in looking for functions, or curves, that maximized or minimized certain functionals.[8] In the 1740s one of these mathematical minds, P. L. M. de Maupertuis (1698–1759), introduced the Principle of Least Action.[9] The principle states that the behavior (or the trajectory in generalized coordinates) of any mechanical system can be found by minimizing (or maximizing, as the case may be) an integral of a certain function of time, a function that, according to Maupertuis, was the kinetic energy T, and according to the (correct) formula of Joseph-Louis Lagrange (1736–1813), $T - V$, where V is the potential energy of the system.[10] Lagrange's formulation is taught to this day in courses of classical mechanics as well as in introductory economics (for solving constrained optimization problems). We count Bentham among the many brilliant philosophers who gave in to the temptation to apply this type of principle to humanity, seeking to maximize or minimize hypothetical quantities of "good" and "bad" and even (in the words of a later economist) the "quantity of feeling of pleasure or pain."[11]

The "rise of statistical thinking" in the mid-nineteenth century provides yet another important example of the intimate relationship between the study of mathematics and that of humanity. One of the protagonists of that romance, Adolphe Quetelet (1796–1874), was mathematician and astronomer and the founder of Brussel's Royal Observatory. What makes his example so revealing to us is his application of the laws of "celestial physics" to the social world. He achieved that application via the notion of "statistical law": the idea that statistics uncover general and stable truths about societies and the individuals that compose them. We will read no further than the titles of a few of his many influential tomes: *On the Human and the Development of Its Faculties, or Essay on Social Physics* (1835); *On the Social System and the Laws that Govern It* (1848); and *On Moral Statistics and the Principles That Should Provide Their Foundation* (1848). In these and many other works Quetelet derived statistical laws and applied them to questions about the attributes and behaviors of human populations: questions about intelligence, fertility, and physiology (he invented the precursor of the Body Mass Indicator used today); about suicide, deviance, and crime.[12]

We don't mean to imply unidirectionality of influence. The statistical laws that Quetelet offered students of society were partially inspired by those of physics and astronomy, but the statistical laws of society also inspired new physical laws. The new powers of number diffused across the many questions and disciplines where they could be put to work. The French mathematician and physicist Joseph

Fourier (1768–1830) put it well, some decades before Quetelet, in a paper applying the so-called error law to insurance underwriting (he had already applied it successfully to the population of Paris and to the theory of heat):

> Mathematical analysis unites the most diverse effects, and discovers in them common properties. Its object has nothing of the contingent, nothing of the fortuitous. . . . This science has necessary relations with all physical causes, and with most of the combinations of spirit[,] . . . and its true advances always recur to two fundamental points: public utility, and the study of nature.[13]

Laws of nature and of society were, as always, intertwined.

And as always, laws of nature and society were also intertwined with concepts of psyche and soul. Nineteenth-century developments in statistics, for example, reshaped debates about determinism and free will, providing new tools for thinking about causality and uncertainty. Is the future of the world determined by its present state? In a world with better statistics and sufficient powers of calculation, could we discover the determinable cause of every event? Would every future event become predictable? Or is there some irreducible randomness in the world, some freedom from complete causality?[14]

In the previous chapter we saw how physicists wrestled with these questions. In this one we are calling attention to a related phenomenon: the attempt by practitioners of the human and social sciences to subject their disciplines to "laws" of reason derived from mathematics and physics. Even historians—today seldom classed among the "social sciences"—were not immune to the lures of determinism, attracted to it, as they so often are, by unexamined yearnings for causality and for a direction to history. In the hands of a Henry Thomas Buckle, for example, the *History of Civilization in England* would stand among "the highest faculties of science": "in the moral world, as in the physical world . . . all is order, symmetry, and law."[15]

Today entirely forgotten, Buckle's popularity in the mid-nineteenth century was something akin to Spengler's in the early twentieth, albeit they rode quite different mathematical and philosophical waves to fame. Even far away Dostoevsky would complain in the same breath about these applications of mathematics to human desire and about Buckle and his deterministic history in his *Underground Man* (1864). Nor was Buckle an anomaly. On the contrary, the young Friedrich Nietzsche thought that a deterministic "science of history" had a hegemonic hold on thinkers of his generation, and it was against them that he discharged his ire in the second of his *Untimely Meditations*, "On the Uses and Disadvantages of History for Life" (1874):

> What, can statistics prove that there are laws in history? Laws? They certainly prove how vulgar and nauseatingly uniform the masses are: but are the effects of inertia, stupidity, mimicry, love and hunger to be called laws? Well, let us suppose that they are: that, however, only goes to confirm the proposition that so far as there are laws in history the laws are worthless and the history is also worthless. But the kind of history at present universally prized is precisely the kind that takes the great mass-drives for the chief and weightiest facts of history.[16]

Again, we do not mean to imply that the advocates of number, axiom, and law ruled unopposed over the nineteenth century or that Nietzsche, Dostoevsky, and their kind were the only critics. One could rank among the very greatest physicists and mathematicians and still be skeptical of excessive faith in the new sciences of certainty. We have already met James Clerk Maxwell (1831–1879), the mathematical physicist whose equations treating light, electricity, and magnetism as manifestations of the same phenomenon laid the foundation for classical electromagnetism. He also established basic principles of thermodynamics, for example, by applying Quetelet's "error law" to the distribution of molecular velocities in a gas. His work would also shape the future of thought about the directionality of time and causality ("time's arrow," as it is sometimes called). But he disagreed with those who believed that logic or empiricism could provide ultimate foundations for knowledge.

Maxwell expressed that criticism both in scientific writings and in verse:

Go to! Prepare your mental bricks,
Fetch them from every quarter,
Firm on the sand your basement fix
With best sensation mortar.
The top shall rise to heaven on high—[17]

These five lines of rhyme contain a great deal of scientific wisdom despite their underlying piety (Maxwell's Christian commitments were well known). He is insisting that no matter what stable ideational "brick" of sameness and apatheia human knowledge might find with which to build an edifice of certainty in the world, there is always flux, patheia, and difference to which that edifice is subject. Such is the condition of the human in this world, he suggests, and to think otherwise is to fall into the same hubris and blasphemy as the builders of the Tower of Babel.

The warning has not dissuaded new architects and new building projects. With every new potential "brick" of sameness—the atom, the gene, and so forth—new claims of certainty and determinism arise in our ways of thinking about the human and the world, along with new protests against them. As Nietzsche wrote in his own protest against the extremists of Darwinian naturalism in the social discourses of his day, "this is an ancient, eternal story: what formerly happened to the Stoics still happens today, too, as soon as any philosophy begins to believe in itself. It always creates the world in its own image; it cannot do otherwise."[18]

Economics: "The Science of Human Action"

Ours is not to belittle or diminish the enormous achievements of the many social sciences built out of "sameness." We seek only to demonstrate the limits of these sciences, limits imposed by the inescapably pathic sands of humanity on which they are built. By way of example we will focus on just one of these sciences, that of economics. The choice is justified not least by the pretensions of the field itself. The Austrian American Ludwig von Mises (1881–1973) was far from alone in claiming that economics is "the science of human action that strives for universally valid knowledge." Economics, he insisted, "like logic and mathematics, is not derived from experience; it is prior to experience."[19] Whatever that may mean, it is certainly the case that much of modern economics has built towering claims to knowledge on empty lots of math and logic. Since we cannot attempt to deconstruct every skyscraper in the crowded skyline of this thriving and important discipline, we will concentrate our fire on just one example, spectacularly built on what we all agreed in the last chapter *is* a "brick" of eternal sameness: namely, ø, the empty set.

The *Theory of Games and Economic Behavior* (1944), coauthored by Ludwig von Mises's student Oskar Morgenstern and John von Neumann, is a massive attempt (632 pages deep) to build new foundations for economics. In the process of construction, the attempt gave rise to tools—known today as game theory—put fruitfully to work in many other fields, from nuclear strategy to artificial intelligence. We should not be surprised by the scale of the authors' ambitions. "Johnny" has already loomed large as an impresario of set theory and physics in previous chapters. Morgenstern could have appeared in chapter 1 as well. He, too, was a product of the collapsing Austro-Hungarian empire (his mother may have been the illegitimate daughter of Emperor Friedrich III) and orbited in the vicinity of the Vienna Circle. He was not, however, primarily a logician, mathematician, or philosopher but rather an economist whose principal professional and intellectual concern, during his career in Vienna, was with the possibility of economic prediction.

The question may have felt particularly urgent during the Great Depression, and Friedrich von Hayek (1899–1992; another economist of enduring fame) hired Morgenstern to explore it at the newly founded Institute for Business Cycle Research. Is accurate economic forecasting of business cycles possible? In 1928 Morgenstern had published a book arguing that it wasn't on the grounds that— unlike, for example, phenomena like wind currents or cold fronts of the type increasingly well predicted by meteorologists—humans are active agents, constantly gathering information about and interpreting predictions and changing their behavior in response to what they learn, thereby rendering the prediction false.[20] Economic predictions, like other forecasts of human behaviour, are in this sense reflexive and self-undermining. We might call this the Jonah effect, since it forms the basis for the prophet Jonah's complaint to God about being sent to Ninevah: the people will change their actions because of my prophecy, and you, God, will forgive them, making me look like a false prophet. (Later in this chapter we will touch on some later twentieth-century responses to this problem, such as Nash equilibria.)

Morgenstern preferred mathematical analogies to biblical ones: he is said to have compared his first book's conclusions about the impossibility of predicting business cycles to the "incompleteness theorem" that his Vienna colleague Kurt Gödel famously developed for mathematics. Much later (in 1976), Morgenstern would recall that his early work was undertaken while under the influence of a number of figures we have already encountered—Hermann Weyl, Bertrand Russell, Alfred North Whitehead (too metaphysical for his taste)—and some others long forgotten, such as F. Y. Edgeworth, the author of *Mathematical Psychics: An Essay on the Application of Mathematics to the Moral Sciences* (1881). That title—which stands "Psychics" cheek by jowl with mathematics—fascinated Morgenstern, who advocated for its reprinting. Perhaps he was already yearning for a science capable of delivering, to borrow the title of a paper he published in 1935, "Perfect Foresight and Economic Equilibrium."

Morgenstern reprised that particular paper on a number of occasions for various arcs of the Vienna Circle, and it was on one such occasion that a mathematician in the audience suggested that a paper published by yet another mathematician might prove helpful. That paper was John von Neumann's "On the Theory of Parlor Games" of 1928, which outlined a set-theoretical approach to such games as poker. Could theories of strategic choice by players in a game be extended to the choices of agents in an economy? Still in Vienna, Morgenstern started "to read a lot of logic and set-theory" and to write papers with titles like "Logic and the Social Sciences" (1936). But it was not until after he immigrated to the United States in 1938 that he finally met his future coauthor, whom he would later call "a gift from heaven."[21] Now began the intense conversations (all in German) that produced a book that the authors first thought of calling *General Theory of Rational Behavior* before settling on the slightly more modest *Theory of Games and Economic Behavior*.

The goal of the book was lapidary: "We hope to establish satisfactorily . . . that the typical problems of economic behavior become strictly identical with the mathematical notions of suitable games of strategy."[22] "Strictly identical": difficult to imagine a more total claim to sameness. What makes initially plausible the assertion that economic (or rational) behavior is strictly identical with mathematical notions? One reason alleged by our authors is "the simile with physics." Over and over again they offer examples of how "mathematics has been applied with great success" in that field, which they offer as elder and successful sibling to economics. Why not emulate that success in the study of the human? The authors are aware that many social scientists "object to the drawing of such parallels on various grounds, among which is generally found the assertion that economic theory cannot be modeled after physics since it is a science of social, of human phenomena, has to take psychology into account, etc."[23] They don't think much of those objections. On the contrary, what they think would demand explanation—what "would itself constitute a major revolution"[24]—would be if it were discovered that the principles that led to progress in other sciences *did not* lead to progress in economics.[25]

But they do recognize that compared with physics, the science of economics is in its infancy and should begin with baby steps. "The free fall is a very trivial physical phenomenon, but it was the study of this exceedingly simple fact and its comparison with the astronomical material, which brought forth mechanics." In the same fashion, von Neumann and Morgenstern resolved to begin with the simplest unit, "the behavior of the individual, and the simplest forms of exchange."[26] They called this "the Robinson Crusoe model," "an economy of an isolated single person or otherwise organized under a single will." From there economics can proceed, as physics had, to greater complexity (in this case larger groups), until it reaches a scale where it can begin to play on "the field of real success: genuine prediction by theory."[27] As von Neumann put it a few years later, "in modern empirical sciences it has become more and more a major criterion of success whether they have become accessible to the mathematical method or to the near-mathematical methods of physics."[28] The *Theory of Games* was meant to put the study of human behavior on that road to success.

This promise enraptured some early readers of the book. The then young future Nobelist Herbert A. Simon proclaimed (in the pages of the *American Journal of Sociology* of 1945) that "the student of the Theory of Games . . . will come away from the volume with a wealth of ideas for application and for development of the theory into a fundamental tool of analysis for the social sciences."[29] These were years of high optimism for the social sciences, an optimism fueled in part by their important contributions to planning and strategy in the massive military and societal mobilization demanded by the Second World War. It was against that optimism that the poet W. H. Auden railed in his "Reactionary Tract for the Times" of 1946:

> . . . Thou shalt not sit
> With statisticians nor commit
> A social science.[30]

But social scientists and their funders preferred the promise of prediction to the prohibitions of poetry. Entire institutions—such as the RAND Corporation, founded in 1948 to provide research to the defense establishment of the United States—were built on the prophetic promise of game theory in economics, political science, and other sciences of human behavior.[31]

Such optimism proved misplaced: von Neumann and Morgenstern's gold standard of "prediction by theory" remains almost as far out of reach of economists and social scientists today as it was when they wrote some seventy years ago. This is true not only of game theory (whose role in modern economics should not be exaggerated) but of economics more generally.[32] And of course we don't mean to imply that our authors invented the yearning for a social science as predictive as physics, which long predated (and survived) them. The point is only that this desire seduced them into making deeply misleading analogies, for example, between the physics of thermodynamics and the concept of utility in economics.[33] Others have written about these analogies. Their history is instructive and their critique is revealing, but it does not go deep enough.[34] After all, the simile between thermodynamics and utility is just that, a simile, not a claim to strict identity. We must take aim at something more fundamental: the axioms of sameness being applied to both disciplines. We have seen that not even physics has been able to subject the universe to strict axioms of sameness. Before we expect better from a science of human desire and social life, such as economics, we must ask its practitioners, What are the assumptions of sameness on which you seek to establish your science?

The Economist's Robinson Crusoe: Transitive Man

We will stick a little longer with von Neumann and Morgenstern not because their contribution to the axiomatization of demand theory was so pioneering (other economists made more fundamental contributions) or because their particular mathematical approach was especially enduring (it saw limited application) but because their assumptions are so clearly stated and so symptomatic of what became common commitments in the field. Recall their first assumption, that in economics, as in physics, the macro is built out of the micro; that the behavior of a social group is a function, or the result, of the decisions (choices) taken by its individual members; and that the science of economics, like that of physics, should start with the simplest form, an economy organized under a single will, an individual. Following a convention among Austrian economists, they called that single will "Robinson Crusoe."[35]

There are mathematical reasons why we might want to treat an economy with a single player as the basic case and paradigm of a full social economy. It permits (and follows from) admitting the axiomatic validity of set theory, where a set is uniquely determined by its elements, the elements, however, are supposed to be unaffected by their belonging to any set. But is it actually the case that our desires and choices are unaffected by the people, objects, and experiences with which we are thrown together in life? Our authors pose a version of this basic question:

> The chief objection against using this very simplified model of an isolated individual for the theory of a social exchange economy is that it does not represent an individual exposed to the manifold social influences. . . . These factors certainly make a great difference, but it is to be questioned whether they change the formal properties of the process of maximizing. Indeed the latter has never been implied, and since we are concerned with this problem alone, we can leave the above social considerations out of account.[36]

We hope that you aren't satisfied with this masquerade of an answer. Stating that social factors that "certainly make a great difference" do not alter "the formal properties of the process of maximizing" because it "has never been implied" that they do is logically equivalent to concluding, for example, that eating parsnips is good for the skin since you have never heard anybody imply that it isn't. Clearly von Neumann and Morgenstern were not interested in interrogating seriously their most basic assumptions. We must do so for them.

If the behavior of economies is built out of the desires and choices of individuals, then economics becomes dependent on a description of psychology, and one of the "chief difficulties" becomes "properly describing the assumptions that have to be made about the motives of the individual."[37] We've received fair warning: assumptions have to be made! But please don't forget: those assumptions are not about the behavior of something alien. They are about *us* humans, about our motives, our desires, our dreams. So we each can (and ought to) ask, Do we recognize ourselves in the assumptions that "have to be made" about our alter ego, Robinson Crusoe?

Here come more assumptions. The second is that the individual agent's goal is to "obtain a maximum of utility or satisfaction" of their various desires and wants within their given constraints.[38] Many—like Dostoevsky's Underground Man—have railed against this maxim: "When they prove to you that in reality one drop of your own fat must be dearer to you than a hundred-thousand of your fellow creatures, ... then you have just to accept it, there is no help for it, for twice two [makes four] is a law of mathematics. Just try refuting it."[39] It is true that self-interest is not our only value, that where some may seek to maximize their own profit, others may value some other notion of the good, such as family, communal solidarity, or free time. This is not, however, a fundamental difficulty for our mathematical economists, for they need not deny that the utility or satisfaction an agent seeks to maximize can take many different forms, including a desire for the welfare of others. The more basic difficulty is this: in order to be maximized, utility and satisfaction should be quantifiable or at least rankable. But why should we think that desires are quantifiable or rankable either by us as human agents or by economists studying them?

This serious difficulty requires yet more assumptions. The rational human subject must be "an individual whose system of preferences is all-embracing and complete, i.e., who for any two objects or rather *for any two imagined events*, possesses a clear intuition of preference."[40] The scope of the assumption is breathtaking: the subject's preferences extend not only to commodities in a market but to anything that may imaginably happen to him or her. Moreover, those preferences are well informed and computationally sophisticated: "we cannot avoid the assumption that all subjects of the economy under consideration are completely informed about ... the situation in which they operate and are able to perform all statistical, mathematical, etc. operations which this knowledge makes possible."[41]

To put this axiom of "completeness and comparability" in slightly more formal terms, given any two objects of desire u and v, the subject can always say which one she prefers, or else that she is indifferent, that is, that she has no preference for either u or v. But what about when there are more than two options on the table, as there so often are? For that we need yet another axiom: for any three or more commodities, objects, or imagined events—call them a, b, and c—all rational agents who prefer a to b and b to c will also prefer a to c. In economics this crucial axiom is called the "transitivity of preference." In *Theory of Games and Economic Behavior* it is axiom 3:A:b. Their justification for it? "Transitivity of preference [is] a plausible and generally accepted property."[42]

Within the magic circle of these assumptions, von Neumann and Morgenstern set out to treat humans and their desires like mathematical objects: transitive, apathic, quantifiable. They were neither the first nor the last to attempt this. What was unusual about their approach was its scope and brilliance, as well as the mathematical bricks they choose for their "creation of a conceptual and formal mechanism." In their desire to establish human behavior on the same firm foundations as one of them (von Neumann) had constructed mathematics, they advocated a turn "away from the algorithm of differential equations which dominate mathematical physics" and embraced "combinatorics and set theory."[43]

The fifteen pages (60–75) that our authors dedicate to providing a set-theoretical axiomatization of games must rank among the greatest monuments to hubris in the history of humanity. By using only mathematical concepts, our authors imagine they have achieved absolute clarity of thought. "We have even avoided giving names to the mathematical concepts . . . in order to establish no correlation with any meaning which the verbal associations of names may suggest. In this absolute 'purity' these concepts can then be the objects of an exact mathematical investigation." Thus, they imagine, we can first develop "sharply defined concepts" that can then be applied without danger of confusion to "intuitively given subjects."[44] And now comes the crowing: "this may serve as an example of the truth of a much disputed proposition: That it is possible to describe and discuss mathematically human actions in which the main emphasis lies on the psychological side." Describe and discuss? That sounds reasonable enough. But they claim more: "a primarily psychological group of phenomena has been axiomatized."[45]

Once again, we should insist: we are not against mathematics or its application to the study of human behavior. On the contrary, it is a powerful tool that has had a vast influence on human knowledge and (not only human) life. Our criticism of von Neumann, Morgenstern, and all who think like them is simply that they have forgotten a basic truth about what mathematics can offer. In the words of an earlier logician and philosopher of astounding talent, Charles Sanders Peirce (1839–1914),

> An engineer, or a business company . . . or a physicist, finds it suits his purpose to ascertain what the necessary consequences of possible facts would be; but the facts are so complicated that he cannot deal with them in his usual way. He calls upon a mathematician and states the question. . . . It frequently happens that the facts, as stated, are insufficient to answer the question that is put. Accordingly, the first business of the mathematician, often a most difficult task, is to frame another simpler but quite fictitious problem . . . which shall be within his powers, while at the same time it is sufficiently like the problem set before him to answer, well or ill, as a substitute for it.[46]

Von Neumann and Morgenstern have forgotten to ask themselves: how "sufficiently like" is the Robinson Crusoe of their axioms to a human one? And how do we decide whether the difference is for well or ill?[47]

The Novelist's Robinson Crusoe: Nontransitive Man

As a path into these questions about the economists' "Robinson Crusoe," let us open again Daniel Defoe's book by that name, published in 1719, and sometimes awarded the title (by the type of critic who thirsts for origins) of "first English novel." (Among Defoe's inspirations was a work we, too, were nourished by in chapter 4: Ibn Ṭufayl's *Ḥayy ibn Yaqẓān*, translated into Latin as *Philosophus autodidacticus*, "the self-taught philosopher," just a few years before Defoe took up his quill.[48])

In the book's opening pages we learn that Crusoe was born in 1632 at York, that when the time came for him to choose a path in life, his father having destined him for the law, he felt inclined, instead, to go to sea. In vain were his father's expostulations, in vain Crusoe was told that he "had a prospect of raising [his] fortune by application and industry, with a life of ease and pleasure," remaining in that "middle state, which he [Crusoe senior] had found . . . the most suited to human happiness, not exposed to the miseries and hardships, the labor and sufferings of the mechanic part of mankind, and not embarrassed with the pride, luxury, ambition, and envy of the upper part of mankind." The young man was briefly persuaded by his father's arguments and "resolved not to think of going abroad any more, but to settle at home according to [his] father's desire. But alas! a few days wore it all off."[49]

Robinson fled his house on the first of September 1651 and went on board a ship at Hull bound for London. He was nineteen, and already in the above extract we detect a basic trait of his character, namely, that his will was like a weathervane that moved with the latest wind.

That pattern will be repeated throughout the book. Overwhelmed by fear during his first storm at sea, for example, "I made many vows and resolutions that if it would please God to spare my life in this one voyage, if ever I got once my foot upon dry land again, I would go directly home to my father. . . . Now I saw plainly the goodness of his observations about the middle station of life, how easy, how comfortably he had lived all his days." But the storm abated, and the next day as Crusoe watched the sun rise over a smooth sea he found the sight "most delightful." That night a party plied with sailor's punch further "drowned all my repentance, all my reflections upon my past conduct, all my resolutions for the future." "My fears and apprehensions of being swallowed up by the sea being forgotten, and the current of my former desires returned, I entirely forgot the vows and promises that I made in my distress."[50]

We might want to attribute these rapid reversals of desire to youthful flightiness to be replaced over the years by a more mature steadfastness of will. The novel has no truck with such nonsense. Years of shipwrecked self-reflection on his desert isle did not erase the fluctuating nature of Crusoe's preferences and aversions so much as heighten his own awareness of those fluctuations, as in this striking paragraph near the end of the novel:

From this moment I began to conclude in my mind that it was possible for me to be more happy in this forsaken, solitary condition than it was probable I should ever have been in any other particular state in the world; and with this thought I was going to give thanks to God for bringing me to this place. I know not what it was, but something shocked my mind at that thought, and I durst not speak the words. "How canst thou become such a hypocrite," said I, even audibly, "to pretend to be thankful for a condition which, however thou mayest endeavour to be contented with, thou wouldst rather pray heartily to be delivered from?"[51]

This literary moment feels familiar and true to our experience, and perhaps to yours as well: a moment in which one suddenly becomes aware of the inadequacy, contradiction, even untruthfulness of one's own convictions about one's happiness. Such insights about conflicts, competing desires, and even contradictions within ourselves is a kind of knowledge that reflection on literature often produces. And one of the innumerable lessons that we can draw from such knowledge is simply this: in so many important aspects of his thoughts, desires, and being, Defoe's Robinson Crusoe is not a transitive man.

Return for a moment to his choice of profession. The great seventeenth-century mathematician Blaise Pascal believed that in the all-important matter of choosing one's profession, "chance rules" (*le hasard en dispose*). Economists today would agree that this choice, alongside other related decisions, such as the choice of college major or educational commitment, is among the most important in any individual's "maximization of utility" (a utility that economists too often measure entirely in terms of future earnings). So does Crusoe's behavior in making this most important of choices conform to the necessary assumptions of rational actor theory and modern economics? Or is the case closer to that other extreme hypothesized by Pascal?[52]

Imagine translating the options Crusoe was considering into von Neumann and Morgenstern's notation of transitive utility. Name three utilities: u, v, and w, as follows: u = become a sailor; v = go back home and take up a settled profession; w = preserve his life. Originally, for Crusoe, $u > v$, and that's why he goes to sea, contrary to his father's wishes. As for w, one would a priori assume that (if asked) the young man would offer $w > u$ and $w > v$, since if he does not preserve his life he can achieve neither u nor v. But w seems not to be in the focus of RC's attention, until he experiences the first storm of his sailing career. At that point, while w is still $> u$ and $> v$, the relation $u > v$ is inverted and becomes $v > u$. When the focus on w in RC's consciousness diminishes, the original $u > v$ again becomes the case. Could one juggle this difficulty by assigning probabilities to w, say, by marking a difference between αw with α near 0 and with α near 1? We don't see how. What seems to be the case is that a greater or lesser focus on w (in other words, a greater or lesser awareness of death) changes the relation between u and v into the opposite, thus violating the axioms.

What can this extended comparison of our two Robinson Crusoes teach us about the limitations of economic rationality for the understanding of human behavior? One could evade the question by insisting that Defoe's Robinson Crusoe is merely a character in a novel, not a "real" rational being, and so can teach us nothing about human behavior. But which Robinson Crusoe feels truer to your own experience, the one in the novel or in the one in the treatise on game theory? Has it happened to you that a sudden awareness (e.g., of your own finitude or mortality) or shift in attention or experience (a feeling of love, the birth of a child) transforms your sense of priorities? Or that, for however fleeting a moment, you are gripped by a feeling—hypocrisy, Crusoe called it—that the form of happiness you act as if you are so committed to achieving is in fact not what you truly desire?

You could also object that our criticisms of rational actor theories are not new. Defoe's contemporary Jonathan Swift, author of *Gulliver's Travels* and much else of devastating wit and insight, was already attuned to the limitations that human psychology imposes on aspirations to rational choice. "When we desire or solicit anything, our minds run wholly on the good side or circumstances of it; when it is obtained, our minds run wholly on the bad ones," he noted in his *Thoughts on Various Subjects* (1711–1726). We are not far from the biases that psychologists and "behavioral economists" such as Amos Tversky, Daniel Kahneman, and Richard Thaler have recently been filing in evidence against the advocates of rational choice. (Though we doubt any economist today would go quite as far as Swift's assertion that "the bulk of mankind is as well qualified for flying as for thinking.")[53]

And plenty of moralists have debated the relative roles of reason, chance, and the passions in our decision making. Samuel Johnson, for example, staged this very problem in his nowadays too seldom read novel of 1759, *The History of Rasselas, Prince of Abissinia*, in which the inexperienced Prince Rasselas needs to be educated on this score by the older and world-wise poet Imlac. The prince begins with the usual vocabulary of rational action:

> "Whatever be the general infelicity of man, one condition is more happy than another, and wisdom surely directs us to take the least evil in the CHOICE OF LIFE."
>
> "The causes of good and evil," answered Imlac, "are so various and uncertain, so often entangled with each other, so diversified by various relations, and so much subject to accidents which cannot be foreseen, that he who would fix his condition upon incontestable reasons of preference must live and die inquiring and deliberating."
>
> "But, surely," said Rasselas, "the wise men . . . chose that mode of life for themselves which they thought most likely to make them happy."
>
> "Very few," said the poet, "live by choice. Every man is placed in the present condition by causes which acted without his foresight, and with which he did not always willingly co-operate, and therefore you will rarely meet one who does not think the lot of his neighbour better than his own."[54]

For these early novelists and many others, it was clear that we humans—in our preferences, hopes, desires, and aversions—fail miserably to conform to the requirement of "an individual whose system of preferences is all-embracing and complete."

Can Sameness Save the Social Sciences?

We have let von Neumann and Morgenstern serve as primary example of a powerful intellectual current: the desire to build a mathematical theory of rational decision making. We know, of course, that their particular formalization was not the only one. Their special fondness for the logic of sets, for example, was not widely shared even by the future giants of game theory. (Kenneth Arrow, Robert Aumann, John Harsanyi, Leonard Hurwicz, Eric Maskin, Roger Meyerson, John Nash, Thomas Schelling, and Reinhard Selten all won Nobels in part for their contributions to the discipline.) Other mathematical approaches (such as the famous Nash Equilibrium) would be developed and put to work by game theorists and economists seeking to address multiple challenges to models of rational decision making. But all these approaches share basic assumptions, assumptions pithily summarized (and famously applied) by the great Chicago economist Gary Becker: "all human behavior can be viewed as involving participants who maximize their utility from a stable set of preferences and accumulate an optimal amount of information and other inputs in a variety of markets."[55]

In terms of these assumptions, every economist we have met thus far more or less conforms to the description provided already in 1860 by the English artist, art critic, and social theorist John Ruskin (1819–1900):

> "The social affections," says the economist, "are accidental and disturbing elements in human nature; but avarice and the desire of progress are constant elements. Let us eliminate the inconstants, and, considering the human being merely as a covetous machine, examine by what laws of labor, purchase, and sale, the greatest accumulative result in wealth is attainable. Those laws once determined, it will be for each individual afterwards to introduce as much of the disturbing affectionate element as he chooses, and to determine for himself the result on the new conditions supposed."

As we might expect from that great Victorian critic of industrial capitalism and utilitarianism, Ruskin did not find these assumptions adequate:

But the disturbing elements in the social problem are not of the same nature as the constant ones; they alter the essence of the creature under examination the moment they are added; they operate, not mathematically, but chemically, introducing conditions which render all our previous knowledge unavailable.[56]

Humans, Ruskin is saying (albeit without our vocabulary), are pathic. They are not like the mathematical objects of set theory, whose sameness persists regardless of the sets they are placed in or the operations to which they are subject. Hence, the "social problem" is not amenable to mathematical law.

This objection would not have come as a surprise to economists in Ruskin's day, or ours. Since John Stuart Mill (1806–1873) at the very least, economists have themselves been well aware of the importance of "sameness assumptions" in making their science possible.[57] They call these sameness assumptions *ceteris paribus* clauses, meaning "other things being equal," "other variables staying the same." It is by the assiduous use of these "sameness clauses" that economics produces its "laws" and claims of causality. Alfred Marshall, one of the nineteenth-century founders of what came to be called neoclassical economics, put it with unusual philosophical clarity in his *Principles of Economics* of 1880:

It is sometimes said that the laws of economics are "hypothetical." . . . Almost every scientific doctrine, when carefully and formally stated, will be found to contain some proviso to the effect that other things are equal; the action of the causes in question is supposed to be isolated; certain effects are attributed to them, but only on the hypothesis that no cause is permitted to enter except those distinctly allowed for.[58]

We might want to say—with Marshall and against Ruskin—that economics is "like every other science" in that it uses provisional claims of "sameness" in order to produce laws that make increasing (perhaps even eventually predictive) sense of the empirical world. "In committing oneself to a law qualified with a ceteris paribus clause, one envisions that the imprecision of the predicate [C] one is picking out will diminish without limit as one's scientific knowledge increases."[59] But what is striking about economics and the social sciences in comparison to "every other science" (such as physics) is just how little the "imprecision" of ceteris paribus clauses (such as "everybody's preferences are transitive") has in fact diminished, and conversely, how little their predictive power has advanced.[60]

This is not for lack of trying. Enormous mathematical effort has been invested in increasing the purchase of formal economic models on the empirical world. The Cowles Commission at the University of Chicago in the 1950s, for example, was tremendously productive in posing the theoretical problems at stake, but by the 1960s the limited empirical reach of their formal models was already recognized. The ensuing search for "policy invariant" structural parameters (i.e., a sameness that stays the same), championed by Leonid Hurwicz with the sharpest of mathematical weapons, foundered on the fact that "economic data, both micro and macro, have not yielded many stable structural parameters."[61]

Yet another increasingly popular approach has taken randomization as a benchmark and tries to identify causal parameters by looking for "natural experiments" within societies (such as a sudden change in law or regulation), specific social experiments (such as the introduction of a childcare program with randomized eligibility), or even laboratory experiments (such as participant behavior in role-playing games).[62] And today, as more and more human behavior from daily life "in the wild" is translated into computational data by smartphone, search engines, wearable devices, and other technologies, formal models and their "sameness" clauses can be tested against (or derived from) databases of exponentially increasing size and quality, driving theory and the empirical into an ever-more-intimate embrace.

Each step in this history has been accompanied by the hope that it might further refine the "sameness clauses" assumed by the theory. For example, in "What Do Laboratory Experiments Measuring Social Preferences Reveal about the Real World?," Steven Levitt and John List write that "the allure of the laboratory experimental method in economics is that, in principle, it provides ceteris paribus observations of individual economic agents, which are otherwise difficult to obtain." But as they then go on to note, the preferences revealed in a lab are not influenced merely by monetary calculations but by "at least five other factors: 1) the presence of moral and ethical considerations; 2) the nature and extent of scrutiny of one's actions by others; 3) the context in which the decision is embedded; 4) self-selection of the individuals making the decisions; and 5) the stakes of the game."[63] Even in a laboratory, desire is not independent of context, nor are preferences necessarily transitive. A subject may, for example, choose monetary profit alone and in private, when they think that their behavior is unobserved, but in the company of another prefer to act otherwise. Perhaps this is what Ruskin meant in his declaration above, that in humans the "disturbing elements . . . operate, not mathematically, but chemically."

Economists and other social scientists have long been aware of these difficulties, and the history of their disciplines is in a real sense the history of their (often successful) efforts to overcome them. Today it is widely acknowledged that preferences are not perfectly transitive; that (except in trivial cases) we are rarely fully informed; that we are not always maximizers or minimizers; that (unlike the objects of set theory) our desires are context dependent; that interpretations of rationality that may seem satisfactory in individual decision making do "not transfer comfortably to *interactive* decisions." Economists and their collaborators in allied disciplines (such as psychology) are producing an enormous literature and coining new fields (such as "behavioral economics" and "psychological game theory") to address all of these questions.[64] But what they are not willing to do is call into question our conformity to the basic laws of thought they take to be necessary for a "science": namely, the Principles of Identity and of Non-Contradiction. And so—like so many before them—they fragment the human into ever smaller particles, hoping to find one of enduring sameness that can be perfectly subjected to these laws.

As a concluding example of this strategy and its limits, let us return to a basic principle of the modern sciences of human desire with which we began: that humans act in pursuit of their happiness, maximizing pleasure and minimizing pain. This "'economic' point of view," as Sigmund Freud called it in his *Beyond the Pleasure Principle* of 1920, underpins an enormous amount of theorizing about human behavior. That theorizing rests on the assumption that we humans know, without contradiction, what it is that we desire, what choices and actions would make us happy.

But is that so? Are our desires oriented toward our happiness? And are we capable of knowing them without contradiction? Even our briefest of excursions into early English novels has already sown doubt about these questions. And we all know countless cases (including ourselves) in which humans who supposedly know what they desire for their well-being and what's best for their happiness nevertheless choose and act in ways that seem contrary to their best interest. Why do we procrastinate? Smoke? Persistently overeat and fail to exercise? Remain in an unhappy relationship? Neglect a loved one in favor of some trivial pursuit?[65]

It is painfully obvious that much human misery is self-inflicted. But how can this be, if our desires are informed, noncontradictory, rational? For the rational theorists the question is urgent, since it seems to threaten their fundamental assumptions, and the very best among them have offered answers to it (we think of Gary Becker and Kevin Murphy's "theory of rational addiction").[66] Among economists the question remains fresher than the grass, but it is also older than the hills. "No one can crave what, in the last analysis, harms him." Plato has Socrates say something along these lines in the *Protagoras*, though we are quoting here from the *Aphorisms* of the novelist Franz Kafka (1883–1924).

Let's stay for a moment with Kafka. Thus far his aphorism has not said anything that Plato, or the most committed economist of the "Chicago School," could not subscribe to. But now the great writer gives us something unusual:

> If in the case of some persons it appears to be so—and perhaps it always does—, the explanation is that someone inside the person craves for something that is useful to *this* someone, but very noxious to *another* someone who has been brought along partly to judge on the matter. Had the person taken the side of the *other*

someone from the outset and not only when the time came for the judgment, *this* someone would have been suppressed, and with him the craving.[67]

"Perhaps it always does!" Kafka puts this contradiction at the center of his psychology. And he explains it by dividing us from a unified subject to multiple personalities.

In order to see just what is so bold here, remember Plato again. He, too, explained deviation from the rule that people choose what they know to be better by introducing difference within the soul, as in his famous metaphor of the psyche as chariot with reason the charioteer and the horses as passions and appetites. In Plato's case the division of the soul was in the service of a hierarchy: reason, the "best part" of the soul, holds the reins.[68] Kafka gives us no hierarchy. Moreover, his multiple personalities coexist in the same subject and in the same temporality, although their relative persuasiveness varies with time. Kafka is here refusing yet another philosophical escape route, the Principle of Non-Contradiction, that Aristotle called "the most certain of all principles": "It is impossible that contrary attributes should belong at the same time to the same subject . . . this is naturally the starting point even for all the other axioms" (*Metaphysics*, 1005b).

Kafka, in short, is defying the logical laws of thought. There is no greater symptom of the power of those laws than the enormous effort expended across the long history of ideas by those who, unlike Kafka, seek to strengthen their grip over our psyche and its desires. Some of that effort has followed well-worn paths. For example, two millennia of philosophers have been building on Aristotle's *akrasia*: the idea that we fail to act according to what our judgment tells us to be good because of weakness of will. Among recent philosophical treatments, perhaps the most influential was that of Donald Davidson (1917–2003), who saw the problem as one of reconciling the following apparently inconsistent triad:

· If you believe A to be better than B, then you want to do A more than B.

- If you want to do *A* more than *B*, then you will do *A* rather than *B* if you only do one.
- Sometimes you act against your better judgment.

We've phrased the problem in this way so that you can see its similarity to the preference and utility problems of the economists. Davidson's solution: when people act in this way, it is because they temporarily believe that the worse course of action is better, not having made an all-things-considered judgment but only a judgment based on a subset of possible considerations.[69] Again, no hint of contradiction.

There are those more willing to seize contradictions by the horns. "We model man as having two sets of preferences that are in conflict at a single point in time." Thus begins Richard Thaler and H. M. Shefrin's "Economic Theory of Self Control," which from its very epigraph invokes a split model of the psyche and embraces the possibility of contradictory desires.[70] The authors invoke important precedent for their positing of a split in the psyche—Adam Smith's "two-self model" in his *Theory of Moral Sentiments* (1759); Thomas Schelling's "Egonomics" (1978); and even Freud's *Beyond the Pleasure Principle*. What they claim as their original contribution is not the introduction of difference and conflict into the soul but its formalization: "to the best of our knowledge our work is the first systematic, formal treatment of a two-self economic man."[71]

The formal treatment they offer has become the foundation for the movement today known as "behavioral economics." The seeming conflict "between short-run and long-run preferences" observed in so many cases of human behavior (smoking, eating, and saving are among the usual examples they offer) can be explained by "viewing the individual as an organization. At any point in time the organization consists of a planner and a doer. The planner is concerned with lifetime utility, while the doer exists only for one period and is completely selfish, or myopic." In other words, by splitting the individual into parts, they seek to make each part consistent in its motivations and preferences. They propose "a formal model of intertemporal choice" that models "man as an organization with a planner and many doers." The conflicts within the individual thus become "fundamentally similar to the agency relationship between the employer and the employee" and susceptible to the same strategies and approaches.

Behavioral economics is sometimes thought of as a revolution against rationality. Nothing could be further from the truth. "Ours is a theory of rational behavior," insisted Thaler and Shefrin, and quite rightly too. What our authors have done is simply divide up the soul in a new way—not along the lines of Plato or Aristotle or Freud, but along those of corporations and organizations—in order to push the rational axioms of economics down to a supposedly deeper level of the psyche.[72]

Others have gone even farther, proposing that we can best be modeled as many selves within ourselves, each of these selves economically rational but applying different discount rates with respect to time and engaged in a massive internal and intertemporal strategic bargaining game. "The orderly internal marketplace pictured by conventional utility theory becomes a bazaar of partially incompatible factions, where, in order to prevail, an option has not only to promise more than its competitors, but to act strategically to keep the competitors from later undermining it." On such a view, our desires and choices are driven by some psychic version of repeated prisoner's dilemmas, and game theory can be used to model our psychology from the bottom up, so to speak, so long as we cut up the psyche into pieces sufficiently small. We move from microeconomics, with its individual Robinson Crusoes, to *Picoeconomics* (as the psychiatrist George Ainslie calls it), which seeks to slice our motivations' orders of magnitude finer, setting each fragment of our desires at war with all the others.[73]

From Plato's chariots and Aristotle's polygons to the corporations and war games of behavioral economics and psychological game theory, these models and metaphors of the psyche all have the goal of preserving within it the axioms of "sameness"—identity and noncontradiction—deemed necessary for science and certainty. But what if the dynamics of our psychic lives are quite otherwise?

The logical laws of thought do not apply in the id, and this is true above all of the law of contradiction.[74]

Sigmund Freud (1856–1939), the author of these words, was just as interested as his younger countryman von Neumann in erecting a psychological science. He, too, drew on the analogy of "other sciences, chemistry or physics, for example." He, too, sought to "establish the laws which [mental processes] obey," and he, too, expected that the "principles of the new science (instinct, nervous energy, etc.)" would "remain for a considerable time no less indeterminate than those of the older sciences (force, mass, attraction, etc.)."[75]

It would be easy enough to show how these scientific ideals sometimes led Freud astray and even easier to show that, as he himself predicted, some of his concepts and principles would prove "indeterminate" (or worse). Without defending any particular part of the Freudian edifice, we simply want to point to two virtues of his teaching. The first was Freud's willingness to confront the possibility that the laws of thought do not apply to much of our psychic lives. The contradictions within us are not simply bargained away or strategically resolved. "When two wishful impulses whose aims must appear to us incompatible become simultaneously active, the two impulses do not diminish each other or cancel each other."[76]

Unlike the other social scientists discussed in this chapter, Freud entertained the possibility that these "incompatible impulses" applied down to our most basic "principles" and drives, such as our preference for pleasure or our desire to live. "No one can crave what, in the last analysis, harms him"? In *Beyond the Pleasure Principle* Freud was even willing to hypothesize that in addition to the drives he had previously incorporated into the psyche—the pleasure principle and the reality principle, mainstays of the "economic view"—there existed as well a drive toward death and therefore a struggle within us between Eros and Thanatos, between the love of life and the desire for self-destruction.[77] But—and here comes the second virtue—even as he posited such bold principles for his new science, Freud insisted on epistemic humility. He was willing to concede that new knowledge might "blow away the whole of our artificial structure of hypotheses." And he did not doubt that, "where ultimate things, the great problems of science and life are

concerned," "each of us is governed by deep-rooted internal prejudices, into whose hands our speculation unwittingly plays."[78]

Nowhere, perhaps, are these lines more important than in our
attempts at human and social sciences such as economics and psychology. For it is in them that speculation about our inner and our
outer lives, about our psyches and societies, meet in ways that
shape the possibilities of our existence. By way of example, consider again that basic question: do our desires conform to the Principles of Non-Contradiction and of Identity, and other laws of rational thought? Much of the political and economic machinery of the
modern world is founded on the assumption that we know what we
desire for our happiness and that the sum of those wants is the aspi-
rational "good" that authorizes our political life. Hence, we have designed systems and sciences that define and measure "the good" as
the freedom to translate certain desires into political and consumer
choices. But if our social sciences have not yet truly interrogated the
nature of our desires and of our happiness—or worse, if they have
built themselves on foundations that confuse that nature—then this
machinery is spinning its dangerously dogmatic wheels at we know
not what risk to our humanity.

We are far from the first to perceive the problem. At more or
less the same time that von Neumann and Morgenstern were publishing their own treatise, Ludwig Wittgenstein was already writing the words that conclude his endlessly inspiring *Philosophical
Investigations*:

> The confusion and barrenness of psychology is not to be ex-
> plained by calling it a "young science"; its state is not comparable
> with that of physics, for instance, in its beginnings. (Rather with
> that of certain branches of mathematics. Set theory.) For in psy-
> chology there are experimental methods and *conceptual confu-
> sion*. (As in the other case conceptual confusion and methods of
> proof.)[79]

If today we have needed to dedicate a chapter to Wittgenstein's
paragraph, it is because in the three-quarters of a century since

Wittgenstein wrote those words, both the confusion and the risk have greatly grown.

Appendix: Formal Definition of Transitive Man

Pace Dostoevsky and others, *Homo œconomicus*, the ideal individual of liberal economic theory, need not be an egoist. Solidarity, or care for the common good, may well be among his or her important concerns. But he must be "transitive." He must be capable of maximizing and minimizing. And in order to be able to maximize or minimize any phenomenon he must introduce in it "the more and the less," that is, the relation of something, *a*, being greater or smaller than something *b*. In this appendix we offer, for those who may be interested, a more formal exposition than we could provide above in prose of what this means.[80]

Let us first examine the formal notion of relation in general, which belongs to set theory and is therefore governed by the Zermelo-Fraenkel axioms, and so, under the imperium of the Identity Principle. Suppose X is a set: one of the Zermelo-Fraenkel axioms states, or implies, that we can form all possible ordered pairs (a, b), where a and b are elements of X, and their forming such a pair does not alter a or b. Then we gather all those pairs into a set, which is usually written $X \times X$ and called "the Cartesian product of X with itself." Now, logicians define a *relation* as being *any* subset R of $X \times X$, and whenever a pair (a, b) belongs to R, we say that a and b are related by the relation R, which we often write aRb. Notice, again, that you can take as R *any* subset of $X \times X$. If R is the empty set, that means that no elements of X are related by R; if R happens to be the whole of $X \times X$, that means that all elements of X are related to all elements of X.

A few more definitions before we give some examples:

1. A relation R is said to be *reflexive* when for all elements a of X we have aRa.

2. A relation R is said to be *symmetric* when aRb implies bRa for any elements a and b of X.

3. A relation R is said to be *transitive* when for any elements a, b, and c of X, if aRb and bRc, then aRc.

EXAMPLES

a. Let X be any set, and R the relation of *being the same as*. Thus, aRb means $a = b$. This relation is reflexive, symmetric, and transitive, as can be easily verified. Reciprocally, any relation that has those three properties is called an *equivalence relation* and defines *some sameness*, as we see in the following example.

b. Let X be the set of piano keys, both white and black. We will say that two such keys are in the relation R if the number of keys between them is any multiple of twelve. Any two keys so related are said to correspond to *the same note* (e.g., two Cs, possibly of different frequencies or octaves). Notice that, indeed, this relation satisfies the three requirements above (it is reflexive since zero is a mutiple of twelve!).

c. Now X is a set of people, say the residents of Columbus, Ohio, and two such people are in the relation R when they have at least a parent in common. R is reflexive since everyone has two parents in common with oneself; it is symmetric since if Jack has a parent in common with Mary, she has a parent in common with Jack; but it is not transitive, since Mary may have, say, her mother in common with Jack, and Jack may have his father in common with Fred, but Mary and Fred may have no parent in common.

d. Let's take X to be the set of natural (counting) numbers 0, 1, 2, 3, ... The relation R will be *greater than*; in other words, nRm means that $n > m$. Here, the relation is not reflexive, since a number is not greater than itself; nor is it symmetric; but it is transitive, since from $n > m$ and $m > p$ we can deduce that $n > p$. If instead we consider the relation *greater than or equal to*, that is, $n \geq m$, it is reflexive, not symmetric, and transitive.

e. X is as in example (d), the set of natural numbers, but we change the relation. A number n will be "greater than" another number m, which we can write $m \angle n$, when both n and m have the same prime divisors and also n is bigger than m in the usual sense as in example (d). Thus, for example, $2 \angle 4$ and $6 \angle 18$, but there is no

relation between 6 and 7 (neither is greater than the other since
6 and 7 do not have the same prime divisors: 6 = 2 times 3, and 7 is
prime). We can see that there are infinitely many pairs that are not
related, and also infinitely many pairs that are.

f. By way of comparison, here is an example of a relation "greater
than" which is not transitive (hence it is not a true order-type
relation), and where there is no maximum or minimum to a finite
set of points, unless you're lucky. Example: let X be the set of the
markings on a clockface, and write $a < b$, b is "greater" than a, if
one can travel from a to b clockwise along an angle less than half
of the circumference ($< 180°$). Diametrically opposite points, like
12 and 6, are not related. If we have the three points 1, 5, and 6, it is
easy to check that 6 is a maximum, but if instead we have the three
points 1, 5, and 9, there is no maximum, for $1 < 5$ and $5 < 9$, but
$9 < 1$. If you try 9, or 10, or 11, or 12 for maxima, they don't work,
since they are all < 1.

The above relation $n \geq m$ in example (d) is an order relation. More
generally, we will call any reflexive and transitive relation a preorder; if, in addition, $a \geq b$ and $b \geq a$ imply that $a = b$, then we call
it an order relation. In example (d), the order relation of the natural numbers is a *total order*, which means that given any two natural
numbers, one must be greater than or equal to the other; but in general, in other situations this is not the case, and we can have an order
relation where two different elements need not be related, that is,
neither need be greater than or equal to the other, as in example (e).

In cases such as example (e), when the relation need not obtain
with all possible pairs of elements of X but only with some, we say
that we have a partial order or preorder, as the case may be.

Our last example (f) contains the definition of transitive man.
The set X will be the preferences of an individual as in *"The Economist's Robinson Crusoe: Transitive Man"* above: "an individual whose
system of preferences is all-embracing and complete, i.e., who for
any two objects or rather *for any two imagined events*, possesses a
clear intuition of preference." This includes the possibility that for
certain pairs of objects or events, the individual, let's call him Fred,
feels indifference, that is, equal preference. In symbols, we'll write

$a \geq b$ when our individual prefers object/event a to object/event b. Notice that this relation need not be an order sensu stricto: if Fred is indifferent to two different cleaning powders, a = Ajax Telamon, and b = Ajax Oileus, then we have $a \geq b$ and also $b \geq a$, although a and b are different. But if we impose on Fred the fundamental restriction that his preferences be transitive, that is, that whenever he prefers a to b and b to c, then he must also prefer a to c, then the set X of Fred's preferences becomes a partially preordered set, and Fred becomes what we call a transitive man.

By way of conclusion, let us demonstrate the close relation between the associativity of the sum (and the apatheia underlying it), which played the stellar role when we examined why and when two plus two is four in chapter 6, and the transitivity of preferences described here. Suppose a binary relation is given between numbers (integers, rationals, or real), a relation noted by $a \propto b$. We do *not* assume that it is transitive. We will assume, however, that this relation is linked to the operation sum by the following two logically independent conditions obtaining between $\propto$ and +:

(*) $0 \propto a$ and $0 \propto b$ imply $0 \propto a + b$ (the sum of two "positive" numbers is "positive"),

and

(**) given $a \propto b$, and c any number, we have $(a + c) \propto (b + c)$ (we can add the same number to both sides and the relation $\propto$ is maintained).

Now we will prove that the associativity of + implies the transitivity of $\propto$.

Suppose $a \propto b$ and $b \propto c$. This means by (**) that $0 \propto (b - a)$ (adding $-a$ to both sides), and, similarly, $0 \propto (c - b)$. Then we use (*) to put those two together and we get $0 \propto ((c - b) + (b - a))$ (#). The right-hand side, *by the associativity of the sum*, is equal to $c + (-b + b) - a$. Since the parenthesis is 0, we just get $c - a$, and so by (#), $0 \propto c - a$, and again by (**), adding a to both sides, we finally get $a \propto c$, which is what we set out to prove.

9

KILLING TIME

My mission is to kill time, and time's mission is to kill me. Assassins are entirely at ease with each other.

EMIL CIORAN[1]

Moreno, I will say
So far as I can infer:
Time is only the delay
Of what is about to occur.

EL GAUCHO MARTÍN FIERRO[2]

We have left to the last a knowledge that divides humanity as painfully as any other, a knowledge that has for all the centuries covered in this book (and presumably many more) split us between being and becoming, sameness and difference, eternity and death. We mean our knowledge of time, a knowledge so powerful and so painful that it was presented by some influential ancients as the original punishment for having eaten from the tree of knowledge: "For dust you *are*, and to dust you shall return." The fall from innocence, according to this view, is the awareness that the cosmos is ruled by the Second Principle of Thermodynamics. Time and entropy flow in one direction. We are condemned to die.

No knowledge of the future could seem more certain than that, yet our species has not easily reconciled itself to it. Already in our earliest mythical records we find heroes like the Sumerian Gilgamesh attempting to conquer mortality. Christianity and Islam, today the two most populous religions, have long appeased our anguish by promising the survival of some part of the self somewhere or even a return to full life at a future time, a resurrection.

And why not? Shouldn't a truly all-powerful god be able to run the film backward, causing a jug broken into a thousand pieces to reconstruct itself into a whole and even cutting the scene in which it shattered, replacing it with whatever suits? Or does logic limit the powers over time of even the greatest god? Circa 1065, around the time that William the Bastard was planning the Norman conquest of England, the monks of Monte Cassino asked Cardinal Bishop Peter Damian this question. Can God restore a woman's virginity once lost? (Yes.) Can God cancel the foundation of Rome? (Perhaps He could but He would not: God works only for good.) But if God can undo what has been done in the past, does this not violate the Principle of Non-Contradiction? "For of what has been it cannot be truly said that it has not been. . . . For contraries cannot coincide in one and the same subject."[3]

Flashback to Aristotle's statement of the Principle of Non-Contradiction cited in our introduction: "it is impossible for the same thing (or attribute) to belong and not to belong *at the same time* to the same thing and in the same respect; and one must add all further such [samenesses] as may be necessary to meet logical objections." The *sameness of time* is already here, an axiom serving as the foundation for this "law of thought," which can then be used to constrain even an (otherwise?) omnipotent deity. But we must ask, How can time serve as a foundation, when its "sameness" is just as problematic as any other?

Forget the past and future: even this very instant, the "now" of your immediate present, is not easily reduced to "one," as we could have learned many chapters ago from Plato, who puzzled over "this strange instantaneous nature, something interposed between motion and rest, not existing in any time."[4] From certain logical points of view it seems that the now cannot exist, since it is never equal to itself, and produces so many contradictions. Yet from the perspective of our experience, its existence seems indubitable. It is in the now that we are at liberty to think, choose, and act. Now is the moment of decision. "Only the present is mine, and the present is all that I live," wrote a modern philosopher, channeling an ancient one.[5]

Oh no, you groan. If this chapter on time has to recapitulate every previous chapter, an already long book will become eternal. You are right to fear. From Parmenides and Heraclitus, with their debates over being and becoming, to the physicists, philosophers, economists, and psychologists of the present: every period and protagonist we've encountered thus far has been deeply concerned with our knowledge of time. All of them have been debating, sometimes explicitly, sometimes not, whether time is one and apathic or multiple and mutable.

The positions available in those debates have been many, and they have certainly changed over . . . time. But fortunately for purposes of our sketch, those arguments have also clustered along a few common themes, which have retained their family resemblance over millennia. Here, for example, are a few words from a letter of consolation that Einstein wrote to the family of a deceased friend, just a few months before his own death: "For us physicists of faith, the separation between the past, the present, and the future holds nothing more than the value of an illusion, however strong it may be."[6] We do not seem so far from Aristotle's *Physics* (bk. 4), with its claim that time is not an agglomeration of instants but a continuum, and that it is only the psyche that arbitrarily isolates segments in the continuum of time and declares them "nows" separating past from future.[7] So rather than recapitulate our entire history, we propose to study just one debate and let the querulous representatives on that stage stand more generally for some of the perpetually feuding clans.

Einstein's Worries/The Time of Physics

Three of those clans are visible in a conversation that the logician Rudolph Carnap reported having had with Einstein.

> Once Einstein said that the problem of the Now worried him seriously. He explained that the experience of the Now means something special for man, something essentially different from the past and the future, but that this important difference does not and cannot occur within physics. That this experience cannot be grasped by science seemed to him a matter of painful but inevitable resignation.[8]

Carnap didn't see a problem. As far as he was concerned, "all that occurs objectively can be described in science." We have physics for "the temporal sequence of events" and psychology for "the peculiarities of man's experiences with respect to time, including his different attitude towards past, present, and future." But Einstein insisted that "these scientific descriptions cannot possibly satisfy our human needs; that there is something essential about the Now which is just outside the realm of science." Both agreed "that this was not a question of a defect for which science could be blamed, as Bergson thought." But they disagreed, and profoundly, on the relationship between science and knowledge. "I definitely had the impression that Einstein's thinking on this point involved a lack of distinction between experience and knowledge. Since science in principle can say all that can be said, there is no unanswerable question left."

In the anecdote's three protagonists—Carnap, Einstein, Bergson—we have excellent representatives of three different ways of confronting the split in our knowledge of time. We will have nothing to say about Carnap, since like Parmenides, his form of confrontation is denial. As far as he is concerned there is no meaningful split in the human experience of time, nothing out of reach of logic and its sciences, except in special psychological or emotional situations, which are of little interest. But Einstein will require engagement if we wish to understand why he was so worried about the now and so convinced that it would always remain beyond the reach of physics. And Bergson, too, will detain us, not least because in his effort to protect the human experience of time from the necessities of number, he attacked Einstein so ferociously.

Why did Einstein believe that the time of physics was severed from the now? In order to answer this question, we don't need to take an excursion into the arcana of space-time in his theories of relativity. Yes, it is true, the theory of special relativity captured the attention of the world with its declaration that an infinity of indistinguishable planes of simultaneity pass through any given space-time point. The relativity of simultaneity—that is, the impossibility of objectively defining a "same instant" for all of space—was one of the most provocative conclusions of the special theory presented by Einstein in 1905, especially after it was translated into a geometric theory of space-time by Einstein's former mathematics professor, Hermann Minkowski in 1908.[9]

Here is an example of the provocation: a space fleet is launched against earth by aliens from the Andromeda galaxy roughly two million light years away. Two earthlings take leave of each other and start walking in opposite directions, one toward and the other away from Andromeda. In keeping with the relativity of simultaneity in special relativity, for the neighbor walking toward Andromeda, a "space fleet has already set off on its journey, while to the other, the decision . . . has not yet been made." Roger Penrose developed this particular example, building on earlier philosophical inquiries with titles like "A Rigorous Proof of Determinism Derived from the Special Theory of Relativity."[10]

This relativity of simultaneity is indeed fascinating, but Einstein did not need relativity in order to worry about the relation of the time of physics to the now. For that, his awareness of a much older problem would suffice, a problem we first encountered among the Greeks, when we discovered that the number line was shot through with "nonbeing," such as the $\sqrt{2}$. We've just encountered the same problem again, in Aristotle's insistence that time is not a series of distinguishable instants but a continuum in which "neither will point be next to point, or instant to instant, in such a way that length or time will be composed of these. For things are next to each other when there is nothing of the same kind between them, whereas between points there is always a line and between instants a time."[11] If time is a continuum, like a geometrical line, then all points in that line are indistinguishable from each other—any point P can be interchanged with any other point Q and no one the wiser. In chapter 7 we discussed some of the logical, physical, and philosophical difficulties that this undistinguishable nature of geometric points presents to physics before and regardless of any consideration of time. And now we add another conundrum: for our consciousness, the now is uniquely special, but if we treat time as a continuum, the now becomes indistinguishable from any other moment.

Fine, you say, don't treat time as a continuum, and have done with it. But this is not an option for the physicist, at least, not for the physicist after Galileo and Newton. In earlier chapters we have talked about the nature of Euclidean space and on the long struggle, beginning with the ancient Greeks and their discovery of the incommensurability of the $\sqrt{2}$, to ensure that the *space* of reason is not poked with holes of not being in order to guarantee such truths as, for example, that a straight line through the center of a circle *does* meet the circle at two opposite points.[12]

Newtonian physics sharpened similar questions with respect to time. Think, for example, of how we calculate fundamental kinematic quantities such as the velocity and the acceleration of a moving object. That calculation involves the quotient of two numbers, one a space interval, the other a time interval, with the calculation of the limit of the quotient as the interval of time gets closer and closer to zero. Notice the requirements: (1) that we have numbers we can divide and (2) that we are able to calculate those limits.

The general requirement that time and space intervals must be numbers already implies that our time and space must both be apathic. This is already a fragmentation of our knowledge, insofar as it amounts to a decision to ignore our pathic experiences of time and space. But the alienation, so to speak, increases when we start to ask what kinds of numbers we need in order to satisfy the requirements of the physicists. Satisfying the first demand (that we have numbers we can divide) is not so difficult: for that we have fractions, that is, rational numbers. But to ensure the second condition, that there actually *is* a calculable limit rather than a "hole," we must enlarge our collection of numbers enormously and treat time in terms of the continuum.

One of the great achievements of nineteenth- and early twentieth-century mathematics was the rigorous definition of this continuum, also known as *the real number system*, an intricate, purely logical construct of sets of sets of sets. In chapter 6 we saw how John von Neumann built the (infinite) set N of natural numbers 0, 1, 2, 3, . . . on the empty set. We can also build the integers, incorporating the negatives of 1, 2, 3, ... Look at the set of all possible *ordered* pairs of natural numbers (pairs, of course, are sets of two elements[13]) and call it $N \times N$. For example, to define the integer -2, we collect together all those ordered pairs of natural numbers where the second element is equal to the first one plus two: $(0, 2), (1, 3), (2, 4), . . .$

and so on forever. That collection, which is an infinite set, is what we call minus 2. Similarly, we define all the other negative integers: each will be an infinite set of ordered pairs of natural numbers. The set of all integers (positive, zero, and negative) is traditionally called Z (from the German *Zahlen* = number). Finally, we order the integers in the familiar way, . . . −3, −2, −1, 0, 1, 2, 3, . . . , and we define how to add them, multiply them, and so forth.

And then we can build the fractions, also known as rational numbers, called Q. Consider all the ordered pairs of integers, a set called $Z \times Z$, but omit those pairs where the second element is zero. Suppose we want to define the fraction ⅔. We take the infinite subset of $Z \times Z$ consisting of the ordered pairs: (2, 3), (−2, −3), (4, 6), (−4, −6), (6, 9), (−6, −9), . . . and so on. Notice that those are all the ordered pairs (p, q) such that $3p = 2q$. All those ordered pairs form an infinite set, and that set is what we call ⅔. Then we order the elements of Q and define their addition, multiplication, and so on. It is worth noting that N is part of Z (namely, the nonnegative integers) and Z is part of Q (−3 is the fraction −³⁄₁, or −⁶⁄₂, etc.).

We will go on, but first we need to comment on infinity. You might notice that we've already used the word multiple times in ways that are different from each other. Thus far, we have started with a set, taken its set of pairs, then defined on the latter an equivalence relation and gathered the so-called equivalence classes into a new set: that's how one goes from N to Z, and from Z to Q. Both N and Z are infinite sets: given any natural number n, we can get a bigger number, $n + 1$; given any negative integer m, we can get a smaller, or "more negative" number, $m − 1$. But the rational numbers Q are infinite, one may say, in a different, additional way. It is easy to see that between two different rational numbers there is always a third, different from both. You can see this in many ways, but the easiest is to add the numerators of the two fractions and take the sum as the numerator of the third fraction (and do the same with the denominators). For example, given ½ and ⅔, the fraction ⅗ lies between them. As a consequence, there are infinitely many rational numbers between any two different rational numbers. Q, therefore, is infinite in extension as well as in intention; it extends indefinitely toward the positive side and toward the negative side, but, also, it is infinitely crowded everywhere. Such sets are called "everywhere dense."

Now let's proceed to define the real numbers R, a set that contains Q (hence, of course, N and Z), but much, oh so much more. We can start, if we wish, with the ancient Greek discovery that there is no fraction whose square is equal to 2. Take the set A to be the set of all fractions whose square is less than two, and throw in, for good measure, all the negative fractions, and call it set A. A is an infinite set contained in Q, and we can find in it fractions whose square is very close to 2, as close as we please, but, of course, never exactly 2. This set A is infinitely extended to the negative side, but it contains no element that's bigger than all the others. To see this, suppose that x is any fraction whose square is less than 2, then take y to be the fraction $2x + 2$ divided by $x + 2$: with algebra one can show that y is larger than x, but that the square of y is still smaller than 2. Of course, if there were a fraction whose square is exactly 2, it would be the largest element of A, but there isn't one, and no other candidate, say x, will do, since there always is, as we showed, another fraction, y, bigger than x and still in A.

Similarly, we take the set B to consist of all positive fractions whose square is larger than 2.

Just like set A, set B is infinite and contained in Q. It is quite easy to see that all the elements of B are larger than all the elements of A. This pair of sets (A, B) is called a *Dedekind cut*, because it was Richard Dedekind (1831–1916) who first proposed those cuts as a way to construct the real numbers.[14] The set of all Dedekind cuts can be ordered in a natural way, and they can be added, multiplied, and so forth; they are, from a purely logical point of view, a perfectly legitimate construction of the *continuum R*.

What does *continuum* mean? It means that if we make a Dedekind cut not, as before, in the set Q, but now in the new set R of all Dedekind cuts in Q, we will find no holes: R is whole.[15] Which also means that we will be able to find the limit of those quotients of space intervals divided by time intervals that, as we saw, are used to define the velocity of a continuously moving object and so find its instantaneous velocity and its acceleration—and the rest is physics.

The continuum R, then, was taken up to mend and perfect Euclidean geometry, to make space whole and free of holes. The Euclidean space of n dimensions is usually thought by mathematicians and physicists alike as provided with a system of coordinates and defined as the set of all n-tuples $(x_1, x_2, \ldots, x_n)$, where each x_i is a real number.[16] Once that decision is made about space, the physicist needs to think of time, too, as a one-dimensional continuum, that is, R. Otherwise, velocity and acceleration and all the functions having to do with motion could not be defined, and physics as we know it would be impossible. So we empower physics by treating time as a continuum, but at the expense of the now. No wonder Einstein was so worried.[17]

Theological Excursus: The Diabolical Diagonal

The other day a distinguished theologian of our acquaintance politely but imprudently asked us what we were working on, so we shared the previous pages. The following conversation ensued.

DT: Are you saying that your Dedekind cut, I mean that pair of sets A and B, A to the left and B to the right of the hole where the square root of 2 would be located if it existed—are you saying that *that* turns out to *be* the square root of 2?

US: Exactly. In Q, the set of fractions, $\sqrt{2}$ does not exist, but in R it does, and it is precisely what you just said.

DT: What I find astounding is that the Dedekind cut in question is an ostension and testimony, the most complete and eloquent possible, of an absence—the absence of the square root of 2. And this very testimony of an absence, by some algebraic abracadabra, ends up being the absent square root of 2, now become present.

US: Ostension? But if you are going to put it in such terms of mystery and testimony, then shouldn't you call it a hocus-pocus instead of an abracadabra?

DT: [*Laughing*] You are right. *Hoc est enim corpus meum.* Jesus enjoined his disciples to partake of the unleavened bread "for the sake of my memory." Every time that bread is baked anew, broken, and passed around, again and again, every time, it evokes the memory of the crucified and absent master.

US: So the combination of absence and presence in Dedekind's construction of the real numbers reminds you of the Eucharist?

DT: Yes, but in the Eucharist, the memory of Jesus and his passion is evoked by a material memento. These Dedekind cuts don't have to go through any intermediary of matter to create in the mind—note that I say create, not re-create—the number that was not. The very lack of it creates it! Add to that, if I have understood you correctly, the fact that the fractions, the integers, the natural numbers, even these real ones, can all built out on another sort of hole, perhaps the paradigm of a hole, the empty set! This creation out of nothing, my friends, is the work of a god, not a human.

US: A god? But of what sort? It is hard for us to imagine, for example, that this god of the real numbers could be omniscient.

DT: It is foolhardy to try to tell what a god can or cannot do.

US: Granted. But didn't the Hebrew God, the God of Genesis, look at His complete creation and find it very good? And doesn't the Qur'an tell us that God keeps strict count of everything?

DT: So it is written.

US: But how could such a god be able to inspect or count the real numbers if by that we mean not a general look on the whole but a review of each and all of them? Georg Cantor, who discovered the many paradises of infinity for mathematicians and whose work still fascinates theologians, revealed a "diagonal procedure" relevant to this question.[18] Have you heard of it?

DT: If I did I have forgotten. Please explain.

US: We'll need a few sheets of paper. To keep it simple, let us use only the digits 0 and 1, and let us imagine we have a list, possibly infinite, of unending sequences of 0s and 1s. Here is a random example:

010101 . . .
111001 . . .
101100 . . .
and so on down, indefinitely.

Cantor devised a very simple way to construct, from a given list, a sequence that is not in that list. We look at the first digit in the first item in the list and change it according to the rule: 0 becomes 1 and vice versa. In our example above, we have a 0, so here, on this other piece of paper, we write, instead, a 1. Now, we look at the second digit in the second item: in our example it is a 1, so we write, instead, a 0, that goes next to the previous 1. Next, go to the third digit in the third item: it is a 1, so we select 0. So, we obtain the three first three terms of a new sequence: 100. We keep going in this way, always down the diagonal: fourth digit in fourth item, fifth in the fifth item, and so on.

The new sequence thus constructed, Cantor claimed, is not in the given list. For if it were, it would have to be in some row, say row number n; but if we look at the new sequence's nth digit, it is different from the corresponding digit in the nth item, because it was chosen that way. Hence, the new sequence is not the same as the sequence in the nth row of the list. Contradiction. Therefore, the sequence constructed by the "diagonal procedure" is *not* in the list. You give me a list of such sequences, and I'll find a sequence that is not in it. In other words, there can be no list that contains all possible sequences of zeroes and ones.

Now, again for the sake of simplicity, consider the real numbers between 0 and 1: each of them can be identified to an infinite sequence of 0s and 1s.[19] So Cantor's capital result is that there is no list containing all those real numbers. Even though the rational numbers Q can be ordered in an infinite list (try it!), the real numbers cannot.

Which returns us to our original theological question: could an omniscient god survey all the real numbers between 0 and 1? Cantor's answer is no if by surveying we mean putting them down on a list (an infinite list, of course, but that in and of itself should not pose a problem for a god).

DT: Again, I don't presume to pronounce on divine capabilities, but I can imagine that Cantor's result would have caused quite a stir, even in the seminaries.

US: Yes, for and against. His infinitely many powers of infinity—which are another result of the diagonal procedure—are not admissible to the champions of human intuition we met in chapter 1. Perhaps after our diagonal exposition you would permit us to say that they should also pose challenges for champions of divine intuition, as well? And this devilish diagonal trick was not done with its troublemaking. In the 1930s it also shook up logic when Gödel used it to prove his Incompleteness Theorems.

Bergson's Worries/The Time of Mind

Perhaps we spent too much time with Spengler and other Germans in chapter 1 and too little in France, Germany's strategic rival in those years between the Franco-Prussian War and World War II. After all, in those years the French were just as concerned as the Germans about what they perceived to be a widening gulf between the mathematicizing sciences and human intuition. The most famous mind worrying the topic in France, the philosopher Henri Bergson, burst on the scene with his doctoral thesis in 1889, some thirty years before the appearance of *The Decline of the West*, and he still held sway over the philosophical firmament of France (and not only France) in 1922, when he and Albert Einstein debated in Paris the nature of time.

We will get to the debate, but first, the doctoral thesis. *Essai sur les données immédiates de la conscience* (translated into English in 1910 as *Time and Free Will: An Essay on the Immediate Data of Consciousness*) sought to open a new front in the ancient battle between time and logic. According to Bergson, the enormous power of causality and determinism that so dominated the world was due to the basic error of seeking sameness where it is not to be found, namely, in number. Modern science was increasingly and inappropriately applying measures of quantity and magnitude to feelings and inner states. As an example, he opened his book asking the reader to imagine how

an obscure desire gradually becomes a deep passion. Now, you will see that the feeble intensity of this desire consisted at first in its appearing to be isolated and, as it were, foreign to the remainder of your inner life. But little by little it permeates a larger number of psychic elements, tingeing them, so to speak, with its

own colour and lo! your outlook on the whole of your surround-
ings seems now to have changed radically. How do you become
aware of a deep passion, once it has taken hold of you, if not by
perceiving that the same objects no longer impress you in the
same manner?

"The fact is," Bergson tells us, "that the further we penetrate into
the depths of consciousness, the less right we have to treat psychic
phenomena as things which are set side by side."[20]

And yet, he laments, this is precisely what people do all the time,
assigning magnitudes to feelings, greater thans and less thans to
desires, as if mental phenomena could be measured and assigned
causes in the same way as physical ones. To some degree he grants
that this error is part of the human tool kit, built into our "common sense." But it has gotten worse and worse, because number has
proven so powerful in the physical world. Physics "encourages and
even exaggerates the mistake which common sense makes on the
point. The moment was inevitably bound to come at which science,
familiarized with this confusion between quality and quantity, between sensation and stimulus, should seek to measure the one as
it measures the other: such was the object of psychophysics" (71).
It is as a result of this confusion that all the seeming paradoxes of
determinism and free will arise. "By invading the series of our psychic states, by introducing space into our perception of duration, it
corrupts at its very source our feeling of outer and inner change, of
movement, and of freedom. Hence the paradoxes of the Eleatics,
hence the problem of free will" (74).

Not only in the early *Essai* but throughout all of his career, Bergson proclaims the two related distinctions/oppositions: space and
time, reason and intuition. He seems to have used the former as ful-
crum for the latter, making space the domain of reason, and time
that of intuition, a double slash into wholeness that enjoyed a pro-
digious success for more than thirty years. Even the Americans William James and John Dewey celebrated the work as the beginning of
a new era, "a true miracle," "a Copernican revolution."[21]

In order to understand how the young Frenchman proposed to

restore our freedom, let's start with a general definition of causal determinism, plucked from an encyclopedia of philosophy:

> The *world* is *governed by* (or is *under the sway of*) determinism if and only if, given a *specified way things are at a time t*, the way things go *thereafter* is *fixed* as a matter of *natural law*.[22]

Beneath every word we have italicized in this definition we see cargoes of cans of worms. But the only crawling creature that concerned Bergson was time. A physicist reading "a time *t*" in the above definition must understand that *t* is a *real number*, and that "thereafter" means "at or for all real numbers *t'* > *t*." Bergson's book was dedicated to proving that this time of number could not be applied to the time of mind.

We have called the physicist's logical construct "apathic time," since it is unruffled by whatever happens along it or flows in it. Bergson called it mechanical, or extensional, or material, or clock time, and he claimed that the mind is not regulated by it but by a radically different intuition of time he called *durée*, or duration. "True duration" has nothing to do with space. Unlike numbers or sets (which Bergson insists are produced from our intuitions about space) duration cannot be divided into parts of which one could say that a part is contained in another part. A duration is not an apathic collection of moments or points in time. In other words, no numbers here, no possibility of counting our thoughts or measuring our psychic states. We have called this a "pathic" point of view. He called it a "wholly dynamic way of looking at things." How could we not be sympathetic?[23]

Alas, Bergson achieved his freedom in the same way we have seen so many others do, by severing our knowledge, plucking out the offending part, and demeaning, like a good dualist, a goodly portion of the human. The distinctions he put to work in order to achieve this were well-worn philosophical ones: the distinction between extension or matter and mind or soul, postulated by Descartes; and the Kantian distinction between an external intuition, called space, and an internal intuition called time. In order to counter the physical determinists' reduction of mental phenomena to material ones, that is, in order to distance mind from matter, Bergson sharpened the distinction between the "internal" intuition called time and the supposedly "external" and material one of space.[24]

Bergson's surgery is easily summarized (or at least caricatured). "The soul," psyche, and inner life is "wholly dynamic." Space, the outer world, is clear and "well defined." But to part of our consciousness, the dynamism of our inner life is unbearable. "This wholly dynamic way of looking at things is repugnant to the reflective consciousness, because the latter delights in clean cut distinctions, which are easily expressed in words, and in things with well-defined outlines, like those which are perceived in space." The reflective consciousness will therefore take every occasion to impose the clarity of space onto the inner life. "It will assume then that, everything else remaining identical, such and such a desire has gone up a scale of magnitudes, as though it were permissible still to speak of magnitude where there is neither multiplicity nor space."[25]

Mistaken uses of identity, ceteris paribus clauses erroneously applied, measure and magnitude applied to desire, the pathic and the apathic confused: we don't seem so far from our own critique of the social and psychological sciences. But notice the divisions Bergson is building within the mind. What is this "reflective consciousness" that "delights in clean cut distinctions"? It cannot be the same as *la conscience de soi*, the consciousness that says *I*. For how could that mental mirror, if at all a true one, delight only in those distinctions whose existence can be asserted and decreed by a set of postulates (like apatheia, or the empty set) and never be verified in actual experience? And worse, notice the deep dualist cuts—every bit as deep in their own way as those of Descartes—that Bergson's scalpel makes in our experience of the world: our intuitions about external space are clear and clean cut, susceptible to measure and number. Our intuitions about internal time are indistinct, unclear, wholly dynamic, untouchable by axiom or mathematics.[26]

Dualism is not a crime. But does it capture a truth? Bergson's distinction between space (apathic) and time (pathic) is not supported by everyday, prescientific experience, and the implications he drew from that distinction even less so. The element of surprise, for example, the sudden appearance of the new, resides only in time for Bergson, not in space. But imagine lying under a clear night sky, in the season of the Aquariids or the Perseids, looking for shooting stars. They can appear anywhere and at any time. Our attention is focused, like that of a hunter, both on the spot and on the moment. As a feeling, it is different in quality from what it would be if we expected the light or the prey at a fixed spot and any time or at a fixed time but anywhere.

Bergson's error is a common one: eyes fixed on one danger, he walks into the jaws of another. His goal is to save humanity from the mechanical time of modernity—time he understands as identified and measured by the regular motion of hands across the space of a clock's face. That metaphor itself could have inspired Bergson to hold clock and human in close proximity, as we saw the poet Rilke do ("And with tiny steps the clocks advance / alongside the day we truly live"). But instead he chose to increase the distance between the space and time of the world and the movement of our inner life. "When I follow with my eyes on the dial of a clock the movement of the hand . . . , I do not measure duration, as seems to be thought; I merely count simultaneities, which is very different."[27]

Bergson seems not to see that if pressed to its conclusions, this choice would make our experience of both space and time unidentifiable and unutterable, as he could have learned from a countryman and contemporary he so much admired, Henri Poincaré:

If I am at a definite point in Paris, at the Place du Panthéon, for instance, and I say, "I will come back *here* tomorrow;" if I am asked, "Do you mean that you will come back to the same point in space?" I should be tempted to answer yes. Yet I should be wrong, since between now and tomorrow the earth will have moved, carrying with it the Place du Panthéon, which will have travelled more than a million miles. . . . In fact, what I meant to say was, "Tomorrow I shall see once more the dome and pediment of the Panthéon," and if there was no Panthéon my sentence would have no meaning and space would disappear.[28]

Why was Bergson so insistent on distancing our inner life from space and time? George Santayana asked that question in 1913. After a few gentile anti-Jewish remarks, he provided a psychological diagnosis: "cosmic agoraphobia." "M. Bergson is afraid of space, of mathematics, of necessity, and of eternity; he is afraid of the intellect and the possible discoveries of science; he is afraid of nothingness and death." Perhaps all that is true, we do not know. But what we want to stress is that this attitude was not Bergson's alone. On the contrary, it constituted a worldview that captured the allegiance of an enormous audience, as Santayana himself conceded: "The most representative and remarkable of living philosophers is M. Henri Bergson. Both the form and the substance of his works attract universal attention."

The popularity of that worldview was not diminished by Santayana's criticism or by the sneers of those who pointed to the enormous female following of Bergson's teachings in order to taint them as irrational. It was still a powerful force when it came face-to-face with Albert Einstein in a crowded Parisian lecture hall on April 6 of 1922.[29]

Bergson versus Einstein

Looking back from 1934, Paul Valéry called the Paris encounter between Bergson and Einstein "the great event" (*grande affaire*) of the twentieth century.[30] From our perspective a century later, that claim seems vastly inflated. But we can better understand the inflation if we think back to the historical context we outlined in chapter 1: a European world convinced that it was caught in a clash of civilizations. That clash was imagined through any number of powerful oppositions: between causality and freedom, reason and immediacy, technology and life, math and feeling, logic and intuition, head and heart, male and female, and yes, also between French and German, Jew and non-Jew, America and Europe, and many others.

The debate between Einstein and Bergson could be (and was) experienced and understood as a staging of these contests in one great boxing match between two heavyweight champions: the one of immediacy, intuition, and the heart (as Santayana had put it, evoking Pascal); the other of causality, mathematics, and the brain. The staging is even reflected in the relics of the two thinkers that would circulate after their deaths: locks of Bergson's hair; Einstein's brain in a jar.

We do not mean that the two thinkers would have recognized themselves in such polarity. Bergson, for example, clamored constantly against the (frequent) accusation that he was an enemy of science just as much as Einstein denied (with more justice) that he was antiphilosophical. Nor do we mean to endorse the polarities ourselves. On the contrary, we agree here with Susan Sontag: "The kind of thinking that makes a distinction between thought and feeling is just one of those forms of demagogy that causes lots of trouble for people by making them suspicious of things that they shouldn't be suspicious or complacent of. . . . For people to understand themselves in this way seems to be very destructive."[31] Like Sontag, who saw herself as waging "a crusade" against the separation of intuition and reason, one of our goals in this book is to remind you that all of these splits and separations are contingent ideological commitments in the long history of humanity's struggle with its own knowledge. But however contingent and dogmatic these ideologies about knowledge might be, they were also resonant and powerful. It was that resonance that amplified the debate and turned the debaters into representatives of, on the one hand, the sovereignty of reason and, on the other, an insurgency against it.[32]

Debate is perhaps not the right word: it was rather a series of speeches posing as questions delivered by a number of eminent savants from slightly differing bandwidths of the intellectual spectrum and occasionally receiving a short response from Einstein, whose limited French may have been strained. Bergson was only one of those savants. (Émile Meyerson, whom we met in chapter 7, was another, and his engagement with Einstein at this event produced an enduring intellectual relation.[33]) The gathering was intended as an international rapprochement, a meeting of the greatest minds of Germany and France, two nations as divided, in those years between world wars, as nations could be, bringing them together in "an affirmation of the universality of the human spirit."

For the first half of the evening, that possibility still held. Einstein responded to questions with a relative lack of dogmatism. "Geometry is an arbitrary conception." "The relationship between the real continuum and the imaginary space of geometry is not univocal, and one cannot say that one way of speaking is preferable to the other." "One can always choose the representation one wishes if one believes it is more adequate than another for the question one is proposing; but that does not have an objective sense." For the first half of the evening, one might almost have thought — except for the accent — that one was hearing the French Poincaré as much as the German Einstein.[34]

Then a Bergson acolyte, Édouard Le Roy, took the floor. "Deep down I have nothing to say," he began. But predictably he did not stop there. In a few short sentences he proceeded to refortify the frontiers between intuition and reason, philosophy and physics, time and space. For although he might have focused on the similarities between the theories of Einstein and of Bergson (both theories, after all, recognized a difference of nature between lived time and the time in our equations of measurement), he did quite the opposite. "The point of view of the philosopher and that of the physicist are both legitimate, but they lead to posing, in terms that seem the same, two problems that are in reality completely different." Specifically, he stated, the "problem of time is not the same for Mr. Einstein and Mr. Bergson." And then he called on Bergson to take the floor.

Bergson did. He began by saying that he had come to listen, not to speak. He then launched into a peroration about eight times longer than those of the other participants. (It could have been much longer, since Bergson had just sent to press a 245-page book expounding his critique of Einstein's theory of relativity.) In his speech he made many of the points we have already learned from his work: about the difference between duration and measurable time; about the imperviousness of the intuition of simultaneity to any attack from the theory of relativity or experiments with clocks or mathematical conventions. All this in the service of his conclusion, that philosophy remained sovereign over physics when it came to determining the philosophical implications of the concepts introduced by physics, which is also to say when it came to determining the boundary between the "real" (such as the intuition of simultaneity and duration) and "convention" (such as measured time).

Einstein's reply was brief, as were all his interventions in this séance. First he restated the question: "Is the time of the philosopher the same as that of the physicist?" And then he answered it. Philosophical time is a mixture of physical and psychological time. The primitive notions of simultaneity in psychological time helped humans make sense of many aspects of their world. From there, humans instinctively took the short step of creating a temporal order in events. But none of these mental constructions allow us to conclude that there really is a simultaneity of events. His last sentence is perhaps the most often cited from this debate: "Il n'y a donc pas un temps des philosophes." "There is therefore no such thing as a time of philosophers; there is only a psychological time that differs from the physicist's."

No one would speak today of a "Bergson-Einstein debate" if it were not for what came after. Bergson published his long critique of Einstein, *Durée et simultanéité: À propos de la théorie d'Einstein*.[35] The physicist read the book on October 8 (as we know from his diary entry), while sailing to Japan: "Yesterday I perused Bergson's book on relativity and time. Strange that time alone is problematic to him but not space. He strikes me as having more linguistic skill than psychological depth."[36] His unhappiness with Bergson seems only to have grown, and soon he was collaborating with Meyerson and other philosophers in order to counter a parallel campaign mounted by the French sage's many allies, which included many prominent scientists.

Over time the polemical fruits produced by both camps became productive for future generations seeking to stage anew scenes of conflict between humanity and science. In the 1950s, for example, Maurice Merleau-Ponty restaged the debate as a defense of "living reason" against the dangerous "rationalists."[37] And in 1965, at the height of Cold War fear of nuclear Armageddon, an art historian translated Bergson's *Duration and Simultaneity* for an American audience with an introduction by the British physicist Herbert Dingle, who warned of global danger if the world's scientists did not waken from their dogmatic Einsteinian slumber and heed once more Bergson's teachings about intuition.[38]

Our purpose in revisiting the debate is not to adjudicate in favor of Einstein's time or Bergson's but only to show how both cut off important parts of human knowledge in their efforts to proclaim the primacy of their approach. Why did Bergson ignore space? Why did he demean clock time? And why did he see in the theory of relativity an assault on the human rather than an allied approach to the conundrum of time? To answer that would require (as Santayana suggested) an approach as open to psychology and history as it is to philosophy and physics.

And why did Einstein deny "objectivity" to vast domains of the human perception of time? Why did he come to insist on the "reality" of certain mathematical conventions and the arbitrariness of others? Why was he so insistent that some constant (in this case the speed of light) must serve as the absolute foundation of knowledge? In his famous paper of 1916 proposing the general theory of relativity, Einstein had only proposed that his theory was "psychologically the natural one." By 1922 he was insisting that it was not only *psychologically* but *necessarily* the natural one, the only one with objective reality. The shift was not demanded by new experimental evidence. Was his stance hardened (as some historians have proposed) by the explicitly anti-Semitic attacks on his work and person that erupted in Germany in 1920?[39] We do not know. We only seek to convince you that neither Einstein's nor Bergson's sharp cuts in time were *necessary*. Both were the product of a choice about what is most important in human knowledge, a choice conditioned by specialization, context, personal history, psychology, and who knows what other contingencies.

The Time of Fiction

To see that other approaches were possible, one need do no more than turn to the fiction of the age. Science fiction might seem a logical place to start, since it took up many of the chronological conundrums we've been discussing. The problem with science fiction for our purposes is that in it the relationship between science and fiction is often slavish rather than reflective, as in the immensely popular writings of the astronomer Camille Flammarion (1842-1925), whose many volumes (beginning with *The Plurality of Inhabited Worlds* of 1862) give him some claim to the title of "founder" of the genre. Consider his *Lumen* of 1872, from the collection *Narratives of Infinity*. *Lumen* is the story of a man (called Lumen, Latin for light) relating his experiences after death. Notice that this newest of literary genres seems hatched from the fossilized egg of the most ancient: we find such narratives already in Sumerian myths.[40]

Lumen is based, its author insists, on a scientific fact: the maximum velocity of light is not infinite but about three hundred thousand kilometers per second. It is also based on a conceit (although its author claims that this, too, is science rather than theology or philosophical dogma), namely, that living beings consist of three parts—body or matter, vital energy, and soul—of which only the immaterial soul subsists after death. Once freed of its fleshy fetters (the conceit continues), the soul can travel through space at any speed, even speeds much greater than the speed of light.[41]

Flammarion imagines that were we capable of flying away from earth at the speed of light, we would keep before our eyes the image that we last saw on our planet, since we are traveling alongside that image and at the same speed. That is equivalent to saying that time has stopped for us: we continue to live at the time of our departure. By the same token, if we are traveling *faster* than light, we will be going backward in time, viewing older and older images. For Flammarion, let us recall, the soul is not subject to the limitations of matter; it can travel at any speed, and it happens to possess, even though immaterial, pretty good eyesight (and apparently no other sense). Lumen's soul witnesses scenes that happened on earth, mostly in or near France, in times gone by. He watches the battle of Waterloo. He determines that the victorious Prussians did not have justice on their side in the Franco-Prussian War of 1870. He sees himself as a school child among other school children at the Place du Panthéon and watches his mother rushing to his rescue when he falls to the ground, hit by another boy.[42]

Lumen is an excellent example of the proximity of science fiction to science both in the sense that the fiction derives its plausibility from the science and that science can be inspired by the fiction. Henri Poincaré read many of Flammarion's works and even derived a paradox on the relativity of time from the novel. (He called it "Flammarion's paradox": time going backward producing more difference, going forward more sameness.)[43] Einstein's childhood encounters with Lumen's time-stopping trip on a beam of light may have influenced the physicist's later thought experiments. Regardless of their role in relativity's origins, fantasies about time travel and immortality certainly affected how the theory was received, not only by the general public but also by physicists like Paul Langevin, an organizer of the Paris gathering in 1922. At a prestigious conference eleven years earlier, Langevin had breathlessly described how relativity could permit an individual to travel into the future and even "to age less rapidly."[44] Such claims help explain why Bergson (who had been present there) later dismissed relativity as one of those "fantastical searches for the fountain of youth."[45]

This type of science fiction continues to thrive. (In *Le Théâtre quantique*, by the mathematician Alain Connes, CERN's particle accelerator is used to achieve the perfect translation of the protagonist's consciousness into a computer network even as her body is destroyed in the process.[46]) For our purposes it is not very interesting, in part because it tends to embrace rather than interrogate Pythagorean fantasies of eternity enabled by the newest science while ignoring any losses these might entail. Lumen, for example, is entirely an optical creature. Sight is his only sense, light his only vehicle. Such travels into the past can only satisfy a reader with no experience or real interest in the business of remembering or reliving one's life, and whatever eternity they can promise feels pale indeed.[47]

But focus for a moment on a very different type of fictional time travel exemplified by the work of a novelist who was, among other things, best man at Bergson's wedding, Marcel Proust (1871–1922). The seven volumes of Proust's *À la recherche du temps perdu* (translated as *Remembrance of Things Past* or *In Search of Lost Time*) were written in the midst of all the debates we have touched on above and were well informed by them. Their author quickly became both celebrated and loathed as the great literary impresario of a modern sensibility about time. (Wyndham Lewis was among the more famous loathers, casting Proust as a homosexual retailer of a cheap Bergsonian Jew-modernist psychology.)[48]

We will dwell on two of the most famous examples of time travel in that massive work, one from the first volume, the other from the last. In the first, the narrator is an adult long installed in Paris. He has forgotten virtually everything about his childhood home in the village of Combray. One cold winter day he dips a petite madeleine in tea and "mechanically" takes it to his lips. The instant the mouthful touches his palate, he is aware that something extraordinary is happening within him. He is invaded by a sweet pleasure without "any notion of its cause." He suddenly ceases to feel himself mediocre, contingent, mortal, but feels only himself. A second mouthful, a third, with similar but diminishing effects. What is the source of virtue in this beverage? Pondering the question, he moves from examining the cup to examining his spirit, reinforcing its élan by focusing his attention in a heroic effort of introspection. "Et tout d'un coup le souvenir m'est apparu." That combination of tea and madeleine is what his aunt had given him those Sunday mornings when, as a small child, he would visit her in her bedroom. And with that sudden recollection, the old house, the garden, the entire town with the streets in which he had played, all that emerged into his memory "from my cup of tea."[49]

And then there is the episode in the last volume, *Time Regained*, when the narrator jumps out of the way of a rushing vehicle, stumbles, and lands a foot on an uneven paving stone. In that instant, "all my fear vanished before the same felicity given me, at different periods of my life, by the view of trees that I felt I recognized . . . , the view of the bell tower of Martinville, the taste of a madeleine dipped in tea, and so many other sensations. . . . Like that moment when I tasted the madeleine, all doubts about the future, all intellectual doubts, were dissipated. . . . Even those [doubts] about the reality of literature found themselves lifted as if by enchantment."[50]

In these moments, Proust's narrator is surprised by memory, seized by an unwilled resurgence of an old experience that transfigures both time and space. We might treat this as an example of

Bergson's dictum that sensation is always saturated in memory. But note that there is no Bergsonian separation here. A spatial experience (stepping on an uneven paving stone) transforms both time and space; a sensory recollection (the taste of madeleine and tea) makes present an entire village. Just as remarkable, this sudden merging of past, present, and future is not experienced (as Einstein might fear) as an anxiety-provoking alienation from lived time but rather as a moment of wholeness. For that moment, the narrator feels coherent unto himself, as the simultaneity of time in memory stills the tension between having been, being, and becoming. Kierkegaard, too, as we will see in the next chapter, wrote about these powers of repetition.[51]

Have you ever experienced such moments? If not, reading fiction may help you cultivate them. You could begin with Borges's short story from 1946, "Nueva refutación del tiempo" (A New Proof That Time Is Not).[52] The title is an ironic oxymoron—how can something be new if there is no time?—and Borges refers to the story explicitly as an "ever-so-slight mockery." But his attitude toward time (which he tells us informs not only this essay but all of his books) is that of a master, for his is "a life in whose course repetitions abound." So let us attend to the story's combination of two arts, which in fact go together: one (the "inner art" Bergson might have called it, with his penchant for dividing time from space) of reflection and speculation, and the other the "spatial" art of walking randomly about a city—in Borges' case, Buenos Aires.

To call the latter activity an art might seem eccentric, yet the word is precise: De Quincey, Dickens, Baudelaire, Marie Bashkirtseff, and Virginia Woolf are only a few of the many artists who have cultivated it. In his story, after a summary of the philosophers (idealist and others) who had sought to refute time before him, Borges tells us he will adopt "a more direct method" and takes to the streets, recounting a walk he titles "To Feel Oneself in Death."

He tells us that one night in 1928 he set out "to walk and to remember," "without destination," "randomly," with "a maximum latitude of probabilities," "accepting the most obscure invitations of chance." Despite that, a "kind of familiar gravitation" drew him toward a certain area of the city, "a zone that is familiar and mythological at once": not the neighborhood familiar to him since child-hood (Palermo) but its periphery, its penumbra.

Borges arrived at a corner. It was a moonlit night. The street be-fore him "was both very poor and very lovely," an unpaved street "of elemental mud." "Over the muddy, chaotic earth a red pink wall seemed not to harbor moon glow but to shed a light of its own. There is probably no better way to name tenderness than that red pink." The feeling that this place evokes in our poet-errant is ec-static. The moment he is experiencing now, he feels, is indistin-guishable from the moment someone experienced at this corner before his birth, "in the 1880s":

> That pure representation of homogeneous facts—calm night, limpid wall, rural scent of honeysuckle, elemental mud—is not merely identical to the scene on that corner so many years ago; it is, without similarities or repetitions, the same. If we can intuit that sameness, time is a delusion: the impartiality and insepara-bility of one moment of time's apparent yesterday and another of time's apparent today are enough to make it disintegrate.

Borges's story claims to teach us the experience of biographi-cal time's collapse into sameness. Given "an ever so-slight mock-ery," we can't know how seriously to take the claim, which is in any case philosophically unjustified.[53] Just who is being mocked? The title points toward philosophers of the logical persuasion, like John McTaggart (whose "The Unreality of Time" we encountered in note 9 of this chapter). Borges's preface mentions philosophers of several stripes: Berkeley, Hume, Schopenhauer, and yes, Bergson. His irony applies even to himself: his first paragraph tells us that he "does not believe" in his own refutation, although it informs all of his writings and comes upon him "in the tired dusk with the illusory power of an axiom."[54]

In Borges's repeated emphasis on the "elemental mud" we sense a mockery aimed at even deeper wellsprings of the philosophy whose history has run like a river through this book. The corner that elicited the vividness of eternity, Borges tells us, is by the river Maldonado. The basin of that stinky polluted stream gave refuge to knifers, ruffians, and prostitutes, and its frequent overflowing produced the elemental mud of those streets, whose sidewalks stood high to protect dwellings from the floods. (The river was entombed in a massive public works project that began in 1928 and concluded in 1937, but continued flooding required new works that began in 2005.) In Plato's *Parmenides* Socrates told us that he could not envision things such as hair and mud participating in *eternal* forms. In Borges's story, mud is the most conspicuous element in the narrator's ecstatic experience of eternity.[55]

If mud is not to your taste, there are many other ways to become conscious of time's conundrums. Borges dedicated an unflattering essay to the apostle of a different technique, J. W. Dunne, author of works like *Nothing Dies* and *An Experiment with Time*.[56] This last work, first published in 1927, engaged the imagination of many writers in addition to Borges: H. G. Wells, James Joyce, W. de la Mare, T. S. Eliot, Aldous Huxley, J. B. Priestly, J. R. R. Tolkien, C. S. Lewis, and Vladimir Nabokov, to name just a few. The experiment, first performed on himself, consisted in waking up immediately after a dream and writing it down in particular detail, then being alert while awake, in the following couple of days or perhaps longer, for any event that was the *same* as something in a dream.

Details are important: if you dream, for instance, that a book falls from a shelf by itself and startles you and this happens in actuality on the following day, one may say that it is no more than a coincidence; but if the book in question was a Czech dictionary that falls in such a way that it opens at the letter *H*, and the following day you are startled, while awake, by the fall of the same kind of book opening at the same letter, it will be hardly believed to be a coincidence. More than half of Dunne's book is dedicated to a list of the latter sort of time-inverting dreams, including those of other people who volunteered for the experiment. They constitute an impressive dossier of mental phenomena that currently resist scientific explanation.

Regrettably, Mr. Dunne did not resist the temptation of providing one. He noticed that English common expressions such as "time flies," "as time goes by," "the flow of time," and so forth, logically require, if taken literally, the existence of an undertime. For time to flow, to go, or to fly, it must do so in a time independent of and underlying the flier, the goer, or the flowing one; this latter time is the undertime. But by the same token, the undertime necessitates an under-undertime . . . and so indefinitely on, from which Mr. Dunne deduced, among other things, our immortality. The endless regress (*mise en abîme*) is a common enough philosophical tool, but in this case it persuaded few of his many readers, and certainly not Borges, about the nature of time. Still, his experimental technique for experiencing sameness inspired many and proved useful to some remarkable writers, the case of Nabokov most remarkable among them.

You can read about Nabokov's engagement with dreams in his *Insomniac Dreams: Experiments with Time*. The subtitle is no accident. Nabokov had read Bergson and perhaps Borges as well, but this did not dissuade him—between October 14, 1964, and January 3, 1965—from endeavoring to replicate Dunne's experiment on himself. In that period, Nabokov filled 118 Oxford index cards with dreams and events. The exercise influenced his late fiction, such as the novel *Ada* (1969), but Nabokov's interest in dreams was evident long before. Just one stunning example from 1916, when the future novelist was seventeen. That year young Vladimir had inherited a valuable estate from his uncle Vasily. He would lose it all a year later with the Russian October Revolution. Shortly after his uncle's death (and before the revolution), Nabokov recorded in his diary that he had dreamt of his uncle, who was telling him in the dream "I shall come back to you as Harry and Kuvyrkin." Forty-two years later Harris and Kubrick Pictures bought from Nabokov the film rights for *Lolita*, thereby restoring much of the fortune he had lost. As if to eliminate the last resort of the skeptic, the book's editor adds that the transformation from Kuvyrkin to Kubrick follows "the correct sound shift from the Slavonic to the Latin."[57]

None of these experiments and experiences made Nabokov think that the subject of time was an easy one to think about. Quite the contrary, as he put it in *Ada*, the question remains a basic and a frustrating one for the human: "Why is it so difficult—so degradingly difficult—to bring the notion of Time into mental focus and keep it there for inspection? What an effort, what fumbling, what irritating fatigue!" (*Ada*, 537). By now, you may be irritated and fatigued as well. So let us provide a conclusion to our engagement with time. That conclusion will not surprise you. The times of physics, the times of philosophy, the times of fiction, the times of dreams . . . these and many others are all part of human time. We can and should debate among them, and for some purposes we may need to choose one and ignore others. But that choice will remain one that must be made carefully and is never absolute. No experiment lurks just over the horizon to put an end to all our questions.

It is not the case that—as the physicist Richard A. Muller recently announced—physics has solved Einstein's problem of "the now" or is at most an experiment or two away from doing so.[58] We are certainly interested in whatever new truths physics will offer us about time, but there are many aspects of human being in time that we hope we have convinced you they cannot address. We are similarly fascinated by our colleagues in neuroscience, with their discoveries that the olfactory bulb plays a key role in critical aspects of learning, memory and behavior in mammals, or that the posterior parietal cortex, an area of the brain that controls spatial awareness, is also causally involved in decision making, such as deciding what images should be in the field of view.[59]

Such findings may help us understand how a madeleine dipped in tea could bring an entire village to mind or suggest how spatial awareness affects the possibilities for thought. But they cannot tell us how to live in such a way that we are able to recognize both repetition, coherence, wholeness, and the knowledge that we have gained by dividing ourselves. If anything, and at their best, such researches can only confirm that when it comes to creatures whose senses, perceptions, and life histories are as intertwined as those of a human (or even a rat), no axiom can decide whether we experience sameness or difference.

10

ETHICAL CONCLUSIONS

*The ethical is and remains the highest task
assigned to every human being.*

SØREN KIERKEGAARD[1]

*Mathematics cannot cast off any prejudice,
it cannot alleviate obstinacy or assuage partisanship;
in the entire sphere of morals it can do nothing.*

JOHANN WOLFGANG VON GOETHE[2]

Does how we think about sameness or difference have any impli-cations for how we should live our lives? Many people in the past have thought so. We've seen Jesus, Muhammad, and their followers translate ideas about some essential, enduring, necessary and true sameness at the foundations of the cosmos into a path to eternal life. From Plato to Kant, we've heard philosophers promise that atten-tion to certain kinds of sameness helps the human achieve, if not life eternal, at least the happiest and most virtuous one possible.

Physicists like Einstein did not produce self-help books, but if he really believed that "everything is determined, . . . human beings, vegetables, or cosmic dust, we all dance to a mysterious tune, in-toned in the distance by an invisible player," then his advice would presumably be like that of the ancient Stoics: be guided by the invisible piper or be dragged by him.[3] Even the claims of today's economists, sociologists, and psychologists about how we should organize human life are often built on implicit values grounded on foundational assumptions of sameness.

We've been advocating a different view: that there is no neces-sary sameness or difference; no entirely stable foundation to our knowledge; no axiom, law of thought, or mathematical analogy that determines how we must choose between the One and the Many. Does *this* view of human knowledge tell us anything about how we should live? Again, there have been philosophers and theologians in many traditions who have thought so. Perhaps the most successful teachers in this vein have been those of the Buddhist traditions. The "Great Abbot" of Tibet Śāntarakṣita (ca. 725–788 CE) laid out such a path in his poetic treatise *Embellishment of the Middle Way*: "using Only Thought, it can be realized that there are no external entities; using our system, it can be realized that even thought has no essence at all."[4]

From the school of "only thought," Śāntarakṣita explained in his own commentary to his poem, we've learned the important lesson that things external to thought, "starting with 'I' and 'mine,' object and subject," have no essence. But his teaching goes one gigantic step farther. Even "thought has no essence; for by ascertaining this Middle Way (which is free of all extremes), one fully realizes that thought is neither essentially unitary nor essentially manifold, and that it thus has no essence at all." There is no essential sameness, no essential difference. The Great Abbot's "Middle Way" is, among other things, a way of life intended to help its followers free themselves from the false attachments produced by the error of believing that there are necessary laws regulating sameness and difference.

Stretching from Sri Lanka through Tibet to northern China, the arc of Buddhism produced many profound meditations on sameness and difference and myriad instructions about how to live in accordance with the resulting wisdom. We mention these not by way of endorsement but simply to establish the point: regardless of their perspective, some of the greatest teachers of humanity have been convinced that how we think about sameness or difference has the deepest implications for how we should want to live our lives. Today, on the other hand, few professors (apart perhaps from theologians and economists) pause to ponder what relationship there should be between our disciplines of knowledge and our form of life, between our epistemology and our ethics.[5]

Ours is not a self-help book, nor are we attempting to establish a religion. But as Søren Kierkegaard suggested in his *Concluding Unscientific Postscript*, "it may also be required of a devotee of scholarship . . . that he continue to understand himself ethically in all his labor, because the ethical is the eternal drawing of breath and in the midst of solitude the reconciling fellowship with every human being."[6] So we will attempt to distill out of the many previous pages some explicit psychological and ethical lesson, that is, a lesson about the self and about relations between the self, the other, and the world.

That lesson will not come as a surprise: strive to cultivate within yourself the simultaneous mysteries of sameness and difference. We have tried to show in the previous chapters that cultivating those mysteries, whatever your intellectual discipline—mathematician or poet, mystic or philosopher, physicist, biologist, economist, psychologist, or historian—will reveal new questions at that discipline's foundations. Here in our conclusion, we will presume to suggest that we may want to cultivate these mysteries not only as scholars, specialists, or intellectuals, but also as psyches, as subjects, as human beings.

Transfigured Night

How can a person strive to cultivate sameness and difference at the same time? The task seems absurd from the point of view of any psychology based on "Laws of Thought" like the Identity Principle ("I am I"), the Principle of Sufficient Reason ("our psychic states can be predicted from their cause"), the Principle of Non-Contradiction (we cannot be at once happy and sad). The Polish writer Olga Tokarczuk, who won the Nobel Prize for literature in 2019, described her own experience studying that sort of psychology in Warsaw. "We were taught that . . . in its essence the world was inert and dead, governed by fairly simple laws that needed to be explained and made public—if possible with the aid of diagrams." From all of this, she drew one simple lesson: "steer clear of psychology altogether. . . . The psyche is quite a tenuous object of study." That, presumably, is why she transferred her own efforts from the scientific to the literary study of the psyche.[7]

Literature has often explored the simultaneity we are advocating. Consider Richard Dehmel's "Transfigured Night" (*Verklärte Nacht*), the German poem that in turn inspired Arnold Schoenberg's sublime sextet of that same title, composed in 1899.[8] In its own day

Dehmel's erotic and poetic experimentation scandalized as well as inspired. Today its sentiments seem almost conventional, but when *Woman and World*, the book containing "Transfigured Night," was published in 1896, it was condemned as blasphemous.

The poem begins, "Two persons walk through a bare, cold wood." The rest consists of a dialogue between those two voices. A woman tells a man that she's with child. She had slept with someone she did not love, desiring motherhood because "I no longer believed I could be happy, and yet I had a strong yearning for something to fill my life." She calls her act a "sin," and "a great offense against myself." "Now life has taken its revenge: Now I have met you, oh you," the man she walks beside while bearing another's child. The man replies that while they are both traveling on a cold sea, he feels in himself her inner warmth and feels his own in her. This warmth, he says, will change, has changed, the child into his own. The child is transfigured, and in the poem's last line—similar to yet different from the first—so is the world: "Two persons walk through high, bright night."

We cannot understand this poem through "pebble" psychologies that teach sameness and impenetrability as the fundamental principle of being. The persons in the poem do not conform to philosophies that insist that "the ultimate characteristic feature of all rational beings is absolute unity, constant self-identity, complete agreement with oneself."[9] The voices participating in the poem's moonlit dialogue *do* have solidity. The woman knows what constitutes offense or fulfillment of her "self," the man speaks of each person's lonely voyage on a cold sea. But their permeability is also exemplary. Each is informed by the other's warmth, as is the fetus within her and the night that enfolds them. Individuality and intersubjectivity, a sense of self and the capacity for transformation through the presence of and dialogue with others: both are simultaneously expressed by the two persons of the poem. We must strive for each of these in their simultaneity.

Unfortunately, the rules of "common sense" and practicality often get in the way. Any number of such rules are embedded in our culture and language. For example, when we say, or point at, "this" or "that," we are already assuming (with Aristotle and so many others) that two *different* bodies cannot occupy the *same* space. The assumption that the thoughts of two *different* persons cannot be the *same* thought is an example of a related practical rule. So is the idea that in order to be an ethical agent, you must have a "practical identity," a description of yourself that has "unity" and "integrity," a description "under which you value yourself."[10] We are asking ethics, the practical part of philosophy, to overcome these practical rules: we cannot expect that to be easy.

Fortunately, our own activities and experiences also provide powerful help. Consider our experience of hearing or making music. When Aristotle took up the subject, he found it challenging to explain what kind of power (*dynamis*) music has on us and what our goals are in engaging in it. Amusement? Relaxation? Education? Eventually he settled on the view that music has an ethical purpose, encouraging in us an imitation of traits of character (*mimēmata tōn ēthōn*). Thus, let us say, martial music may lead, by imitation, to heroic deeds.[11] Such a position is true to Aristotle's psychological and logical commitments, but imitation does not seem to us to account adequately for the "ethical" aspects of the experience of making or being moved by music.

Return for a moment to "Transfigured Night." For string players to perform Schoenberg's piece, each of them must be what we might call a fairly solid musical self, a virtuoso of her instrument. But for the performance to be moving, both for musicians and for audience, it is not enough that they follow the score, not even enough that the detail of their rehearsals be remembered. The players must hear each other, feel each other's inner warmth and interpretive choices, know the others' intentions a moment before they are manifested. In short, they must compenetrate one another, just as the man and the woman of Dehmel's poem. The moving performance is a compenetrating of the artists with one another and with the audience. We could say that in making music we ought to behave like fields or waves as well as like particles, we need to be blue pebbles as well as conventional ones.

What is true of music is also true of dance and of drama and indeed of every muse, including Erato, the muse of love poetry, with whom we have already flirted so much. Rilke, one of her favorites, offered us a similar teaching in one of the *Sonnets to Orpheus*, which we encountered in chapter 7:

> And should the Earthly forget you,
> say to the ground: I flow.
> speak to the swift water course: I am.
> (2.29)

The norm Rilke is articulating here is one with deep roots in the old gigantomachy between being and becoming, between the primacy of sameness or difference. But what we want to stress here is that it also belongs to an ethics of dialogue struggling to be born in our modern era, an ethics demanded by the capacity of *Homo sapiens*'s powers of thought and production to overwhelm the human. The remedy to this madness is to abandon the strict oppositions self/nonself, subject/object, same/other, and adopt a veritable discipline of transformation.

That discipline begins by recognizing that truth is made by us humans. In the words of the great Neapolitan philosopher, "the true and the made are convertible into each other."[12] Even space, that classical paragon of apriorism and mathematical apatheia, can be imagined as made by us. "Feel how your breath increases space," Rilke tells us in his poem's first lines. "The Earthly" that may forget you is the world of truths we humans have made across countless generations of dialogue on the planet we occupy. It is engraved in our sinews, veins, and brains, but it will forget us "in this night of excess," whenever we are overcome by a madness of overreaching (line 9). The antidote to this madness, according to Rilke, is not victimhood or powerlessness but transformation. "Go in the transformation out and in," he tells us. "What is your most painful experience? / If to you drinking is bitter, become wine" (6–8).[13]

Kierkegaard's "Venture in Experimenting Psychology"

It may seem odd for us to insist that cultivating a subjectivity capable of experiencing the simultaneity of sameness and difference and of being open to transformation between them is a difficult discipline. Isn't every human being already a subject? "To become what one is as a matter of course—who would waste his time on that?" The question is Kierkegaard's, and he goes on to answer it as well: "Quite so. But that is why it is already so very difficult, indeed, *the most difficult of all*, because every human being has a strong natural desire and drive to become something else and more."[14]

Kierkegaard paid a great deal of attention to this most difficult task.[15] In 1843 he published a book titled *Repetition: A Venture in Experimenting Psychology*, whose authorship he attributed to a scholar he called Constantin Constantius.[16] The title should attract us, because the topic of repetition is central to the question of how humans (and not only humans) experience and learn from sameness and difference. Repetition requires *both* sameness and difference. Two events that are mathematically identical would be the same event and not a repetition. Two events that are mutually exclusive would constitute simply a new event and also not a repetition. In repetition, these two aspects of sameness and difference are not experienced separately and then conjoined but are coextensive. This is precisely the kind of simultaneity we've said we need.[17]

So how should we cultivate that feeling of repetition? How is it achieved? Both the ironic subtitle of the book and the pseudonym (constancy squared!) of its author already suggest that Kierkegaard does not think experimental psychologies can tell us much about our experience of sameness, difference, or their simultaneity in repetition.

Constantius begins his book by announcing the world-historical importance of his discovery.[18] Repetition, he declares, is what recollection was for Plato, except that it goes in the opposite direction. Recollection gives us the past, repetition the future. Provided it exists—which Constantius recognizes it is his task to establish— repetition would make us happy, whereas recollection (he says) makes us unhappy. Much as psychologists do today, Constantius devises an experiment to establish empirically whether or not repetition and its pleasures exist. Here's the experiment: before he started mulling about repetition, Constantius had taken a trip to Berlin. Now he will do it again, and see if his experience there can be repeated.[19]

Constantius tells us that on his first visit he'd gone to the theater to see a farce starring a famous comic actor, recounting the experience in detail. His retelling of that first experience is full of long asides about theater, actors, and audiences. But in the midst of one of those asides, Constantius in his theater box is suddenly transported. He finds himself in an eclogue, with all its classical frills: nymph, murmuring brook, immense sky. The nymph represents his happy childhood at his father's farm: "You, my faithful comforter, you who preserved your innocent purity over the years, you who did not age as I grew older, you quiet nymph to whom I turned once again, weary of people, weary of myself, so weary that I needed an eternity to rest up, so melancholy that I needed an eternity to forget."

And then just as suddenly he switches again to description, "—Thus did I lie in my theater box, discarded like a swimmer's clothing, stretched out by the stream of laughter and unrestraint and applause that ceaselessly foamed by me. . . . Only at intervals did I rise up, look at [the actor], and laugh so hard that I sank back again in exhaustion alongside the foaming stream."

It is clear to the reader that this eclogue that gripped Constantius in the theater on his first visit to Berlin was itself a repetition, a return of his childhood and his innocence. It also seems that he himself did not grasp this when it happened to him. "By itself this was blissful, and yet I lacked something." What he lacked, it turns out, was a girl to watch. Lo, there sat one, in the third row of a box directly across from him, half hidden by a couple sitting in the first row. How lovely! Now his thoughts turned to this and other beautiful, watchable girls, both in Copenhagen and in Berlin.

In short, on that first trip Constantius had indeed been granted a repetition, but he lacked the subjectivity to notice it. Hence, he devises an experimental approach: can the pleasures of that visit to the theater be repeated? Constantius returns to Berlin, books a ticket for the same seat at the same play. But this time he does not find the comedian funny, nor does he see the beautiful, innocent girl anywhere, and so he concludes angrily that repetition does not exist, oblivious to his own previous experience of it. The most obvious lesson we should take from Constantius's adventure in Berlin (although by no means the only one) is not that repetition does not exist but rather that it cannot be tested by such experiment or achieved simply by willing it.[20]

There are other characters in *Repetition*. There is, for example, a poet whom Constantius seeks to instruct on the practice of repetition, a young man full of self-contradiction, in love but thinking of breaking his engagement. Unlike Constantius, with all his principles and observations, the conflicted poet does indeed discover and recognize the experience of repetition. He also realizes that this experience is neither willed nor easy to explain. "So there is repetition, after all," says the poet. "When does it occur? Well, that is hard to say in any human language. When did it occur for Job? When every *thinkable* human certainty and probability were impossible."[21]

The experience of repetition, the simultaneity in self and time of sameness and difference, is indeed "hard to say." But *Repetition*'s poet, like Kierkegaard, like all of us, has at his disposal many previous attempts to express it. Here he draws on one of those resources, the Hebrew Bible, by invoking the story of Job, a just man who is nevertheless deprived by God of wealth, health, family, and everything else he has in the world before having it all restored to him just as it had been before. Johannes de Silentio, the pseudonymous author of *Fear and Trembling*, the book Kierkegaard wrote and published at much the same time as *Repetition*, drew on a different story from the same biblical tradition. "Once upon a time there was a man who as a child had heard that beautiful story of how God tempted Abraham and of how Abraham withstood the temptation, kept the faith, and, contrary to expectation, got a son a second time."[22]

But we must not forget that in the writing of *Repetition* Kierkegaard also drew on and devoted attention to another important resource: his own subjectivity, his own self. He had himself gone on a trip to Berlin on October 25, 1841, shortly after breaking off his own engagement to Regina Olsen, with the purpose of attending the philosopher F. W. J. Schelling's inaugural course at the university. (The lectures themselves constitute a historical event: in addition to Kierkegaard, Karl Marx, Engels, Feuerbach, and Bakunin were among the many crowding the packed hall.)[23] During his four months in the city he also took German lessons, attended opera (including Mozart's and Da Ponte's *Don Giovanni*), and wrote, amid a storm of feeling and ideas, his book *Either/Or*. He was well aware, as he wrote to a friend, that "this winter in Berlin will always be of great significance to me. I have accomplished a great deal."[24]

Kierkegaard returned to Copenhagen on March 6, 1842. He had not forgotten Regina, and their paths sometimes crossed. On one occasion, at church, the pair exchanged signals, nods and head-shakes. On May 8, 1843, he left on a second trip to Berlin. His letters during that three-week stay are exulting: "Ideas are pouring over me—healthy, happy, thriving, cheerful, blessed children, easily born, yet all bearing the birthmark of my personality."[25] *Fear and Trembling* and *Repetition* are two of those blessed children. How so much, and of such complexity, could be achieved in three weeks is in itself a miracle. Perhaps Kierkegaard viewed his sacrifice of his love for Regina as the enabler of that miracle, just as God had rewarded Abraham's willingness to sacrifice his son with descendants "as numerous as the stars of heaven and the sands of the seashore." Perhaps he was thinking of the pagan advice of the poet Horace: "the more we deny ourselves, the more will we get from the gods."[26]

Did Kierkegaard nurse the secret hope that, on top of his wondrous, blessed children, he would also get back Regina? That, although jilted, she would wait for him? Upon returning from Berlin he found Regina engaged to another. That day in the church, through her nods, she had apparently been trying to confirm that Søren agreed to her engagement, but he had completely misunderstood those gestures. The news hit Kierkegaard hard, and the manuscript of *Repetition* underwent significant changes that made it more complex, traversed by more conflicting currents. Which is as it should be, for it is a book that teaches, among other things, that if we do not attend to our own experience, we can neither feel nor know the moment, perhaps only a fleeting one, in which our lived space-time is transfigured.

This may be but one of the reasons why Kierkegaard is often called the "father of existentialism," but we do wish to underline its importance. Recall from chapter 7 the metaphor of the two shields that Hermann Weyl offered us in his 1954 Columbia University lecture: the sound shield of Knowledge, which for Weyl meant knowledge of "that which is unmistakably demonstrable," and the broken shield of Being, tainted by "man's infinite capacity for self-deception." Many have preferred, like Weyl, to philosophize behind the protective shield of "unmistakable demonstrability," that is, the shield of the fundamental logical principles with which our book has been concerned.[27] Kierkegaard, instead, made his own contradictions, even his own human capacity for self-deception, into an acknowledged theme and subject of his thought. The difference between those two kinds of quests is crucial not least from the ethical standpoint.

Kafka's Trial of the Self

Regardless of era or faith, attending to such moments has never been easy. There are many who maintain that in modernity it has become even harder. We turn to literature again for example, this time to the writings that Franz Kafka produced in his short life (1883–1924), writings with a vivid impact on the literary imagination of the twentieth century. The leap from Kierkegaard to Kafka is not as far as it might seem. Kafka read Kierkegaard, and considered him a kindred spirit, noting in his diary entry for August 21, 1913, "As I suspected, his case, despite essential differences, is very similar to mine. At least he is on the same side of the world. He confirms me like a friend." In 1917 he wrote that "Kierkegaard is a star, although he shines over territory that is almost inaccessible to me." And in the penultimate extant entry of his diaries (dated December 18, 1922), Kafka jotted down, "All this time in bed. Yesterday Either/Or." Which part of Kierkegaard's *Either/Or* was Kafka reading? From his letters we know that he had started reading the book in 1918 but found repugnant the first part (the aesthetic life, with "The Diary of a Seducer"). We infer that now, in 1922, he is reading the second part, concerned with the ethical life.[28]

"Essential differences," yet "very similar": Kafka's comparison of himself with Kierkegaard applies to their views of the ethical life as well. Consider the short story Kafka published in 1915 titled "Vor dem Gesetz" ("Before the Law").[29] The story is quickly summarized: Before the Law stands a doorkeeper. A man from the country comes to the gate and asks to be let in, but the doorkeeper refuses; the country man asks if he will be allowed to enter later on, and he is told, "Perhaps, but not now." The gate of the Law is always open, so the man bends over to look inside. The doorkeeper tells him to go ahead and enter if he dares even though he has forbidden it, but to keep in mind that he, the doorkeeper, is powerful, and that once inside, in front of each hall through which one will pass there is a more and more powerful doorkeeper, so much so that even he, the doorkeeper, cannot bear the looks of the third one.

The country man, who had thought that the Law should be accessible to anyone at any time, decides that it is better to wait until he is granted entry. The doorkeeper gives the man a stool to ease his wait. Years pass; the man is always waiting; the doorkeeper answers his questions but does not grant him entrance. When the man is close to death, a last question overwhelms him, and he asks the doorkeeper, "How is it that through all these years, I have been the only one to beg for admittance to the Law?" And the doorkeeper, seeing that the man is at his last, replies, "No one but you could gain admittance through this door, since this door was intended only for you. I am now going to shut it."

We could not be further from the psychological comforts of the Identity Principle, with its promise that the "I" posits itself, is identical to itself. To the contrary, Kafka's fable, like Kierkegaard's ethics, suggests that our self, our law, is not *given* to us at any age. We have to search for our own law, our particular "identity," and the quest is lifelong. Perhaps—there can be no statistics about this—most human beings never face their law; perhaps only a few, or none, are granted that miracle. But even if we recognize the unknowability of the law or of the self, that is not an excuse for resigning ourselves to never seeking it.

"Before the Law" is a myth, one whose motif is ancient, even as it is itself modern. Like the man from the country, Odysseus, Aeneas, and other heroes of the classical world must confront terrors in order to find a way forward and fulfill their destiny. But they are valorous and determined, whereas the man from the country is not: he is terrified by the words and looks of the doorkeeper, so he decides to wait. Ancient heroes are assisted by gods and their messengers. Odysseus, Athena's protégé, is helped by Hermes, Circe, and Tiresias; Aeneas, the son of Venus, is helped by a seven-hundred-year-old Sibyl. But the man from the country has no one to protect him. He is utterly alone. In this he is typical of our age.

The man from the country is not, however, to be classed among the "many [who] live out their lives in quiet lostness," as Kierkegaard had called those people who are unaware of those moments in which they are presented with a choice or are unable to make one.[30] Though the man from the country never overcomes his fear and succumbs before the gate, he does so seeking for what he knows is, of all things, the most important. The quest is not achieved, perhaps even not achievable, but it is heroic enough.

The heroism of the man from the country is all the more striking in contrast with Joseph K., the protagonist of Kafka's unfinished novel *Der Prozess* (*The Trial*), started in 1914 and published posthumously in 1925. Joseph K. is no hero at all. He feels no guilt, cannot think of any crime he has committed, not even (like the woman in "Transfigured Night") a crime against himself. On the contrary, at every turn he is persuaded of his own innocence, unable to conceive under what law he is being indicted and tried. Hence the interpretation often given to the novel that he is the guiltless victim of a mysterious, pitiless bureaucracy, his condition a portent of tyrannies soon to come. Joseph K. may indeed be a victim of the times. But if we bear in mind the contrast with the man from the country, we can add that he is guilty (if *guilt* is the right word) of neglecting the work of becoming attentive, receptive to intimations of self or other.[31]

K. cannot bear that work. When, for example, his uncle takes him to a lawyer who might help, K. is distracted by lust and rushes off instead to lie with Leni, the lawyer's aide. In the penultimate chapter the prison chaplain relates to him a parable designed to open K.'s eyes to his predicament. That parable is the very same one we've just described: "Before the Law," first published by Kafka years earlier, and inserted here again near *The Trial*'s conclusion. K. responds to the priest's parable by debating with him who is superior and who the subordinate, the man from the country or the doorkeeper, a problem not only otiose but ridiculous insofar as the doorkeeper stands for the man's own fear. The priest tries to explain that the doorkeeper's words should be understood as necessary for the subject rather than true. But K. cannot face necessity or tolerate the conversation. He invents an urgent errand, at which point the priest gives up on him.

Kafka affords Joseph K. a number of moments through which he might step into some awareness of his (and our) dilemma, but he disregards them all, living instead in "lostness." Perhaps for the lost, as for the physicist, all points of time are interchangeable. But for a human being who knows that identity is not a given like an axiom and yet yearns to become a self, space-time is not homogenous. It is punctuated by moments, wormholes through which he or she might slide into despair, ascend to joy, experience repetition, or even simply stand before the door. Those wormholes are not physically or even psychically objective. They pertain to each person, like Kafka's gate of the Law, like the repetition of Kierkegaard's poet. We cannot offer prescriptions other than our mantra: to recognize the wormhole requires a sensitivity to both sameness and difference, a simultaneous solidity and permeability of self.

Dialogue

Much could be written about Kafka's relationship to the literature that came before him and what came after. Often enough, what *has* been written chooses to emphasize Kafka's role in the development of "the absurd" and gives short shrift to his deep exploration of the psychological and the ethical. Already in her 1947 essay "From Dostoevsky to Kafka," Nathalie Sarraute took this path. She understood the movement between Dostoevsky's Underground Man and Kafka's Joseph K. (whom she treats as a continuation of the former) as one in which the Kafkaesque liberated the future of the novel from its earlier psychological bent.[32]

Sarraute was herself a remarkable writer, critic, and "psychologist." Her first book, *Tropisms* (1939), was about the interior movements that precede our conscious actions, and the best seller she published at the age of eighty-three was a psychological autobiography called *Childhood*. Her life spanned nearly the entire twentieth century, experiencing many of its terrors and its glories. (She was born a Jew in czarist Russia in 1900, survived World War II hiding in Paris, and died there in 1999.) We should expect her to have something significant to say about sameness and difference in the self, and indeed she does. Explaining what it is that she sees as the most basic attribute of both Dostoevsky's and Kafka's characters, she tells us that their psychological foundation "is nothing but what Katherine Mansfield called, with some fear and, perhaps, slight distaste: 'this terrible desire to establish contact.'"

We wish we could pause to establish contact with Mansfield (1888–1923), whose short stories would repay our attention and who died even younger than Kafka, at much the same time and of the same disease. But we'll remain with Sarraute in order to stress her point. In her own words,

> It is this continual, almost maniacal need for contact, for an impossible, soothing embrace, that attracts all of these characters like dizziness and incites them on all occasions to try, by any means whatsoever, to clear a path to the "other," to penetrate him as deeply as possible and make him lose his disturbing, unbearable opaqueness; in their turn, it impels them to confide in him and show him their own innermost recesses.[33]

Solidity and penetrability, self and other, same and different. Sarraute is absolutely right to point to the "almost maniacal" desire to overcome these distinctions through contact and dialogue as a fundamental aspect of fictional characters in a literary modernity she called the "Age of Suspicion."

There is, however, an important difference between Dostoevsky's and Kafka's novelistic characters. Both crave to penetrate and be penetrated, but Dostoevsky's are simultaneously solid selves whereas Kafka's tend to be unstable bundles of quivering and quarreling subselves. Joseph K. may crave contact, but he cannot enter in dialogue with the other. This is not because Kafka's novels replace psychology with the absurd but because they are grappling with what he takes to be a fundamental ethical and psychological question of his age, an age in which, as Sarraute put it referring to Kafka's *The Castle*, "one no longer knows whether one has resisted or given in." In such an age, how can a fragmented being gain sufficient awareness to participate in and be transformed by dialogue with the other?

The problem here, we hasten to add, is not with our (inevitable) fragmentation. Our lack of self-unity, our inner diversity, may just as well facilitate dialogue with others as hinder it if we can learn to attend to that diversity rather than ignore it. How to learn to attend? Versions of that question vary across time, place, and person. Sarraute's "Age of Suspicion" is one answer to it, drawing on Stendhal's pronouncement a century earlier that in the aftermath of Napoleon, "the poetic genius is dead, but the genius of suspicion has been born."[34]

Today that genius has matured into a powerful "hermeneutics of suspicion" that has extended its reach over many domains of culture. That hermeneutics has yielded much critical fruit. But it has also corroded any confidence in the possibility of true dialogue except in mathematical languages, where axioms rather than hermeneutics rule. It can now only seem naive to speak (as we are doing) of the solidity and permeability of self, of contact with the other or the transfiguring power of dialogue. The sophisticated reader who stumbles on an eighteenth-century poet like Edward Young will smile not only at the language but also at the archaism of the sentiments:

> 'Tis thought's exchange, which, like th' alternate push
> Of waves conflicting, breaks the learned scum,
> And defecates the student's standing pool.
>
> . . . .
>
> 'Tis converse qualifies for solitude;
> As exercise, for salutary rest.[35]

And yet the need for this simultaneity of solidity and permeability, for the possibility of transformation between sameness and difference through dialogue (among other means), is as great in our age of suspicion as it has ever been. It is difficult to imagine much of a future for self or society if we do not find some way of remembering that need.

The Ephemeral and the Eternal (Borges Again)

There are many moments in our life where, if we attend, we can feel an intimation of the simultaneity of sameness and difference that will help us to remember the need. And there are also resources to help us cultivate our attention: literature and art, liturgy and music, psychoanalysis and philosophy, love and friendship, gurus, meditation, walks in nature, psychotropic medication, and many, many more. Each has, of course, its limitations and its deceptions, but at its best, each recalls us to the question and the task.

What do such moments look like? Borges offered us an example in his "New Refutation of Time" when he described the effects of a nighttime walk in Buenos Aires.

> That pure representation of homogeneous facts—calm night, limpid wall, rural scent of honeysuckle, elemental mud—is not merely identical to the scene on that corner so many years ago; it is, without similarities or repetitions, the same. If we can intuit that sameness, time is a delusion: the impartiality and inseparability of one moment of time's apparent yesterday and another of time's apparent today are enough to make it disintegrate.

Perhaps Borges's experience was more literary than phenomenological, since we know he kept both *Fear and Trembling* and *Repetition* on his shelves. And as we suggested in the previous chapter, his claim that such an experience represents the annihilation of time is not justified philosophically. But we should not quibble with Borges's important point: "if we can intuit that sameness" in such moments, we feel that a time we had lost, which life had taken away from us, a time we had forgotten or hardly remember, is unexpectedly given back.[36]

We can learn a similar lesson from Borges's poetry, namely, from two sonnets he published toward the end of 1963 on a theme closely related to repetition, that of eternity as opposed to oblivion. The first sonnet bears a strange title, "Everness," a word not recorded in the Oxford English Dictionary (let alone in Spanish, the language of the rest of the poem).[37] It starts with a startling line: "Sólo una cosa no hay. Es el olvido" (Only one thing does not exist. It is oblivion). The sonnet's second quartet starts with a no less puzzling and resounding "Ya todo está" (Everything is already present).[38]

The second sonnet bears the German title "Ewigkeit" (eternity, perpetuity).[39] Its two quartets recall, by their style and Stoic reference, the poetry of the Spaniard Francisco de Quevedo (1580–1645).

> Let in my mouth Castilian verse repeat
> what it is always saying
> since Seneca's Latin: the horrendous
> verdict that all is food for worms.

> Let the gray ash proclaim
> the pomp of death and the victory
> of that rhetorical queen trampling
> the banners of vainglory.

But the following two tercets are very different, beginning with a revelatory reversal:

> Not so. That which blessed my mud
> I shall not cowardly deny.
> I know that one thing does not exist. It is oblivion;

> I know that in the eternal it survives and burns,
> the precious plenty of all that I have lost:
> that forge, that moon, that afternoon.

Is this emphatic refutation of oblivion and affirmation of eternal being just a dogmatic pronouncement, inflated by hope but unjustified by argument or experience? Not so. The justification is found in the mud. The poet has not forgotten the "elemental mud" of the Maldonado and his "New Refutation of Time."[40] Nor has he forgotten the calm night, limpid wall, and rural scent of honeysuckle. Those things and their *experienced sameness* ought not to be forgotten: to forget them, or to deny them, is cowardice, the poet says, and in saying this he is agreeing with Kierkegaard, Proust, Rilke,

Kafka, and many others. The experienced sameness cum difference that we have been calling repetition puts us face-to-face with a task, and there are only two alternatives: either we shrink from it and go on living as usual—this is what Borges calls cowardice—or we devote our lives to the task.

We humans have often been called a synthesis of ephemeral and the eternal, but the destiny of those two words has been diverse.[41] *Ephemeral* literally means lasting a single day, and by metaphoric extensions it may mean lasting for one year, a lifetime, or even longer times; that is, it means, in a word, *finite*. There seems to be agreement about that. *Eternal*, however, presents old confusions. It is often taken as a synonym of *perpetual* or *perdurable*. But *eternal*, perhaps as often and with better sense, is taken to mean that time is not or does not pass irretrievably, that it has been transcended. It was to avoid such confusions that Borges avoided using *eternity* or *eternidad* as a title for either of his two sonnets.

Repetition, or experienced sameness cum difference, does not give us eternity in the sense that our souls are granted perpetual hospitality up in heaven or down in the infernal caves; rather, it gives us the conviction, unwarranted though it may be, that our presence in past, present, and future is perhaps arranged differently from our conventional notions. We will avoid the word *eternity* and call this revolution in our sense of sameness and difference an intimation of *transcendence*.

The involuntary nature of those experiences of repetition, of sameness cum difference, make us feel that they are not merely a deception of our fragmented mind. Their power of conviction is all the greater because they seem to be outside our power. They suggest to us—with greater eloquence the greater the detail that is "given back" from the past—that there is no oblivion. How important are they? That depends on the shape of your own psyche. To our mind, these intimations of transcendence, whether through the experience of sameness cum difference *or otherwise*, are a truer lighthouse for human living in time than Heideggerian "determination" in the face of nothingness and death.

We have emphasized the word *otherwise* because there are many other, different ways of persuading ourselves that oblivion is not.

One such was Nietzsche's resolution to live as if all points in time must recur infinitely often. Rilke's affirmation in the ninth of his *Duino Elegies* is yet another:

> Everything *once*, *once* only. Just *once* and no more.
> And we also *once*. Never again. But this
> having been *once*, though only *once*:
> to have been of the earth, seems irrevocable.

Both ways have something to be said for them, but both require faith, and faith requires a split or cleavage in the mind. One part, essential for earthly life, wants to stick to, and act according to, our remembered experience. Another part wants to believe, to take as fact something we have never experienced, or even, as in the case of Nietzsche's eternal return and of Rilke's irrevocability, something that is impossible for humans to experience. Still another part mocks the naive faith of the second part, while another belittles the crass empiricism of the first. It is otherwise with repetition, an intimation of nonfragmentation that we humans can experience on this earth.

There are many lookouts that afford a vision of transcendence. We stress repetition because it seems to us relatively secure from attribution to wishful eyes, free from the torments of faith and from vicious circles of begged questions. Still, repetition is not a panacea for anguish (remember the farcical fate of Constantin Constantius), and in any event we are not licensed to prescribe remedies for the psyche. Are there things we can do, advice we can follow, to cultivate our humanity? We can only give the advice with which we began: live with the mystery of sameness and difference at heart.

We can reduce it to a simple interrogatory mantra. Why, for a given question, or at a given instant, do I attend to one rather than the other? What would I gain and lose if my attention shifted? What knowledge about self, or other, about humanity, or the world, would emerge? Our "laws of thought," our choices about sameness and difference, depend on the task at hand, that is, on what it is we want. Even wanting, as the Daodejing teaches, is not the only option:

So the unwanting soul
sees what's hidden,
and the ever-wanting soul
sees only what it wants.[42]

From dreamers, mystics, and poets to mathematicians and logicians, we all dwell in untold registers of being. Kierkegaard states, "even if a man his whole life through occupies himself exclusively with logic, he still does not become logic; he himself therefore exists in other categories." Our introduction dwelt on Borges's "Blue Tigers," a story about the terror of a logician who wanders into some of those other categories and discovers that (Kierkegaard again) "existence mocks the one who keeps on wanting to become purely objective." But whatever type of knowledge you devote yourself to, a story could be written about the terror that comes from wandering into worlds where that knowledge does not hold. Such stories, at their best, remind us that choices exist even if we have forgotten them.

Regardless of the story we choose to write, it will in every case require empathy, imagination, openness, sensibility: the permeability of the self by the other. Our point is similar to Kierkegaard's: "In order to shed light on logic, it might be desirable to become oriented psychologically in the state of mind of someone who thinks the logical—what kind of dying to oneself is required for that purpose, and to what extent the imagination plays a part in it."[43] But we must also insist that there is also a "dying to oneself" in *refusing* to think the logical, in seeking to dwell in difference, in exaggerating the claims of irrationality.

We are each of us always dying to some parts of ourselves and resurrecting others. If at times in this book we have seemed to focus more on the losses incurred by thinking too much the logical, it is because we feel that the accelerating extension of the rules of sameness and the imperial claims of number threatens many domains of the human and nonhuman world today.

The words inscribed on the Fields Medal, the highest prize awarded in mathematics in our time, boast inadvertently of the danger: *transire suum pectus mundoque potiri* (you are seeking to pass beyond your human limitations and make yourself master of the universe). The words and the ambition are perennial. In fact they were written some two thousand years ago by the Roman astronomer-poet Marcus Manilius (who had the wisdom to continue "nor are such high attainments secured without a price"). But at present, with our earthly home at such risk and our mastery so great, it seems especially urgent that we remember to cultivate rather than transcend those attributes of humanity that cannot be reduced to the glorious power of the Principle of Identity and its attendant axioms.[44]

Where does the greater danger lie, in excesses of sameness or of difference, in the tyranny of axioms, or of unreason? Your own answer will be both personal and social. It will depend on the particularities of your own interests, aptitudes, and fears, on the family, language(s) and culture you grew up in, the disciplines you trained in, the faiths that attracted or repelled you. But whatever your answer, the more you arrive at it attentive to the bottomless mysteries of sameness and difference, the greater will be the adequacy of your particular quest. By adequacy we mean, of course, both the capacity to cultivate confidence and to cultivate doubt, to discover new foundations and the abysses that yawn beneath them. Both are required to avoid what a poet described well as a moment of grave danger to your humanity:

> And the eye, self-satisfied, will be misled,
> Thinking the puzzle solved, supposing at last
> It can look forth and comprehend the world.
> That's when you have to really watch yourself.[45]

ACKNOWLEDGMENTS

Schools, libraries, universities: these constitute the crucial plumbing of the collective intellect. We thank the institutions that taught us and within which we have taught. To these, and indeed to all the communities that contribute to the creation and transmission of knowledge, we and humanity are indebted.

We have also accrued debts more specific to this book. A large one is to Lorraine Daston, who spent a long train ride reading early drafts of our first chapters. Because the course corrections she suggested came at the start of our own voyage, their effect on our destination was all the greater. Alberto Bernabé Pajares, Glenn Most, and Sofía Torallas Tovar often helped us in our excursions among the Greeks. Sarah Stroumsa generously engaged our Islamic philosophy. Edward (Rocky) Kolb lent us his astrophysicist's eye. Richard Robb read chapter 8 as a skeptical economist. As we neared the end, Matthew L. Jones, Yitzhak Melamed, Peter Pešić, Chad Wellmon, and Brook Ziporyn read the whole and provided sharp criticism, much of which we have attempted to implement and all of which has improved the result. As for our editor Alan Thomas, we join the many authors (and readers!) who have so much to thank him for, including for his rashness in taking on a project such as ours.

David is grateful to Daniel Watling, for assistance with research and expertise in Arabic philosophy, and to Alexandra Montero

Peters, for shepherding the bibliography. He thanks Brenda Shapiro for believing in ideas despite much evidence to the contrary.

But first and last we must acknowledge the chances of nativity and love, and all the more in a book authored by father and son. One does not choose those determinants of fate: where, when, and to whom one is born. Ricardo's father, Guillermo Nirenberg, aspired to get to the heart of all things even when that heart is as unreachable as the bottom of the soul. By the example of a long life well lived and honestly departed, Ricardo's father-in-law, Emilio Lida, taught him the meaning of Socrates's last words, "We owe a cock to Asclepius." Isabel Lida Nirenberg, Ricardo's wife and David's mother, has read and improved every chapter; what we have gained from her cannot be contained in any learning or philosophy. And from a chance transatlantic encounter with Sofía Torallas Tovar has come philology and laughter, wisdom and love, and an absurd desire to dance.

It is to Isabel and to Sofía that we dedicate this book.

NOTES

Introduction

1. William James, *Pragmatism: A New Name for Some Old Ways of Thinking* (London: Longmans, Green, 1907), 129.
2. Jorge Luis Borges, "Tigres azules," in *Cuentos Completos* (Barcelona: Random House Mondadori, 2011), 521–31.
3. Erwin Schrödinger, "July 1952 Colloquium," in his *The Interpretation of Quantum Mechanics: Dublin Seminars (1949–1955) and Other Unpublished Essays*, ed. Michel Bitbol (Woodbridge, CT: Oxbow Press, 1995), 32. The passage continues: "If you happen to get 1000 [or] more records of a proton, as you often do, then notwithstanding the greatest psychological urge to say: it is the *same* proton, you must remain aware, that there is no absolute meaning in this statement. There is a *continuous* transition between cases where the sameness obtrudes itself to such where it is *obviously* meaningless" (emphasis in original).
4. This principle—or perhaps better, family of principles—has been defined in different ways, as have most of the "laws of thought" we discuss. On Leibniz' formulations, see Vincent Carraud, *Causa sive ratio: La Raison de la cause, de Suarez à Leibniz* (Paris: Presses universitaires de France, 2002), and the excellent introduction by Y. Melamed and M. Lin, "Principle of Sufficient Reason," *Stanford Encyclopedia of Philosophy*, https://plato.stanford.edu/archives/spr2018/entries/sufficient-reason/.
5. Aristotle, *Metaphysics*, 4.3.1005b–1006a.
6. William James, *The Principles of Psychology* (Mineola, NY: Dover, 1950), 2:459–60. We are not accusing James of having been especially dogmatic: quite the contrary. We are simply pointing out that he could have made much the same point by stressing the mind's ability to discover or intend difference, since every assertion of sameness is also an assertion of difference. That he did not, but instead preferred to represent the human mind in terms of its capacity for "identity" and "sameness," is itself a symptom and product of the dogmatic divisions we are studying.
7. Søren Kierkegaard, *Concluding Unscientific Postscript*, ed. and trans. H. V. Hong and E. H. Hong (Princeton, NJ: Princeton University Press, 1992), 1:117.

8. On regularity and anomaly in ancient Mesopotamian attempts to make sense of their world, see Francesca Rochberg's *Before Nature: Cuneiform Knowledge and the History of Science* (Chicago: University of Chicago Press, 2016), 120.

9. Sir James G. Frazer went so far as to formulate principles of magical thought based on sameness, which he called "the law of similarity": *The Golden Bough*, abridged ed. (New York: Macmillan, 1922), 11. As we will see, Frazer's law of similarity does not sound so different from David Hume's basic principles of association (which Hume did not restrict to human "reasoning" but saw as in many ways reducible to animal rules of conditioning). It is also one of the principles of Gestalt psychology.

10. Pliny, *Natural History*, 2.9.54–55. Our translation varies from that of H. Rackham, Loeb Classical Library (Cambridge, MA: Harvard University Press, 1938), 202–5.

11. On Egyptian cultic intervention to "support" the sun's "life-giving beneficent powers with hymns and gifts," see Jan Assmann, *The Search for God in Ancient Egypt* (Ithaca, NY: Cornell University Press, 1984), 64–65. On the Aztec/Mexica cosmos, "filled with living entities who must eat other entities if they are not to die before their time spans are completed," a cosmos whose motion was ordered "through controlled acts of sacrificial feeding," see Kay Almere Read, *Time and Sacrifice in the Aztec Cosmos* (Bloomington: Indiana University Press, 1998), xix, 28–29.

12. In the eighteenth century, Thomas Bayes and Richard Price, whose "Bayesian" probabilistic approach to epistemic confidence underpins much machine learning today, used the example of the sun to demonstrate exactly that point, and David Hume explored the question in less mathematical terms. See Stephen M. Stigler, "Richard Price, the First Bayesian," *Statistical Science* 33 (2018), 117–25.

13. Erwin Panofsky, *Three Essays on Style*, ed. William S. Heckscher (Cambridge, MA: MIT Press, 1997), 190.

14. The phrase, from the conclusion of Kant's *Critique of Practical Reason*, was partly inscribed on the philosopher's tombstone. See Immanuel Kant, *Gesammelte Schriften*, ed. Akademie der Wissenschaften, vol. 5, *Kritik der praktischen Vernunft, Kritik der Urtheilskraft* (Berlin: De Gruyter, 1913), 161; *Critique of Practical Reason*, trans. Mary Gregor (Cambridge: Cambridge University Press, 2015), 129.

15. We hasten to add that many ethics have gotten along well without free will. The problem is sharp, however, within Western traditions derived from Christianity and Islam, where it is linked to theodicy and divine justice. Spinoza provides an excellent but rare example of an early modern European philosopher willing to develop an ethics of determinism.

16. Kant, *Critique of Practical Reason*, 78; *Gesammelte Schriften*, 5:96.

17. Descartes, Hobbes, Spinoza, Leibniz, and other early modern philosophers explored notions such as that of a "Spiritual Automaton" and debated whether the human was reducible to artificial mechanism. On the pre-Babbage tradition see Matthew L. Jones, *Reckoning with Matter: Calculating Machines, Innovation, and Thinking about Thinking from Pascal to Babbage* (Chicago: University of Chicago Press, 2016). For a now classic (and a bit outmoded) statement of the problem as

posed by modern computing, see Hubert L. Dreyfus, *What Computers Still Can't Do: a Critique of Artificial Reason* (Cambridge, MA: MIT Press, 1992).

18. Paul Erickson et al., *How Reason Almost Lost Its Mind: The Strange Career of Cold War Rationality* (Chicago: University of Chicago Press, 2013); Cathy O'Neil, *Weapons of Math Destruction: How Big Data Increases Inequality and Threatens Democracy* (New York: Broadway Books, 2016).

19. Friedrich Nietzsche, *The Gay Science*, trans. Walter Kaufmann (New York: Vintage, 1974), §121, 177.

20. The complex history of this phrase will be discussed in chap. 2.

21. Adelard of Bath, *Conversations with His Nephew*: On the Same and the Different, Questions on Natural Science, *and* On Birds, ed. and trans. Charles Burnett (Cambridge: Cambridge University Press, 1998), 47; Alfred North Whitehead, *An Introduction to Mathematics* (New York: Henry Holt, 1911), 9.

22. Pedro Salinas, "La voz a ti debida" (1933), in *Poesías Completas*, ed. J. Marichal (Madrid: Aguilar, 1961), 133: "Y súbita, de pronto, / porque sí, la alegría."

23. Daniel Bernoulli (1738); Daniel Bernoulli and Johann Bernoulli, *Hydrodynamics and Hydraulics*, trans. Thomas Carmody and Helmut Kobus (Mineola, NY: Dover, 1968), containing Daniel Bernoulli's treatise and also Johann Bernoulli's *Hydraulica* of 1743, falsely dated 1732.

24. We are here loosely extending an opposition articulated in Aristotle's Greek in *On the Cosmos*, 392a, between that which is *atrepton, anheteroiôton kai apathê*—unchangeable, unalterable, and impassive—and that which is *pathêtê, te kai treptê*—liable to change and alteration. Aristotle, *On Sophistical Refutations, On Coming-to-be and Passing-Away, On the Cosmos*, trans. E. S. Forster and D. J. Furley, Loeb Classical Library (Cambridge, MA: Harvard University Press, 1955), 352–53. The Greeks did not apply the word *apathê* to number or mathematics. In Greek philosophy the term applied to the soul in the sense of impassive, not suffering from passions.

25. As we will see, to the extent that we can do this with physical objects, such as a clock (which we can take apart piece by piece and reassemble without change), we say that they are machines or mechanisms. To do this with humans is one of the dreams of science fiction (as in Star Trek's "beam me up, Scotty").

26. We are certainly far from the first to ask this question. Within the European philosophical tradition, Spinoza's Letter 12, to Lodewijk Meyer, on the subject of infinity, represents an early attack against those who forget that "Measure, Time and Number are nothing other than modes of thinking, or rather, modes of imagining" that cannot be applied to all things: *Spinoza: Complete Works*, ed. Michael L. Morgan and trans. Samuel Shirley (Indianapolis, IN: Hackett, 2002), 789.

27. Sir David Brewster, *Memoirs of Sir Isaac Newton* (New York: Johnson Reprint, 1965), 2:407.

28. The story of Hobbes and Buckingham is reported by John Aubrey; see *Aubrey on Education. A Hitherto Unpublished Manuscript by the Author of Brief Lives*, ed. J. E. Stephens (London: Routledge, 2012), 160.

29. Leonhard Euler, *Lettres à une Princesse d'Allemagne sur divers sujets de Physique &*

de Philosophie. Saint Petersbourg, 1768); slightly revised and expanded by Condorcet as *Lettres de M. Euler à une princesse d'Allemagne sur différentes questions de physique et de philosophie, Nouvelle Édition, Avec des Additions, par MM. le Marquis de Condorcet et De La Croix* (Paris, 1787).

30. We cite from L. Euler, *Letters of Euler on Different Subjects in Physics and Philosophy Addressed to a German Princess*, trans. H. Hunter, 2nd ed, (London: Murray and Highley, 1802), letters 69 & 70 of October 1760, pp. 263–70, emphasis added. Newton himself had taken impenetrability to be an experimentally justified property of all bodies in the third of his philosophical rules in the *Principia*. The Cartesians, on the other hand, had held that body is extension, a notion Euler dismissed as inadequate, offering the princess the examples of space, or a vacuum, that are extension but are not bodies.

31. *The Book of the Pheasant Cap Master/He guan zi*, chap. 5.9. The Chinese text can be found at Chinese Text Project, https://ctext.org/he-guan-zi. We thank Professor Farquhar for bringing this cryptic and fascinating text to our attention and providing her translation, made in collaboration with Bruce Rusk. On the work generally (though without specific attention to our topic or this passage), see Carine Defoort, *The Pheasant Cap Master (He guan zi): A Rhetorical Reading* (Albany: State University of New York Press, 1997).

32. We cite Ebü-s-Suʿūd's *fatwā* from Shahab Ahmed's *What Is Islam? The Importance of Being Islamic* (Princeton, NJ: Princeton University Press, 2016), 288–89. Ebü-s-Suʿūd died in 1574. His chronology overlaps suggestively with another remarkable skeptic about the more expansive claims of legal logic, Michel de Montaigne (1533–1592).

Chapter One

1. Max Scheler, *The Human Place in the Cosmos*, trans. Manfred S. Frings (Evanston, IL: Northwestern University Press, 2009), 5, first published in German: *Die Stellung des Menschen im Kosmos* (Darmstadt: Reichl, 1928).

2. Richard Rorty, *Philosophy as Poetry* (Charlottesville: University of Virginia Press, 2016), 23–24.

3. We have used the (unabridged) edition in one volume: Oswald Spengler, *Der Untergang des Abendlandes: Umrisse einer Morphologie der Weltgeschichte* (Munich: C. H. Beck, 1990). Citations are of the translation by Charles Francis Atkinson, *The Decline of the West: Form and Actuality* 2 vols. (New York: Alfred A. Knopf, 1926–1928). On the book's reception history, see *Spengler ohne Ende: Ein Rezeptionsphänomen im internationalen Kontext*, Schriften zur politischen Kultur der Weimarer Republik, vol. 16, ed. Gilbert Merlio and Daniel Meyer (Frankfurt am Main: Lang, 2014).

4. Spengler, *Decline*, 1:424.

5. Spengler, *Decline*, 1:59. The point is not as original as Spengler claimed. Spinoza treated numbers as "entia imaginationis," beings of the imagination, and to the extent that we do not all share the same imagination, relative. And as we will see, the Islamic philosopher Ibn Ṭufayl made a similar claim in the twelfth century.

6. Spengler, *Decline*, 1:60. This originality, he added, should not count against him:

"we must not forget that unanimity about things that have not yet become problems may just as well imply universal error as universal truth." We do not subscribe to Spengler's view of mathematics, but with this last sentiment we entirely agree.

7. *Umrisse einer Morphologie der Weltgeschichte.* Though Spengler was not trained as a historian, he was a pioneer in the development of what we today call "world history." See for example his indictment of the "West-European conceit" in *Decline*, 1:17-18. The historical establishment largely rejected his methodology, but Arnold Toynbee was an exceptional admirer.

8. Faustian soul: Spengler, *Decline*, 1:75, 90. Spengler's judgment that Western mathematics had reached its end was both dogmatic and incorrect, but that is beside the point.

9. Spengler, *Decline*, 1:6.

10. Spengler, *Decline*, 1:4, 70. Spengler imagined the birth of alienation as the moment humanity first stumbled upon number: "As soon as the primitive's astonished eye perceives the dawning world of ordered extension . . . there arises in the soul— instantly conscious of its loneliness—the root feeling of longing (Sehnsucht)." *Decline*, 1:78. Compare Plato's *Timaeus* and *Epinomis* in chap. 3 of this book.

11. See, for example, St. Paul, Romans 7:5-6.

12. The novel was set circa 1913 and published in 1930. Ulrich, the novel's protagonist, is himself a mathematician, and Musil was a mathematically well-trained PhD. Robert Musil, *The Man without Qualities*, trans. Sophie Wilkins and Burton Pike (New York: Vintage International, 1996), 36-37. Compare Musil's prewar essay of 1913, "Der mathematische Mensch" (1913), Projekt Gutenberg, http://gutenberg .spiegel.de/gutenb/musil/essays/; English translation in Robert Musil, *Precision and Soul: Essays and Addresses*, trans. Burton Pike and David S. Luft (Chicago: University of Chicago Press, 1995), 39-43.

13. Musil's review is titled "Geist und Erfahrung: Anmerkungen für Leser, welche dem Untergang des Abendlandes entronnen sind." The first word of the title could better be translated as "Spirit" or "Reason," but we follow here the choice of the English translators, "Mind and Experience," in Musil, *Precision and Soul.* The quote is from 139.

14. Adorno gave a conference on Spengler in 1938, reworked it as an English text in 1941, and lastly published the German essay "Spengler nach dem Untergang," in Adorno, *Gesammelte Schriften in 20 Banden*, vol. 10, bk. 1, *Kulturkritik und Gesellschaft 1: Prismen. Ohne Leitbild* (Frankfurt: Suhrkamp, 1977), 47-71. English translation "Spengler after the Decline," in T. Adorno, *Prisms*, trans. Samuel and Shierry Weber (Cambridge, MA: MIT, 1981), 51-72.

15. Paul Forman, *Weimar Culture, Causality, and Quantum Theory: Adaptation by German Physicists and Mathematicians to a Hostile Environment*, Historical Studies in the Physical Sciences 3 (Philadelphia: University of Pennsylvania Press, 1971), 1-115; "*Kausalität, Anschaulichkeit*, and *Individualität*, or How Cultural Values Prescribed the Character and Lessons Ascribed to Quantum Mechanics," in *Society and Knowledge*, ed. Nico Stehr and Volker Meja (Brunswick, NJ: Transaction Books, 1984), 333-47.

16. We cite from Max Weber, "Wissenschaft als Beruf," in *Schriften 1894-1922*, ed. Dirk

Kaesler (Stutgart: Kröner 2002), 488, 489. English translation from *Max Weber: Essays in Sociology*, ed. and trans. H. H. Gerth and C. Wright Mills (New York: Oxford University Press, 1946), 129–56. For a critical edition see *Max Weber Gesamtausgabe*, ed. W. J. Mommsen, W. Schluchter, and B. Morgenbrod, vol. 1 (Tübingen: J. C. B. Mohr, 1992). For a recent critique of Weber's notion of disenchantment, see Jason Ā. Josephson-Storm, *The Myth of Disenchantment: Magic, Modernity, and the Birth of the Human Sciences* (Chicago: University of Chicago Press, 2017).

17. Max Weber, "Wissenschaft als Beruf," 490.

18. Max Weber, 506, 511.

19. Spengler, *Decline*, 1:84, writing "Theory of Aggregates" rather than set theory. The reference here is presumably to Frege's, Russell's, and Whitehead's project to provide a foundation for mathematics, on which, see below. We wonder what Spengler would have made of John von Neumann's articles published a few years later: "Eine Axiomatisierung der Mengenlehre," *Journal für die Reine und Angewandte Mathematik* 154 (1925): 219–24; and "Die Axiomatisierung der Mengenlehre," *Mathematische Zeitschrift* 27, no. 1 (1928), 669–752.

20. Bertrand Russell, *My Philosophical Development* (London: Routledge, 1993), 74.

21. Bertrand Russell, "On Denoting," *Mind* 14, no. 56 (1905), reprinted in Russell, *Logic and Knowledge: Essays 1901-1950*, ed. Robert Charles Marsh (Nottingham: Spokesman, 2007), 41–56. "Given that the words": the formulation is Rorty's, not Russell's: *Philosophy as Poetry*, 26–28. Curiously enough, the poet T. S. Eliot praised precisely this part of the project: "how much the work of logicians has done to make of English a language in which it is possible to think clearly and exactly on any subject. The *Principia mathematica* are perhaps a greater contribution to our language than they are to mathematics." See T. S. Eliot, "A Commentary," *Monthly Criterion* 6, no. 1 (October 1927): 291.

22. It may be unnecessary to point out that the goal remains out of reach. In the century since Russell, analytic philosophers have scarcely dared to propose an axiomatization of philosophy.

23. "Every philosophical problem": Bertrand Russell, *Our Knowledge of the External World as a Field for Scientific Method in Philosophy* (Chicago: Open Court, 1914), 33. "Unimportance of time": Bertrand Russell, "Mysticism and Logic," in *Mysticism and Logic, and Other Essays* (London: Longmans, Green, 1919), 22: "Both in thought and in feeling, even though time be real, to realize the unimportance of time is the gate of wisdom." Compare Samuel Alexander, *Space, Time and Deity* (New York: Dover, 1966), 36, "To realize the importance of Time is the gate of wisdom."

24. For an energetic narrative of the events involving Adler and Neurath, see Karl Sigmund, *Exact Thinking in Demented Times* (New York: Basic Books, 2017), 86–88, 91–99. Neurath's treatise can be found in his *Gesammelte philosophische und methodologische Schriften*, ed. R. Haller and H. Rutte (Vienna: Hölder-Pichler-Tempsky, 1981). See also P. Neurath and E. Nemeth, eds., *Otto Neurath oder die Einheit von Wissenschaft und Gesellschaft* (Vienna: Böhlau, 1993).

25. Johann Nelböck, "Die Bedeutung der Logik im Empirismus und Positivismus" (PhD diss., University of Vienna 1930). On the murder, see Friedrich Stadler,

"Documentation: The Murder of Moritz Schlick," in *The Vienna Circle. Studies in the Origins, Development, and Influence of Logical Empiricism*, ed. Friedrich Stadler (Vienna: Springer, 2001), 866–909; Sigmund, *Exact Thinking*, 312–20, citing "conflicts of a philosophical nature" at 317.

26. Readers interested in an account of these controversies that is equally attentive to the mathematical and the politico-cultural issues may turn to Herbert Mehrtens, *Moderne — Sprache — Mathematik. Eine Geschichte des Streits um die Grundlagen der Disziplin und des Subjekts formaler Systeme* (Frankfurt am Main: Suhrkamp, 1990), and to his article "Modernism vs. Counter-Modernism, Nationalism vs. Internationalism: Style and Politics in Mathematics, 1900–1950," in *L'Europe mathématique: Histoires, mythes, identités*, ed. Catherine Goldstein, Jeremy Gray, and Jim Ritter (Paris: Éditions de la Maison de l'homme 1996), 518–29.

27. For a relatively recent overview, see José Ferreirós, "The Crisis in the Foundations of Mathematics," in *Princeton Companion to Mathematics*, ed. I. Leader, J. Barrow-Green, T. Gowers (Princeton, NJ: Princeton University Press, 2010), 142–56.

28. In the 1930s Kurt Gödel and Gerhard Gentzen proved that the Dedekind-Peano axioms of arithmetic are consistent relative to formalized intuitionistic arithmetic ("consistent" is a type of sameness: if a statement is true in one arithmetic, it will be true in the other, and if it is false in one, it will be false in the other). Similar proofs for set theory or for systems of analysis have not been found. See Kurt Gödel "Zur intuitionistischen Arithmetik und Zahlentheorie," in *Ergebnisse eines mathematischen Kolloquiums*, ed. Karl Menger (Leipzig: F. Deuticke, 1933), 4:34–38, translated as "On Intuitionistic Arithmetic and Number Theory," in *The Undecidable: Basic Papers on Undecidable Propositions, Unsolvable Problems and Computable Functions*, ed. Martin Davis (Mineola: Dover, 1993), 75–81; G. Gentzen, "Die Widerspruchfreiheit der reinen Zahlentheorie," *Mathematische Annalen* 112 (1936), 493–565, English translation as "The Consistency of Arithmetic," in *The Collected Papers of Gerhard Gentzen*, ed. M. E. Szabo (Amsterdam: North-Holland, 1969), 132–213. For more recent explication, see the work of William Tait, "Gödel on Intuition and on Hilbert's Finitism," in *Kurt Gödel: Essays for His Centennial*, ed. Solomon Feferman, Charles Parsons, and Stephen Simpson (Cambridge: Cambridge University Press, 2010), 88–108. Later efforts to raise the problem of foundations, such as that of the "constructivists," have had little resonance. For an example, see Gabriel Stolzenberg, "Can an Inquiry into the Foundations of Mathematics Tell Us Anything Interesting about Mind?," in *The Invented Reality: How Do We Know What We Believe We Know? Contributions to Constructivism*, ed. Paul Watzlawick (New York: W. W. Norton, 1984), 257–308.

29. John Dewey, *The Quest for Certainty: A study of the Relation of Knowledge and Action* (New York: Minton, Balch, 1929), 30.

30. Spengler, *Decline*, 1:418.

31. Spengler, *Decline*, 1:6.

32. Spengler, *Decline*, 1:4, 118, 421. On the distinction between organic and mechanical, which corresponds to the distinction between chronological and mathematical number, see pp. 4, 6, and 70 of the English.

33. Forman, *Weimar Culture*, 71, citing Einstein, letter of January 27, 1920, from Albert

Einstein, Hedwig Born, and Max Born, *Briefwechsel 1916-1955*, ed. M. Born (Munich: Nymphenburger, 1969), 42-45. Like Musil, Einstein attributes Spengler's obsession ("the whole monomania") to "schoolchild mathematics."

34. Einstein, "On the Present Crisis," and Moritz Schlick, "Naturphilosophische Betrachtungen über das Kausalprinzip," *Naturwissenschaft* 8 (1920), 461; both as cited by Forman, *Weimar Culture*, 64-65. Compare Spengler: "Causality is coextensive with the concept of law. There are only causal laws."

35. See, for example, C. S. Peirce's brilliant and brief "The Doctrine of Necessity Examined," *Monist* 2 (1892), 481-86. On the impact of probability on deterministic explanations of the world, see especially Ian Hacking, *The Taming of Chance* (Cambridge: Cambridge University Press, 1990).

36. Max Planck, "Dynamische und statistische Gesetzmässigkeit," *Physikalische Abhandlungen und Vorträge*, ed. Verband Deutscher Physikalischer Gesellschaften und der Max-Planck-Gesellschaft zur Förderung der Wissenschaften e. V. (Braunschweig: F. Vieweg, 1958), 77-90, cited in Forman, *Weimar Culture*, 67. Planck was certainly correct that one did not have to make an absolute choice between the two conceptions. As physicists and philosophers would soon accept, weaker postulates of lawfulness that did not require every detail of every natural process to be determined unambiguously could still be satisfied. But many thinkers in this period tended to extremes, seeing, for example, in any relaxation of determinism the collapse of causality.

37. Wolfgang Pauli, "Quantentheorie," in *Handbuch der Physik*, vol. 23, *Quanten* (Berlin: Springer, 1926), 1-278, here 11; cited by Forman, *Weimar Culture*, 96.

38. See Forman, *Weimar Culture*, 105-7.

39. Erwin Schrödinger, "Was ist ein Naturgesetz?" *Naturwissenschaft* 17 (1929), 9-11; translated as "What Is a Law of Nature?" in *Science, Theory, and Man* (New York: Dover, 1957), 133-47. Though unpublished until 1929, the lecture was delivered December 9, 1922. Forman discusses the lecture in *Weimar Culture*, 87-88.

40. Erwin Schrödinger, "Quantisierung als Eigenwertproblem (Zweite Mitteilung)," *Annalen der Physik* 79 (April 1926), 509. See Forman, *Weimar Culture*, 104. Schrödinger seems to us remarkable for his desire to reconcile sameness and difference.

41. Max Dehn, *Über die geistige Eigenart des Mathematikers: Rede anlässlich der Gründungsfeier des Deutschen Reiches am 18. Januar 1928*, Frankfurter Universitätsreden 28 (Frankfurt am Main: Werner und Winter, 1928), cited in Forman, *Weimar Culture*, 55.

42. Paul Valéry, "La crise de l'esprit, première lettre," in *Œuvres*, vol. 1 (Paris: Gallimard, 1957), 994.

43. Hendecasyllabic: in French, a language deprived of tonic stresses, it becomes *décasyllabe*, which in this poem is divided by a caesura after the fourth syllable. The poem was published in 1920 by Émile-Paul Frères, then collected in *Charmes* (Paris: NRF, 1922). We cite the translation by Cecil Day-Lewis, *The Graveyard by the Sea* (London: Martin Secker & Warburg, 1946). Valéry stressed often the role of the Mediterranean, which he characterized in his *La Liberté de l'esprit* of 1939 as a "machine for making civilization" (though not the only such machine: he presents the Rhine basin as another). See *History and Politics: Collected Works of Paul*

Valéry, ed. J. Mathews and trans. D. Folliot (Princeton, NJ: Princeton University Press, 1971), 196.

44. Thus *cimetière* (cemetery) joins with *colombes* (doves) through *columbaria* (the ancient Roman cemeteries or dovecotes), with *pins* (pines) by metonymy, and with *tombes* by synecdoche, and again through *colombes* by rhyme; the Latin *pendo* (weigh, value, pay, think, ponder, judge, regard) is at the origin of *récompense* as well as *pensée* and recalls *Midi le juste* and *regard*. And *calme* derives ultimately from late Latin and Greek *kaûma*, the heat of the midday sun, from *kaíō*, I burn.

45. Borges's judgement was expressed to Bioy Casares in 1964. See Adolfo Bioy Casares, *Borges*, ed. D. Martino (Barcelona: Destino, 2006), 1042. For a contemporary critique of Valéry's poem, see Paul Claudel, "Sur le vers français" (1925), in *Œuvres en prose* (Paris: Pléiade, 1965), 8–9. One of us (Ricardo) is fond of Elizabeth Sewell's reading in *Paul Valéry: The Mind in the Mirror* (Cambridge: Bowes & Bowes, 1952).

46. As Valéry put it, "les 'idées' qui figurent dans une œuvre poétique n'y jouent pas le même rôle, ne sont pas du tout des *valeurs de même espèce*, que les idées de la prose." Gustave Cohen and Paul Valéry, *Essai d'explication du "Cimetière marin"*: *Précédé d'un avant-propos de Paul Valéry au sujet du "Cimetière marin"* (Paris: Gallimard, 1933), 29.

47. "Mauvaises pensées," in *Œuvres* (Paris: Pléiade, 1960), 2:794 (written ca. 1940). Compare this *pensée* to Plato's *Philebus*, 15d–e.

48. The German title was *Die Krisis der europäischen Wissenschaften und die transzendentale Phänomenologie: Eine Einleitung in die phänomenologische Philosophie*, published as volume 6 of Husserl's *Gesammelte Werke*, ed. Walter Biemel (The Hague: Martinus Nijhoff, 1954). Published in English as *The Crisis of European Sciences and Transcendental Phenomenology*, trans. D. Carr (Evanston, IL: Northwestern University Press, 1970).

49. Or as John Dewey put it in *The Quest for Certainty*, 30, "Thus philosophy in its classic form became a species of apologetic justification for belief in an ultimate reality in which the values which should regulate life and control conduct are securely enstated."

50. Husserl, *Crisis of European Sciences*, 46. Like Heidegger, Husserl here shares Plato's deprecatory attitude toward *technē* and those who use it. Philosophy once more makes its age-old aristocratic claim to a monopoly of power over truth, at the expense of the laboring classes of mere "technicians" and producers.

51. Husserl, *Crisis of European Sciences*, 13.

52. Just one representative example of the mode of argument: "The obscurity [of the relation between both kinds of existence] was strengthened and transformed still later with the development and constant methodical application of pure formal math. 'Space' and the purely *formally* defined 'Euclidean manifold' were confused; the *true axiom* (i.e., in the old, customary sense of the term), as an ideal norm with unconditional validity, grasped with self-evidence in pure geometric thought or in arithmetical, purely logical thought, was confused with inauthentic [*uneigentliches*] 'axiom'—a word which in the theory of manifolds signifies not judgments ('propositions') but forms of propositions as components of the definition of a

'manifold' to be constructed formally without internal contradiction." Husserl, *Crisis of European Sciences*, 56.

53. Alain Badiou offers similar solutions in our own day, claiming a difference between the supposed cruelties of counting and number, on which modern technology and tyranny are based, and the equally supposed ontological truths of set theory that he offers as firm ground for being. We offer our view of this approach in "Badiou's Number: A Critique of Set Theory as Ontology," *Critical Inquiry* 37 (2011), 583–614, and "Critical Response," *Critical Inquiry* 38 (2012), 362–87.

54. "Philosophy and the Crisis of European Man," in Edmund Husserl, *Phenomenology and the Crisis of Philosophy: Philosophy as Rigorous Science and Philosophy and the Crisis of European Man*, trans. Quentin Lauer (New York: Harper & Row, 1965), 153.

55. "Truly invariant": the description of Einstein's thought is from Ernst Cassirer's 1921 essay on the theory of relativity, translated in *Substance and Function and Einstein's Theory of Relativity*, trans. William Curtis Swabey and Marie Collins Swabey, (Chicago: Open Court, 1923), 379, cited in Peter Gordon, *Continental Divide: Heidegger, Cassirer, Davos* (Cambridge, MA: Harvard University Press, 2010), 17.

56. Cassirer provides a definition of his concept in "Der Begriff der Symbolischen Form im Aufbau der Geisteswissenchaften," in *Vorträge der Bibliothek Warburg*, ed. Fritz Saxl (Leipzig: B. G. Teubner, 1921–1922), 1:15. Cassirer was well aware of the fact that in the crisis precipitated by the German defeat in World War I, many of his contemporaries understood "form" itself as a type of tyranny allied with reason and that moreover they considered that tyranny to be in some way "Jewish." It was against that view that he had published *Freedom and Form: Studies in German Intellectual History* in 1916.

57. As when, in a lecture course from the 1930s, he suggests that modernity has inverted the priorities of the Greeks in speaking of mathematics: "what does τὰ μαθήματα mean? When we speak of 'the mathematical,' we run the *risk of misinterpreting* the Greek concept. For with the 'mathematical' we initially and exclusively think of number and numerical relations, of the point, line, plane, solid (spatial elements and forms). But all this is called mathematical only in a derivative sense, insofar as it satisfies precisely what *originally* belongs to the *essence of the* μαθήματα; the μαθήματα are not to be explained through the mathematical, but vice versa." Martin Heidegger, *Being and Truth*, trans. Gregory Fried and Richard Polt (Bloomington: Indiana University Press, 2010), 26.

58. Heidegger writes of Descartes: "his ontology of the world is not primarily determined by his leaning towards mathematics . . . but rather by his ontological orientation in principle towards Being as constant presence-at-hand, which mathematical knowledge is especially well suited to grasp. In this way Descartes explicitly switches over philosophically from the development of traditional ontology to modern mathematical physics and its transcendental foundations." Heidegger, *Being and Time*, trans. John Macquarrie and Edward Robinson (New York: Harper Perennial, 2008), 1.3, §20, 95–96 (128–29) (section and page numbers are to the standard German edition, page numbers in parentheses are to the translation). Heidegger remained preoccupied with these issues. See, for example, his 1957 lec-

ture on "Die Satz der Identität," in Martin Heidegger, *Identity and Difference*, trans. Joan Stambaugh (Chicago: University of Chicago Press, 1969), English on 23–41, German on 85–106.

59. On the past's failure to properly interrogate the condition of "notness," see Heidegger, *Being and Time*, 2.2, §58, p. 286 (331–32).

60. Heidegger, *Being and Time*, 1.3, §18, p. 88 (121–22). A fine example of the type of work Heidegger was attacking here, published a year later, is Rudolf Carnap's *Der logische Aufbau der Welt* (The logical structure of the world) (Berlin: F. Meiner, 1928).

61. Martin Heidegger, *Kant and the Problem of Metaphysics*, trans. Richard Taft (Bloomington: Indiana University Press, 1997), 192.

62. Philipp Lenard, *Deutsche Physik* (Munich: J. F. Lehmann, 1936), 1:6. For many more examples, see Klaus Hentschel, ed., and Ann M. Hentschel, trans., *Physics and National Socialism: an Anthology of Primary Sources* (Basel: Birkhäuser, 1996). We thank Robert Gordon for bringing this last to our attention.

63. Johannes Stark, "The Pragmatic and the Dogmatic Spirit in Physics," *Nature* 141 (1938), 772. On the mathematicians see Herbert Mehrtens, "Ludwig Bieberbach and 'Deutsche Mathematik,'" in *Studies in the History of Mathematics*, ed. Esther R. Phillips (Washington, DC: Mathematical Association of America, 1987), 195–241; "Mathématiques, sciences de la nature et national-socialisme: Quelles questions poser?" in *La Science sous le Troisième Reich: Victime ou alliée du nazisme?*, ed. Josiane Olff-Nathan (Paris: Seuil 1993), 33–49; "Irresponsible Purity: On the Political and Moral Structure of the Mathematical Sciences in the National Socialist State," in *Scientists, Engineers, and National Socialism*, ed. Monika Renneberg and Mark Walker (Cambridge: Cambridge University Press 1994), 324–38; and Sanford Segal, *Mathematicians under the Nazis* (Princeton, NJ: Princeton University Press, 2014). For a preliminary approach to the large question of why these debates about mathematics were conducted in terms of Judaism, see David Nirenberg, *Anti-Judaism: The Western Tradition* (New York: W. W. Norton, 2013).

64. Spengler is idiosyncratic in this regard. He advocated the view that the mathematization of physics and the "ruthlessly cynical hypothesis of the Relativity theory" (Spengler, *Decline*, 1:419) were extreme symptoms of the collapse of the West. But because his racial theories ascribed "Magian" rather than "Faustian" roots to the Jews, and he believed that the concept of force was a Faustian idea, he though Jews *incapable* of physics. (See, for example, 414: "It is an astounding proof of the secret power of root ideas that Heinrich Hertz, the only Jew amongst the great physicists of the recent past. . . .") We might say that Spengler's prejudice led him *not to* blame the Jews for the mathematics and physics he was criticizing.

65. Vienna alone could produce many more examples. Christian von Ehrenfels, creator of Gestalttheorie, wrote that the contemplation of the "field of numbers and their eternal, necessary connections" soothed his agonized spirit (*Weltangst*) after World War I: *Das Primzahlengesetz* (Leipzig: O. R. Reisland, 1922), iii. Hermann Broch set mathematics at the center of the spiritual crisis of his age in a 1934 essay, "Geist und Zeitgeist," in *Geist and Zeitgeist: The Spirit in an Unspiritual Age*, ed. and trans. John Hargraves (New York: Counterpoint, 2002), 45–46.

66. Dewey, *The Quest for Certainty*, 10.
67. Dewey, 19.

Chapter Two

1. Hans Blumenberg, *The Laughter of the Thracian Woman: A Protohistory of Theory*, trans. S. Hawkins (London: Bloomsbury, 2015), vii.
2. See Herodotus, *Histories*, 2.109; cf. Aristotle, *Metaphysics*, 1.1.981b21. On the influence of Babylonian mathematics, see note 4 below.
3. Diogenes Laertius, *Lives of Eminent Philosophers*, 1.22–23, 27; Proclus, *A Commentary on the First Book of Euclid's "Elements,"* 250.18–251.2, 298.12–13, 347.13–16; Thomas L. Heath, *A History of Greek Mathematics* (Minneola, NY: Dover, 1981), 1:128–40.
4. Leonid Zhmud stresses Pythagorean innovation in *Pythagoras and the Early Pythagoreans* (Oxford: Oxford University Press, 2012), 245–46, and elsewhere. On Mesopotamian mathematics see especially Jens Høyrup, *Lengths, Widths, Surfaces: A Portrait of Old Babylonian Algebra and Its Kin* (New York: Springer, 2002). For a discussion of the influence of Mesopotamian on Greek, Egyptian, Indian, and Islamic mathematics, see pp. 400–417. See also Høyrup's *Influences of Institutionalized Mathematics Teaching on the Development and Organization of Mathematical Thought in the Pre-Modern Period: Investigations into an Aspect of the Anthropology of Mathematics*, Materialien und Studien. Institut für Didaktik der Mathematik der Universität Bielefeld, vol. 20 (Roskilde: Roskilde University Center, 1980), where Høyrup understands the Greeks to be the first Mediterranean culture to study "pure" as opposed to "applied" mathematics. Annette Imhausen's *Mathematics in Ancient Egypt: A Contextual History* (Princeton, NJ: Princeton University Press, 2016) has little to say about angles or their measurement.
5. Reviel Netz, *The Shaping of Deduction in Greek Mathematics: A Study in Cognitive History* (Cambridge: Cambridge University Press, 1999), 196–97. Netz continues: "On a wider view still, another small class, of relations which are transitive without being equivalence relations, governs much of what remains of Greek proofs: mainly the relations 'greater/smaller.' The combinations of transitive relations and equivalence relations yield the set of legitimate substitutions which is the core of Greek mathematical argumentation." To be sure, Netz is referring to the later (and documented) period of classical Greek mathematics, not to these retrospective stories about mythic founders.
6. Except for trivial, noninformative ones like "if the point P is the same as R and Q is the same as S, then the segment PQ is the same as the segment RS."
7. Andrew Gregory, *The Presocratics and the Supernatural: Magic, Philosophy and Science in Early Greece* (London: Bloomsbury, 2013), 219, citing Simplicius, *Physics*, 23, 29.
8. Aristotle, *Metaphysics*, 1.3.983b; Heraclitus Homericus, *Quaest. Hom.*, 22. Although Thales seems not to have enunciated any principle of persistence or conservation, Heraclitus seems to have thought in these terms in his fragment B31: "Sea pours out, and is measured by the same amount as before it became earth." Compare Lavoisier's formulation of the Law of Conservation of mass in the 1770s, now more

generalized as the First Law of Thermodynamics: matter is neither created nor destroyed, but merely changed in form. (We put these side by side not to assert influence, but merely to highlight the enduring power of sameness claims.)

9. Thales on earthquakes: Seneca, *Natural Questions*, 3.14; Pseudo-Plutarch, 3.15.

10. Andrew Gregory, *Presocratics and the Supernatural*, 43–67.

11. Pseudo-Plutarch 1.25, Stobaeus 1.4.7a, cited by Gregory, *Presocratics and the Supernatural*, 62. We would add that Homer and other poets also spoke of necessity, which they sometimes explained in terms of the gods.

12. Earthquakes: Cicero, *De Divinatione*, 1.50.112. Proportions of the cosmos: Hippolytus, *Refutation of all Heresies*, 1.6.5.

13. Aristotle, *On the Heavens/De caelo*, 295b, lines 10 ff., DK 12A26, emphasis added. W. C. K. Guthrie (Loeb) translates *homoiótēta* as "indifference," comparing with Plato at *Phaedo* 108e; we prefer the more immediate "sameness." *Phaedo* 108e–109a does, however, make clear that Plato should be included among the "some" referred to by Aristotle.

14. For example, Carlo Rovelli, *The First Scientist: Anaximander and His Legacy*, trans. M. L. Rosenberg (Yardley, PA: Westholme, 2011); Daniel Graham, *Explaining the Cosmos: The Ionian Tradition of Scientific Philosophy* (Princeton, NJ: Princeton University Press, 2006). Dirk L. Couprie, "The Discovery of Space: Anaximander's Astronomy," in *Anaximander in Context*, ed. D. Couprie, R. Hahn, and G. Nadaff (Albany: State University of New York Press, 2003), 167–240.

15. See Katherine Brading and Elena Castellani, eds., *Symmetries in Physics: Philosophical Reflections* (Cambridge: Cambridge University Press, 2003), 10.

16. Herodotus, *Histories*, 1.170, reports that Thales counseled the general assembly of the Ionian federation to establish a single seat of government at "the center of Ionia." For the general political notion of *eis meson*, "taking things to the center," see Marcel Detienne, *The Masters of Truth in Archaic Greece*, trans. J. Lloyd (Boston: Zone Books, 1996), 101 ff. On the prestige of "the center" in archaic Greek thought see Jean-Pierre Vernant, *Myth and Thought among the Greeks* (London: Routledge, 1983), 184ff, 190 ff.

17. Aristotle, *Metaphysics*, 4.3.1005b.

18. *De Caelo* 295b, lines 17–19, 31–34.

19. We might call this a μετάβασις εἰς ἄλλο γένος, or slippage into a different category. In his *Theodicy* of 1710 G. W. Leibniz, a modern champion of the Principle of Sufficient Reason, criticized the case of Buridan's ass (and thereby Aristotle) on both of the grounds we have just outlined. See G. W. Leibniz, *Theodicy* (Chicago: Open Court Classics, 1985), 150.

20. Netz, *Shaping of Deduction*, 198, has pointed to the importance of this choice for Greek mathematics: "At some stage, some Greeks—impelled by the bid for incontrovertibility, described in Lloyd—decided to focus on relations in so far as they are transitive, to demand that in discussions of relations of area and the like, the make-believe of ideal transitivity should be entertained." The reference is to G. E. R. Lloyd, *Demystifying Mentalities* (Cambridge: Cambridge University Press, 1990).

21. These words, like all the ones we have attributed to Anaximander, come via poste-

rior media: in this case the writings of Simplicius, who lived a thousand years later and is said to have taken them from Theophrastus, Aristotle's disciple. Simplicius, *Physics* 24.13; Diels-Kranz 12.Bl; Aristotle, *Physics*, 3.4.203b7 f.

22. No boundaries: Imagine *X* to be a plane as in Euclidean geometry, consisting of numberless points but no lines whatsoever: lines have not been invented yet. Since boundaries, the boundaries of figures or things like triangles, rectangles, circles, and so forth, are lines, it follows that since there are no boundaries in *X*, there are no figures or definite things in *X* either. Think of it this way: any *definite* thing requires, by definition, a boundary, or a criterion, to decide what the thing is and what it is not.

23. Nietzsche already demonstrated the error in Friedrich Nietzsche, *Die Philosophie im tragischen Zeitalter der Griechen*, in *Werke in drei Bänden*, vol. 3, ed. Karl Schlechta (Munich: Hanser, 1954), §4. Nevertheless H. F. Cherniss, in "The Characteristics and Effects of Presocratic Philosophy," *Studies in Presocratic Philosophy*, ed. D. J. Furley and R. E. Allen (New York: Humanities Press, 1970), 1:8, construes the *ápeiron* as a mixture of all the elements. Charles H. Kahn, *Anaximander and the Origins of Greek Cosmogony* (New York: Columbia University Press, 1960), 233, understands the *ápeiron* as essentially the cosmos.

24. No mixture: A mixture must have at least two different ingredients; say *A* and *B* are two different ingredients of the *ápeiron*. The part of the *ápeiron* that is different from *A* (not *A*) is then not empty. Hence, the *ápeiron* is divided into two nonempty classes, *A* and not-*A*. This, however, is a limitation or determination, an added boundary, contradicting the definition of the *ápeiron*.

25. Or perhaps we should think of the poet Hölderlin's *Sein* (Being) before the *Ur-Theil*. The German *Ur-Theil* means "primordial cut," and *Urteil* (without the dash and without the obsolete *h*), means "judgment." "The thing *X* is *Y*," or "The thing *X* is not *Y*," are judgments: "being before the judgments" is therefore equivalent to the *ápeiron*. Compare George Spencer-Brown, *Laws of Form* (Leipzig: Bohmeier, 2014), 1-3, who identifies the notion of *form* with the introduction of a *difference* into what is otherwise total *sameness*.

26. Edgar Allen Poe, in the last pages of his short life, engaged in a similar reading of Anaximander: "In the original unity of the first things lies the secondary cause of all things, with the germ of their inevitable annihilation." *Eureka: A Prose Poem* (Amherst, NY: Prometheus, 1997), 5-6.

27. W. H. Auden, "In Sickness and in Health, 1940," *Selected Poems* (New York: Vintage, 1979), 113, lines 74-75.

28. Diogenes Laertius 2.4-6, in *Lives of Eminent Philosophers*, trans. R. D. Hicks, Loeb Classical Library (Cambridge, MA: Harvard University Press, 1925), 1:132-35. Diogenes (3rd century CE) lived centuries after the philosophers whose lives he described, and we cannot take his reports at face value. According to other equally problematic accounts, Pythagoras also studied under Zoroaster (or Zabratus) in Babylonia (Porphyry, *Life of Pythagoras*, 12; Alexander Polyhistor apud Clement's *Stromata*, 1.15; Diodorus of Eritrea, *Aristoxenus apud Hippolitus*, 6.32.2).

29. In his *Commentary on Ptolemy's "Harmonics"* (30.1-4), Porphyry cites the Platonist Xenocrates (fourth century BCE): "Pythagoras also discovered that musical

intervals do not have their origin without number, for they are a comparison/combination of a quantity with a quantity." Of the scholars who hold that the Pythagorean harmonic traditions we do know more about (ca, 400 BCE) may have drawn from a tradition going back to Pythagoras himself, few have been as categorical as Zhmud in *Pythagoras and the Early Pythagoreans*, 290: "Pythagorean harmonics must go back to Pythagoras himself." On the many philosophical and cultural futures of the almost certainly apocryphal hammers, see Daniel Heller-Roazen, *The Fifth Hammer: Pythagoras and the Disharmony of the World* (Cambridge, MA: MIT Press, 2011).

30. The classic treatment is Walter Burkert, *Lore and Science in Ancient Pythagoreanism* (Cambridge, MA: Harvard University Press, 1972). Burkert does not credit early Pythagorean arithmetic with anything beyond playing with pebbles and rejects any early venturing into logical proof (401–47). For a similarly negative take on the mathematical merits of Pythagoras, see Jonathan Barnes, *The Presocratic Philosophers* (London: Routledge, 1982); Netz, *Shaping of Deduction*, 272–76, goes even further, and places the beginnings of Greek mathematics at "440 BC." By contrast, Zhmud, *Pythagoras and the Early Pythagoreans*, insists that Greek mathematics developed into a deductive science between roughly 550 and 450 BCE and that Pythagoras and his disciples most likely carried that out. See also the essays gathered in G. Cornelli, R. McKirahan, and C. Macris, eds., *On Pythagoreanism* (Berlin: De Gruyter, 2013); and A.-B. Renger and A. Stavru, eds., *Pythagorean Knowledge from the Ancient to Modern World: Askesis, Religion, Science* (Wiesbaden: Harrassowitz, 2016).

31. Aristotle, *Metaphysics* 1.5.986a. The translation is by H. Tredennick, Loeb Classical Library (Cambridge, MA: Harvard University Press, 1947).

32. Philolaus frag. 4, in C. A. Huffman, *Philolaus of Croton: Pythagorean and Presocratic; A Commentary on the Fragments and Testimonia with Interpretive Essays* (Cambridge: Cambridge University Press, 1993), 172.

33. This is the introduction of arithmetic into geometrical approaches on which we have already heard philosophers (like Husserl) opining in chap. 1. Add to those Karl Popper, *The Open Society and its Enemies*, vol. 1, *The Spell of Plato*, 5th ed. (London: Routledge and Kegan Paul, 1945), 248; "The Nature of Philosophical Problems and their Roots in Science," in *Conjectures and Refutations: The Growth of Scientific Knowledge* (London: Routledge and Kegan Paul, 1963), 87; M. R. Wright, *Cosmology in Antiquity* (London: Routledge and Kegan Paul, 1995), 54.

34. Aristotle, *Metaphysics*, 1.5.986a; Plutarch, *Isis and Osiris*, 370.

35. We put both Ionians and Pythagoreans in scare quotes to remind ourselves that these were only retrospectively characterized as schools. On the distinction between limited and unlimited, see C. Huffman, "Limité et illimité chez les premiers philosophes grecs," in *La Fêlure du Plaisir: Études sur le Philèbe de Platon*, vol. 2, *Contextes*, ed. M. Dixsaut (Paris: Vrin, 1999), 11–31.

36. Philolaus (frag. B 5) suggests that the Pythagorean oppositions started with even/odd numbers, then were applied to other domains.

37. The relevant passages of Aristotle are *Metaphysics* 1.5.985b27, 986a2, 986a16, 986a18; 1.6.987b11, 987b28; 13.8.1083b. Zhmud, Huffman, and Gregory all doubt

that Pythagoras or any known early Pythagorean stated that "all is number" or held the view that the world is actually constituted from number, a characterization of the early Pythagoreans that Gregory suggests originated with Aristotle. To us Aristotle's characterization seems apt insofar as the Pythagoreans took τὰ ὄντα (beings) and ἀριθμὸν ἔχοντα (having number) to be one and the same. See Gregory, *Presocratics and the Supernatural*, 129; Zhmud, *Pythagoras and the Early Pythagoreans*; *Wissenschaft, Philosophie und Religion im frühen Pythagoreismus* (Berlin: Akademie, 1997); Andrei V. Lebedev, "Idealism (Mentalism) in Early Greek Metaphysics and Philosophical Theology: Pythagoras, Parmenides, Heraclitus, Xenophanes and Others (With Some Remarks on the *Gigantomachia about Being* in Plato's Sophist)," *Indo-European Linguistics and Classical Philology* 23 (2019): 651–704; C. A. Huffman, "The Role of Number in Philolaus' Philosophy," *Phronesis* 33 (1988), 1–30; idem, "The Pythagorean Tradition," in *The Cambridge Companion to Early Greek Philosophy*, ed. A. A. Long (Cambridge: Cambridge University Press, 1999), 66–87.

38. W. Knorr, *The Evolution of the Euclidean Elements: A Study of the Theory of Incommensurable Magnitudes and Its Significance for Early Greek Geometry* (Dordrecht: D. Reidel, 1975), 36–40, reminds us that pre-Socratics never allude to incommensurability.

39. Plato, *Laws*, 7.819d–820d.

40. Diogenes Laertius 9.21, for example, considered Parmenides a disciple of the Pythagorean Aminias. The quote is from W. K. Guthrie, *History of Greek Philosophy*, vol. 2, *The Presocratic Tradition from Parmenides to Democritus* (Cambridge: Cambridge University Press, 1965), 2–3. But, Guthrie concludes, "he certainly broke away from it, as from all other previous philosophical systems." "True creators": Árpád Szabó, *The Beginnings of Greek Mathematics* (Dordrecht: D. Reidel, 1978). Among recent scholars maintaining Pythagorean influence on Parmenides, see Catherine Rowett, "Philosophy's Numerical Turn: Why the Pythagoreans' Interest in Number Is Truly Awesome," in *Doctrine and Doxography: Studies on Heraclitus and Pythagoras*, ed. David Sider and Dirk Obbink (Berlin: De Gruyter, 2013), 22–25, 30; Leonid Zhmud, "Pythagorean Communities: From Individuals to Collective Portraits," in *Doctrine and Doxography: Studies on Heraclitus and Pythagoras*, ed. David Sider and Dirk Obbink (Berlin: De Gruyter, 2013), 45; *Pythagoras and the Early Pythagoreans*, 253. Other scholars still stress Parmenides's distance from such schools: cf. John Palmer, *Parmenides and Presocratic Philosophy* (Oxford: Oxford University Press, 2009), 19.

41. Frag. B 1, lines 28–30. We refer to fragments of the pre-Socratics according to the standard Diels-Kranz numeration but have made use of the edition and translation by André Laks and Glenn W. Most, eds. and trans., *Early Greek Philosophy*, Loeb Classical Library (Cambridge, MA: Harvard University Press, 2016), 5:37.

42. Porphyry, *Life of Pythagoras*, 46.

43. For the sake of comparison, it is a good deal sharper than in the famous figure of the cave through which Plato would later attempt to explicate a version of the problem. In Plato's cave, sense perception remains connected to true being, albeit only as shadows or projections are to solid bodies.

44. For a different translation see Laks and Most, *Early Greek Philosophy*, 5:39. Ours is closer to that provided by L. Tarán in *Parmenides* (Princeton, NJ: Princeton University Press, 1965), 32, and to G. E. M. Anscombe in "Parmenides, Mystery and Contradiction," *Proceedings of the Aristotelian Society* 69 (1968), 125–32: "It is the same thing that can be thought and can be."

45. Frag. B 3. The translation offered by Laks and Most (39), "for it is the same, to think and also to be," is linguistically simpler but makes less philosophical sense in our view. We follow L. Tarán in his translation of this fragment. For a discussion of the issues involved, see Tarán, *Parmenides*, 41–44.

46. Frag. B 4, with its insistence on the continuity of being, offers some (admittedly enigmatic) support for our view: "Nevertheless, keep the absent firmly present in mind: / for you shall not cut off being from holding on to being, / either by dispersing it all through the cosmos / or by bringing it together." If we interpret κατὰ κόσμον in another of its possible senses, as "according to an order," the fragment would acquire a somewhat different meaning.

47. Frag. B 8, lines 3–6, 15–16.

48. Cf. Plato, *Timaeus*, 29b–c, 48d–e. "As being is to becoming, so is truth to opinion"— a pithy mathematical way of writing Parmenides's and Plato's doctrine.

49. Heraclitus, frag. B 40, B 81, B 129. "Evil arts" (frag. 129): The Greek word is *kakotechniên*, a compound of "evil" (κακο-) and "practice, technique, art" (τέχνη). "Chief swindler" (frag. B 81): *kopidôn archêgos*. The translation "bullshitter extraordinaire" is authorized by Gregory, *Presocratics and the Supernatural*, 133.

50. Charles H. Kahn, *The Art and Thought of Heraclitus* (Cambridge: Cambridge University Press, 1979), 204, argued that Heraclitus's notion of harmony was a generalization of Pythagoric numerical ratios; we will argue that it is not. Rowett, "Philosophy's Numerical Turn," argues that Heraclitus refers to number implicitly, since *lógos*, the key Heraclitean term, may mean, among many other things, "ratio." But (a) the meanings of *lógos* are too manifold to draw such a conclusion, and (b) ratios, or proportions, need not be numerical but could be solely geometrical.

51. Plato attributes to Heraclitus the phrase τὰ ὄντα ἰέναι τε πάντα καὶ μένειν οὐδὲν, "all beings move and none remains still," then uses a slightly different phrase, πάντα χωρεῖ καὶ οὐδὲν μένει, "everything moves, and nothing remains," to "explain" the name of the god Kronos (from χωρεῖ = moves, goes) and finally adds the Heraclitean river flow—ποταμοῦ ῥοῇ—to "explain" the name of the goddess Rhea (*Cratylus*, 401d, 402a).

52. As Roman Dilcher points out in "How Not to Conceive Heraclitean Harmony," in *Doctrine and Doxography: Studies on Heraclitus and Pythagoras*, ed. David Sider and Dirk Obbink (Berlin: de Gruyter, 2013), 263–80.

53. Rodolfo Mondolfo, in *Heráclito: Textos y problemas de su interpretación* (Madrid: Siglo 21, 1966), 160, maintains that certain fragments are critical of Pythagorean oppositions.

54. Frag. B 59: γναφείῳ ὁδὸς εὐθεῖα καὶ σκολιὴ μία εστί καὶ ἡ αὐτή. The precise meaning of γναφεῖον is a crux. Laks and Most, *Early Greek Philosophy* (3:163) translate it "carding-comb," where some translators use "screw." Hippolytus, in his *Refutation of All Heresies* 10.3 (Laks and Most, 3:301) explains it thus: "The rotation

of the instrument called the screw in the carding-comb is straight and crooked, for while it turns in a circle, at the same time it moves up."

55. In his *Dissertation* of 1770, section III, §15, Kant built on the fact that a right and a left hand are "intellectually the same," but not congruent by means of rotations and translations, in order to claim that our notion of space is not intellectual but sensitive. This claim would not have escaped Heraclitus's demolition service: in four spatial dimensions the two hands are congruent.

56. Compare Augustine, *On the Morals of the Manichaeans*, 11, who argues against the existence of absolute evil through the example of a scorpion's poison, harmful to the human but salvific for the scorpion.

57. Frag. B12. The dative case of the word ἐμβαίνουσιν (those who step into) can be translated using several English propositions: we have settled on "for" and "on." Also, τοῖσιν αὐτοῖσιν (the/those same) can be referred to ποταμοῖσι (the rivers), or to ἐμβαίνουσιν (the waders), or to both. Aristotle (*Rhetoric* 3.5.1407b = Heraclitus A 4 DK) complained of the difficulty of "punctuating" Heraclitus's text "because it isn't clear whether a word goes with what precedes or follows it." We consider this a deliberate aspect of Heraclitus's writing, very much part of his radical critique of the Principle of Non-Contradiction. Other "river" fragments are B 49a, 91a and b, and A 6.

58. Aristotle, *Metaphysics* 4.3.1005b. For recent disagreements on B 12, see Daniel W. Graham, "Once More unto the Stream," in *Doctrine and Doxography: Studies on Heraclitus and Pythagoras*, ed. David Sider and Dirk Obbink (Berlin: de Gruyter, 2013), 303-20.

59. Frag. B 45.

60. Frag. B 115.

61. Even this is controversial. See the evidence reviewed in Netz, *Shaping of Deduction*, 63-64. Plato himself made the analogy between mathematical arts and "pebble games" (*petteutikē*) in *Laws* 819d-820d and *Gorgias* 450c-d.

Chapter Three

1. Plato, *Plato: Phaedrus*, 265c-266d, trans. A. Nehamas and P. Woodruff (Indianapolis, IN: Hackett, 1995), 65. See also *The Statesman*, 286d-287d. In Plato's *The Sophist* 253b-e the Stranger defines the philosopher as the person who knows how to divide according to kinds, and does not take "the same form for a different one or a different one for the same." Cf. Aristotle, *De anima* 3.6, 430b5-20, with a different but related limitation.

2. Heraclitus, B10DK, in André Laks and Glenn W. Most, eds. and trans., *Early Greek Philosophy*, Loeb Classical Library (Cambridge, MA: Harvard University Press, 2016), 3:161. The Greek word translated here as "conjoinings" is συνάψιες. Guy Davenport prefers "joints" in his loose translation of the fragment: "Joints are and are not part of the body. They cooperate through opposition, and make a harmony of separate forces. Wholeness arises from distinct particulars; distinct particulars occur in wholeness": *Herakleitos and Diogenes*, trans. Guy Davenport (Bolinas, CA: Grey Fox Press, 1979), 30. Socrates's Delian diver is from Diogenes Laertius, *Lives of Eminent Philosophers*, trans. R. D. Hicks, Loeb Classical Library (Cambridge,

MA: Harvard University Press, 1925), 2:22. Nietzsche shares our suspicion of Plato on this point in *Beyond Good and Evil*, trans. Walter Kaufmann (New York: Vintage, 1966), 71: "Here and there one encounters an impassioned and exaggerated worship of 'pure forms,' among both philosophers and artists: let nobody doubt that whoever stands that much in *need* of the cult of surfaces must at some time have reached *beneath* them with disastrous results." Is Nietzsche suggesting that this insistence on form and foundational sameness is what we today might call a response to a "trauma" of indeterminacy?

3. We cite the Loeb Classical Library translation by H. N. Fowler, *Plato: Cratylus, Parmenides, Greater Hippias, Lesser Hippias* (Cambridge, MA: Harvard University Press, 2002).

4. Flummoxed, but critically self-conscious: Socrates at 131d is admirable in recognizing the degree to which his choice about which ideas to busy himself with—the gross or the sublime—is a matter of taste.

5. See Aristotle, *Metaphysics* 990b17 = 1079a13, 1039a2; *Sophistical Refutations* 178b–179a. Just how seriously Plato himself took the argument he makes here remains an open question: see Robert Barford, "The Context of the Third Man Argument in Plato's *Parmenides*," *Journal of the History of Philosophy* 16 (1978), 1–11. We are aware that the theory of Forms is articulated differently elsewhere in Plato's work.

6. We are confident that Plato would *not* agree with our characterization of his method here. He took pains to differentiate his philosophical way of proceeding from that of the "practitioners of geometry and arithmetic" who, he claimed, "make hypotheses . . . as if they knew them to be true. They do not expect to give an account of them to themselves or to others but proceed as if they were clear to everyone. From these starting points they go through the subsequent steps by agreement." *Republic*, 510c–e. Our claim is that Plato, too, has his "axioms," and that they are in some ways more dogmatic (not least because unstated and less constrained in their pretensions to explanatory power) than those of the "practitioners of geometry and arithmetic."

7. We borrow this list from Samuel Rickless, "Plato's *Parmenides*," *Stanford Encyclopedia of Philosophy*, https://plato.stanford.edu/archives/spr2020/entries/plato-parmenides.

8. Greek mathematicians conceived of number (*arithmós*) not in terms of sets (as did the modern mathematicians we encountered in chap. 1 and those we will meet in chap. 6) but as a plurality of units (*monas*). On Greek concepts of number, see, among others, Jacob Klein, *Greek Mathematical Thought and the Origin of Algebra*, trans. Eva Brann (New York: Dover, 1992); and Ivor Grattan-Guinness, "Numbers, Magnitudes, Ratios, and Proportions in Euclid's Elements: How Did He Handle Them?," *Historia Mathematica* 23 (1996): 355–75.

9. 143a–144b. The one and the many is a version of the question of sameness and difference, but note the dialogue's exploration of the difference between "one" and "same." At 139d–e Parmenides asserts that "The nature of one is surely not the same as that of the same" and argues that by adding sameness to the many into the one, the one would no longer be one. This will become of enormous philosophical and

theological import, as (for example) later thinkers try to distinguish between the Oneness of God and equality or sameness.

10. Plato, *The Sophist*, 216a–d. The Stranger's interlocutor in the dialogue, a boy called Theaetetus, is also known to history as a first-rate mathematician. Both were also characters in the *Theaetetus*, a dialogue represented as having taken place the day before. Here, too, we use the translations by H. N. Fowler, *Plato: Theaetetus, Sophist*, Loeb Classical Library (Cambridge, MA: Harvard University Press, 2006).

11. Curiously, Plato does not seem to have been very interested in the question of the "deceptiveness" of the mathematical diagrams that played such an important role in the development of deductive truth claims in his day, though he touches on them in *Republic* 529d–e, contrasting them to drawings by artists and craftsmen like Daedalus. On the place of diagrams in Greek mathematics, see especially Reviel Netz, *The Shaping of Deduction in Greek Mathematics: A Study in Cognitive History* (Cambridge: Cambridge University Press, 1999), 12–86. Conversely, Vitruvius (*De architectura*, 7.2) provides some evidence that the perspectival discoveries of Greek painters may have been underpinned by mathematical theory, but if Plato knew of this, it did not lessen his condemnation.

12. Tertulian, *On Idolatry*, 3.2-4. Origen, *Homilies on Genesis and Exodus*, trans. Ronald E. Heine (Washington, DC: The Catholic University of America Press, 1982), 8.3, p. 321.

13. If we compare "fair Portia's counterfeit" portrait in *The Merchant of Venice* to the living statue in *A Winter's Tale*, we might deduce that Shakespeare shared this view. Origen's argument drew on the fact that the Greek Septuagint uses different terms at Exodus 20:4 and Genesis 1:27: *homoiosis* and *homoioma*. Gerhard Ladner saw in the term *homoiosis* a turning point in the history of ideas: a deliberate likening of humanity to God. The Latin Vulgate of Exodus 20:4 and Genesis 1:27 will use "idolum" and "similitude." On Origen and Tertullian, see Carlo Ginzburg, *Wooden Eyes: Nine Reflections on Distance* (New York: Columbia University Press, 2001), 96–105. "E dicono che la scultura imita la forma vera, e mostra le sue cose, girandole intorno, a tutte le vedute . . . Ne hanno rispetto a dire molti di loro, che la scultura è tanto superiore alla pittura, quanto il vero alla bugia"; Giorgio Vasari, *Le vite de più eccellenti pittori scultori ed architettori scritte da Giorgio Vasari* (Florence: Sanson, 1906), 1:94.

14. By way of beginning, compare the Chinese aesthetics described by François Jullien in *The Great Image Has No Form, or On the Nonobject through Painting*, trans. Jane Marie Todd (Chicago: University of Chicago Press, 2009), with the Western tradition limned by Tom Rockmore in *Art and Truth After Plato* (Chicago: University of Chicago Press, 2013).

15. See Plato, *Cratylus* 386d–e: "But if neither is right, and things are not relative to individuals, and all things do not equally belong to all the same moment and always, they must be supposed to have their own proper and permanent essence: they are not in relation to us, or influenced by us according to fancy, but are independent and maintain to their own essence the relation prescribed by nature."

16. Cf. Plato, *The Sophist*, 259e–260a: "This isolation of everything from everything

else means a complete abolition of all discourse, for any discourse we can have owes its existence to the weaving together of forms." Compare Montaigne: cut things into small enough pieces and everything loses all meaning.

17. The Stranger's explanation of one of the reasons why this is a problem reminds us of Frege's analysis of why the sentence "The present king of France is bald" is not a logical proposition.

18. Simply by discussing the *number* of *archaí*, philosophers like Plato and Aristotle reveal themselves to believe that these can be counted without difficulty. In other words, for such philosophers, as for the Pythagoreans, *number is prior to being*.

19. We use here the translation by Benjamin Jowett in *The Collected Dialogues of Plato*, ed. Edith Hamilton and Huntington Cairns (Princeton, NJ: Princeton University Press, 1961). Note, however, that Plato has Timaeus characterize his account as a "likely account" (*eikôs logos*) or "likely story" (*eikôs muthos*); that is, the account is itself an "image" (*eikôn*). See T. K. Johansen, *Plato's Natural Philosophy: A Study of the* Timaeus-Critias (Cambridge: Cambridge University Press, 2004), 62–64; M. F. Burnyeat, "Eikôs Mythos," *Rizai* 2, no. 2 (2005): 143–65; G. Betegh, "What Makes a Myth *eikôs*? Remarks inspired by Myles Burnyeat's 'Eikôs Mythos,'" in *One Book, The Whole Universe: Plato's "Timaeus" Today*, ed. R. Mohr, K. Sanders, and B. Sattler (Las Vegas, NV: Parmenides, 2009), 213–24. One cosmological implication of 28a–29d that future philosophers explored is that there can be only one universe.

20. *Timaeus* 31c–32a. For those who may prefer a more formal demonstration of how proportion makes of many one: if a and b are nonzero numbers, the ratio of a^3 to a^2b is the same as the ratio of a^2b to ab^2, which is the same as the ratio of ab^2 to b^3. In fact all those ratios are equal to a/b. This equality is what Plato means by "will all be one." On the enormous importance of analogy and proportion in Greek thought, a classic treatment is that of G. E. R. Lloyd, *Polarity and Analogy: Two types of Argumentation in Early Greek Thought* (Cambridge: Cambridge University Press, 1966).

21. How the one and the many, or the same and the different, come to coexist is a core question for Platonic cosmogony. In *Philebus* Plato presumes but does not explain the primal existence of both the one and the many, and attempts to describe how they came to coexist in unity, creating the third "type," or synthesis. A similar attempt is being undertaken here in this account of creation. Aristotle characterized Plato's approach by saying that he embraced the one and the indefinite dyad (e.g., "the great and the small") as his basic principles, the one being the essence of the Forms, and the dyad their matter. Aristotle, *Metaphysics*, 987b14–29; and see the still interesting treatment in Walter Burkert, *Lore and Science in Ancient Pythagoreanism* (Cambridge, MA: Harvard University Press, 1972), 21–22.

22. See also 40a–b, discussing the creation of the fixed stars, made out of a mixture of two movements, the first "a movement always on the same spot after the same manner, whereby they ever continue to think consistently the same thoughts about the same things, in the same respect," and then a "forward movement, in which they are controlled by the revolution of the same and the like."

23. Every sentence of the *Timaeus* has received centuries of commentary. Readers looking for a recent starting point may turn to Mohr, Sanders, and Sattler, *One Book, The Whole Universe.*

24. This relationship was not lost on the *Timaeus*'s future readers. For but one example, see Christopher Gill's study of the implications of *Timaeus*, 86b–87b, for Galenic and Stoic thinking about the relation between physical and psychic illness: "The Body's Fault? Plato's *Timaeus* on Psychic Illness," in *Reason and Necessity: Essays on Plato's "Timaeus,"* ed. M. R. Wright (London: Duckworth, 2000), 59–84. Many of the other essays in that volume are also relevant to the topics we have touched on here.

25. M. F. Burnyeat explores the reasons for the centrality of mathematics in Plato's pedagogical program in "Plato on Why Mathematics Is Good for the Soul," in *Mathematics and Necessity: Essays in the History of Philosophy*, Proceedings of the British Academy 103, ed. T. Smiley (Oxford: Oxford University Press, 2000), 1–81. See also the multiple articles of Ian Mueller, especially "Mathematics and Education: Some Notes on the Platonist Programme," in *"Peri Tōn Mathēmaton*: Essays on Greek Mathematics and Its Later Development," ed. Ian Mueller, special issue, *Apeiron* 24 (1991), 85–104. Plato was aware that math could be studied for more mundane ends such as commerce, and he dismissed those who did so. For him, the proper motives for mathematical study are "for the sake of war and to attain ease in turning the soul itself away from the world of becoming and toward truth and reality" (525c).

26. In the *Republic* Plato classified mathematical reasoning (*dianoia*) not as the highest but as the second highest form of thought, behind understanding (*noesis*), which brought direct participation in the Forms themselves (510–11).

27. Plato, *Laws*, 818–20, here 818a–b; cf. 818d. The translation is by A. E. Taylor in *The Collected Dialogues of Plato*, ed. Hamilton and Cairns. On what Plato means by *necessity*, see Burnyeat, "Plato on Why Mathematics," 21.

28. Plato has Socrates provide a justification for this playful pedagogy in *Republic*, 536e, stressing that free men should not study slavishly and that forced study is not retained in the soul. So not only are mathematics a prerequisite for philosophy and human freedom, but whether they make us slavish or free depends to some degree on how we feel about studying them.

29. Among those who account it spurious because of its excessive claims for mathematical astronomy, see Leonardo Tarán, *Academica: Plato, Philip of Opus, and the Pseudo-Platonic "Epinomis."* (Philadelphia: American Philosophical Society, 1975). Others have judged it genuine on (statistically determined!) stylistic grounds: Charles M. Young, "Plato and Computer Dating," *Oxford Studies in Ancient Philosophy* 12 (1994): 227–50, reprinted in Nicholas D. Smith, ed., *Plato: Critical Assessments*, vol. 1, *General Issues of Interpretation* (London: Routledge, 1998), 35. We cite from the translation by A. E. Taylor in *The Collected Dialogues of Plato*, ed. Hamilton and Cairns.

30. Early moderns, such as Pico della Mirandola, were attracted to the *Epinomis* for this very reason.

31. *Nicomachean Ethics*, 1.6.1.1096a11–15, trans. H. Rackham, *Aristotle: The Nicoma-chean Ethics*, Loeb Classical Library (Cambridge, MA: Harvard University Press, 2003). For a history of the Latin proverb, see Leonardo Tarán, "Amicus Plato sed magis amica veritas: From Plato and Aristotle to Cervantes," in *Antike und Abend-land* 30 (1984): 93–124. Don Quixote's letter comes in part 2, chap. 51, of the novel.

32. Aristotle, *Metaphysics* 1.6.987b10–11, 1.9.992a30. Here and in what remains of this chapter we generally cite the translation edited by Jonathan Barnes: *The Complete Works of Aristotle* (Princeton, NJ: Princeton University Press, 1984).

33. For a good introduction to the context, see D. H. Fowler, *The Mathematics of Plato's Academy: A New Reconstruction*, 2nd ed. (Oxford: Oxford University Press, 1999). See also M. F. Burnyeat, "Platonism and Mathematics: A Prelude to Discussion," in *Mathematics and Metaphysics in Aristotle*, ed. A. Graeser (Berne: P. Haupt, 1987), 213–40.

34. We won't follow him through those critiques. For Aristotle's thoughts on the nature of mathematical objects, readers can turn to books 13 and 14 of the *Metaphysics*, with the excellent commentary provided by Julia Annas, *Aristotle's Metaphysics Books M and N: English Translation and Commentary*, 2nd ed. (Oxford: Oxford University Press, 1988).

35. Aristotle, *Nicomachean Ethics*, 1.1.1094a5–7, 1.3.1094b23–25.

36. Among the many valuable works of Jonathan Barnes on this subject, see "Aristotle's Theory of Demonstration," *Phronesis* 14 (1969): 123–52. Indispensable is G. E. R. Lloyd, "The Theories and Practices of Demonstration," in *Aristotelian Explora-tions*, ed. G. E. R. Lloyd (Cambridge: Cambridge University Press, 1996), 16–19. On demonstration and enthymeme in the *Rhetoric*, see M. F. Burnyeat, "Enthymeme: Aristotle on the Logic of Persuasion," in *Aristotle's Rhetoric: Philosophical Essays*, ed. D. J. Furley and A. Nehemas (Princeton, NJ: Princeton University Press, 1994), 3–55.

37. Aristotle, *Posterior Analytics* 1.71a, translation in Jonathan Barnes, *The Complete Works of Aristotle*. On this attraction see W. Knorr, "On the Early History of Axi-omatics: The Interaction of Mathematics and Philosophy in Greek Antiquity," in *Theory Change, Ancient Axiomatics, and Galileo's Methodology: Proceedings of the 1978 Pisa Conference on the History and Philosophy of Science*, ed. J. Hintika, D. Gruender, and E. Agazzi, (Dordrecht: Reidel, 1980), 1:145–86. Among the at-tractions we have not yet mentioned was the cognitive force of mathematical let-tered diagrams. On Aristotle's own reference to and use of such letters and dia-grams (especially in the *Posterior Analytics*) see Netz, *Shaping of Deduction*, 36–37, 45, 48–49, 61.

38. On starting points in Greek mathematics (including that of Aristotle) see Netz, *Shaping of Deduction*, 171–77; and I. Mueller, "On the Notion of a Mathematical Starting Point in Plato, Aristotle and Euclid," in *Science and Philosophy in Classical Greece*, ed. A. C. Bowen (New York: Garland, 1991), 59–97.

39. Aristotle, *Metaphysics* 4.4.1006a5–9. On the *archaí* of demonstrations, see Lloyd, "Theories and Practices," 27.

40. On *aether* see the cogent introduction of G. E .R. Lloyd, *Aristotle: the Growth and*

Structure of His Thought (Cambridge: Cambridge University Press, 1968), 134-39. Lloyd notes the "religious considerations" that influenced Aristotle here, citing among other passages *Metaphysics*, 12.8.1074a38, and *On the Heavens*, 270b5.

41. Aristotle, *Metaphysics*, 4.4.1006a15.

42. Aristotle, *Metaphysics*, 4.3.1005b. Here and in the next paragraph we follow the translation of Jonathan Barnes in *The Complete Works of Aristotle* (Princeton, NJ: Princeton University Press, 1984), 2:1588.

43. Aristotle, *Metaphysics*, 4.3.1005b9-30.

44. See, for example, *Metaphysics*, 13.1-2.1076a-1076b.

Chapter Four

1. See in general Hans Joas, "The Axial Age Debate in Religious Discourse," in *The Axial Age and Its Consequences*, ed. Robert N. Bellah and Hans Joas (Cambridge: Harvard University Press, 2012), 9-29. See also Benjamin I. Schwartz, "The Age of Transcendence," *Daedalus* 104 (1975): 1-7; Shmuel Eisenstadt, "The Axial Age in World History," in *The Cultural Values of Europe*, ed. Hans Joas and Klaus Wiegandt (Liverpool: Liverpool University Press, 2008), 22-42. "The age of criticism" is Arnaldo Momigliano's felicitous term, from his *Alien Wisdom: The Limits of Hellenization* (Cambridge: Cambridge University Press, 1975), 8-9. See also the essays collected in Johann P. Arnason, S. N. Eisenstadt, and Björn Wittrock, eds. *Axial Civilizations and World History* (Leiden: Brill, 2005).

2. Karl Jaspers, *The Way to Wisdom: An Introduction to Philosophy* (New Haven, CT: Yale University Press, 2003), 98.

3. Søren Kierkegaard, *Concluding Unscientific Postscript to "Philosophical Fragments,"* ed. and trans. H. V. Hong and E. H. Hong (Princeton, NJ: Princeton University Press, 1992), 1:37-38.

4. Plato sometimes uses the Greek word *methexis* to describe the simultaneous sameness and difference of things with respect to Forms. Often translated as "participation," the term appears in the *Phaedo, Parmenides, Timaeus,* and *Sophist.* See, for example, *Parmenides*, 131c-132b, on which, inter alia, see Patricia Curd, "'Parmenides' 131c-132b: Unity and Participation," *History of Philosophy Quarterly* 3 (1986), 125-36.

5. The literature is enormous, but for a clarifying start see Michael Davis, *The Soul of the Greeks: An Inquiry* (Chicago: University of Chicago Press, 2011).

6. *On the Soul*, 404b, in Aristotle, *On the Soul, Parva Naturalia, On Breath*, trans. W. S. Hett, Loeb Classical Library (Cambridge, MA: Harvard University Press, 1957), 22-23: γινώσκεσθαι γὰρ τῷ ὁμοίῳ τὸ ὅμοιον.

7. *De anima*, 430a. Separable, apathic and unmixed: "χωριστὸς καὶ ἀπαθὴς καὶ ἀμιγής." Immortal: "χωρισθεὶς δ' ἐστὶ μόνον τοῦθ' ὅπερ ἐστί, καὶ τοῦτο μόνον ἀθάνατον καὶ ἀΐδιον (οὐ μνημονεύομεν δέ, ὅτι τοῦτο μὲν ἀπαθές, ὁ δὲ παθητικὸς νοῦς φθαρτός), καὶ ἄνευ τούτου οὐθὲν νοεῖ." Our translation here differs from Hett's at 171, not least in retaining Aristotle's vocabulary of pathic and apathic.

8. *De anima*, 414b, 29-31. Again, our translation differs from Hett, 83, and from Aristotle, *On the Soul*, trans. Joe Sachs (Santa Fe, NM: Green Lion Press, 2004), 89.

Compare Spinoza's "Preface" to part 3 of his *Ethics*: "I shall, then, treat of the nature and strength of the emotions, and the mind's power over them, by the same method as I have used in treating of God and the mind, and I shall consider human actions and appetites just as if it were an investigation into lines, planes, or bodies." *Spinoza: Complete Works*, ed. Michael L. Morgan, trans. Samuel Shirley (Indianapolis, IN: Hackett, 2002), 278.

9. This was a common concern of ancient philosophers. For another example, that of Claudius Ptolemy (ca. 100–ca. 170 CE), see Jacqueline Feke, "Mathematizing the Soul: The Development of Ptolemy's Psychological Theory from *On the Kritêrion* and *Hêgemonikon* to the *Harmonics*," *Studies in History and Philosophy of Science* 43 (2012): 585–94.

10. Eusebius, *Praeparatio Evangelica*, 11.10.14 (cf. 9.6.9); citing Numenius, frag. 8.13. Cf. Clement of Alexandria, *Stromata*, 1.22. See also M. F. Burnyeat, "Platonism in the Bible," in *Metaphysics, Soul, and Ethics in Ancient Thought: Themes from the Work of Richard Sorabji*, ed. R. Salles (Oxford: Oxford University Press, 2005), 143–69.

11. Clement of Alexandria, *Stromata*, 2.100.3. See the discussion in David T. Runia, "Why Does Clement of Alexandria Call Philo 'The Pythagorean?,'" *Vigiliae Christianae* 49 (1995) 1–22. The passage is cited at page 2. For an edition of Clement's text, see Clement of Alexandria, *Clemens Alexandrinus Stromata*, 3rd ed., vols. 1–2, ed. O. Stählin, L. Früchtel, and U. Treu, (Berlin: Akademie, 1960–1985).

12. For an excellent introduction to Philo's intellectual context and to the various influences (Roman, Hellenistic, Jewish; Stoic, Platonic, etc.) on his work see Maren Niehoff, *Philo of Alexandria: An Intellectual Biography* (New Haven, CT: Yale University Press, 2018).

13. Philo, *De Vita Mosis*, 38. We thank Sofía Torallas Tovar for the reference. The Babylonian Talmud stresses that seventy-two scholars working independently in separate rooms miraculously came up with the same translation: BT Megillah 9a.

14. David Runia, *Philo of Alexandria and the Timaeus of Plato*, Philosophia Antiqua 44 (Leiden: Brill, 1986). On the influence of Philo on Christianity, see Runia, *Philo in Early Christian Literature: A Survey* (Assen: Van Gorcum, 1993).

15. We use the term *Abrahamic* here to refer to Judaism, Christianity, and Islam, three monotheistic religions that consider themselves heirs of the prophet Abraham. We realize that, though convenient, the term is controversial. For a justification, see Guy G. Stroumsa, "From Abraham's Religion to the Abrahamic Religions," *Historia Religionum* 3 (2011): 11–22.

16. Quoted in Diogenes Laertius (fl. 3rd century CE), *Lives of the Philosophers*, 8.25. Alexander himself was apparently quoting from a text of unknown date he called *Pythagorean Memoirs*. Again, this is citation but with difference: here Plato's dualism is translated into a more Pythagorean monadism. The literature on the *Timaeus* influence is vast. See, inter alia, J. F. Phillips, "Neo-Platonic Exegeses of Plato's Cosmology," *Journal of the History of Philosophy* 35 (1997), 173–97.

17. For example, Psalm 8:5–9: "What is man, that you are mindful of him? . . . You have made him a little lower than god/s." The word is the plural *elohīm*, used in biblical Hebrew to mean both singular God and plural gods or divine beings.

18. For an attempt at reconstruction see Karl Staehle, *Die Zahlenmystik bei Philon von*

Alexandreia (Leipzig: Teubner, 1931). See also Frank Egleston Robbins, "Arithmetic in Philo Judaeus," *Classical Philology* 26 (1931), 345-61.

19. On Nicomachus, see, inter alia, Leonardo Tarán, "Nicomachus of Gerasa," in *Collected Papers (1962-1999)* (Leiden: Brill, 2001), 544-48; and G. Wolfgang Haase, "Untersuchungen zu Nikomachos von Gerasa" (PhD diss., University of Tübingen, 1974), 34-119. Our treatment draws on David Albertson, *Mathematical Theologies: Nicholas of Cusa and the Legacy of Thierry of Chartres* (Oxford: Oxford University Press, 2015), 50-56.

20. "The necessary point of departure": Albertson, *Mathematical Theologies*, 53. "God and the Monad": Nicomachus of Gerasa, *Theologoumena arithmeticae* I, ed. Victor De Falco (Stuttgart: Teubner, 1975), 3, as translated in Albertson, *Mathematical Theologies*, 54.

21. Albertson, *Mathematical Theologies*, 55, citing Nicomachus of Gerasa, *Introductio arithmetica libri I*, ed. Richard Gottfried Hoche (Leipzig: Teubner, 1866), 1.6.1, 12; *Nicomachus of Gerasa: Introduction to Arithmetic*, trans. Martin Luther D'Ooge, Frank E. Robbins, and Louis C. Karpinsky (New York: MacMillan, 1926), 189.

22. The term was coined by Etienne Gilson describing Plotinus's philosophy of the One as distinct from an ontology (or philosophy of being). See his *L'Être et l'essence* (Paris: Vrin, 1948), 42. See also Werner Beierwaltes, *Denken des Einen: Studien zur Neuplatonischen Philosophie und ihrer Wirkungsgeschichte* (Frankfurt am Main: Vittorio Klostermann, 1985); and Egil A. Wyller, "Zur Geschichte der platonischen Henologie: Ihre Entfaltung bis zu Plethon/Bessarion und Cusanus," in *Greek and Latin Studies in Memory of Cajus Fabricius*, ed. Sven-Tage Teodorsson (Göteborg: University of Göteborg, 1990), 239-65.

23. Indeed, it is only because Philo's questions were in some sense shared by early Christians that his writings survive, for they left little trace in the Jewish communities of his day and later but were transmitted to posterity by Christians like Clement.

24. There we cited 1 Cor. 8.4, "an idol [*eidolon*] is nothing in the world." As Albertson points out (*Mathematical Theologies*, 292, 296), in this period Plato's *Parmenides* was often interpreted as a henological manifesto articulated through a negative theology. The dominant argument here remains that of E. R. Dodds, "The *Parmenides* of Plato and the Origin of the Neoplatonic One," *Classical Quarterly* 22 (1928), 129-41. Without supposing that Paul was aware of these interpretations of *Parmenides* we can still suggest that his position here resonates with them.

25. For a more categorical approach, see Geurt Hendrik van Kooten, *Paul's Anthropology in Context: the Image of God, Assimilation to God, and Tripartite man in Ancient Judaism, Ancient Philosophy, and Early Christianity* (Tübingen: Mohr Siebeck, 2008).

26. On these early Christian movements, see especially Joel Kalvesmaki, *The Theology of Arithmetic: Number Symbolism in Platonism and Early Christianity*, Hellenic Studies 59 (Cambridge, MA: Harvard University Press, 2013).

27. J. W. B. Barns, G. M. Browne, and J. C. Shelton, eds., *Nag Hammadi Codices: Greek and Coptic Papyri from the Cartonnage of the Covers* (Leiden: Brill, 1981). Other im-

portant surviving texts of this general school include Ptolemy, *Lettre à Flora*, 2nd ed., ed. Gilles Quispel (Paris: Cerf, 1966).

28. Marius Victorinus, *Le Livre des XXIV philosophes: Résurgence d'un texte du IVe siècle*, ed. and trans. Françoise Hudry (Paris: Vrin, 2009), 150. Compare Augustine, *De doctrina christiana*, ed. and trans. R. P. H. Green (Oxford: Oxford University Press, 1996), 1.12 (v. 5), 16-17; trans. Green, 10. The attribution of the *Le Livre* to Marius Victorinus follows Hudry and is based on shared terms with Victorinus's *Adversus Arium* and other works. According to Hudry, Porphyry's *Vita Pythagorae* is among Victorinus's most important sources in these works (*Le Livre des XXIV Philosophes*, 24-29). The juxtaposition of Augustine and Victorinus is Albertson's, *Mathematical Theologies*, 79.

29. On Augustin and number, see, inter alia, Christoph Horn, "Augustins Philosophie der Zahlen," *Revue des Études Augustiniennes* 40 (1994): 389-415; Alois Schmitt, "Mathematik und Zahlenmystik," in *Aurelius Augustinus: Die Festschrift der Görres-Gesellschaft zum 1500; Todestage des Heiligen Augustinus*, ed. Martin Grabmann and Joseph Mausbach (Cologne: J. P. Bachem, 1930), 353-66; Werner Beierwaltes, "*Aequalitas numerosa*. Zu Augustins Begriff des Schönen," *Weisheit und Wissenschaft* 38 (1975), 140-57; and Albertson, *Mathematical Theologies*, 68-80.

30. *De libero arbitrio* 2.8.20-22, quote from 2.8.21. The translation here and following is from Augustine, *On the Free Choice of the Will, On Grace and Free Choice, and Other Writings*, ed. and trans. Peter King (Cambridge: Cambridge University Press, 2010).

31. *De libero arbitrio* 2.15.39. On number in this treatise, see, among many others, Horn, "Augustins Philosophie," 396-400.

32. It seems to us that Augustine here comes oddly close to the position of the Athenian in the *Epinomis*, namely, that good is number and evil the absence of number.

33. *Soliloquia*, 1.11, beginning at "Although you urge me vehemently and convince me, nevertheless, I do not dare to say that I desire to know God as I know these things [about geometry]. Not only the things, but even the very type of knowledge seems to me dissimilar." We thank Clifford Ando for the reference. See also John Peter Kenney, *Contemplation and Classical Christianity: A Study in Augustine* (Oxford: Oxford University Press, 2013), 74-76.

34. *Confessiones*, 7.6.

35. The formulation is inspired by Albertson, *Mathematical Theologies*, 61. Those interested in discussion of Augustine's change of mind may see Peter Brown, *Augustine of Hippo: A Biography* (Berkeley: University of California Press, 2000), 139-50; Carol Harrison, *Rethinking Augustine's Early Theology: An Argument for Continuity* (Oxford: Oxford University Press, 2006), 14-19.

36. *Confessions*, 6.4: "Volebam enim eorum quae non viderem ita me certum fieri, ut certus essem, quod septem et tria decem sint." Note how he continues: "Neque enim tam insanus eram, ut ne hoc quidem putarem posse conprehendi." To doubt the certainty of $7 + 3 = 10$ would be the ultimate insanity.

37. *Confessions*, 10.12: "The memory also contains innumerable reasons and laws of

numbers and dimensions, which no bodily sense has impressed upon it. . . . I have also perceived with all my bodily senses those numbers that we count, but those numbers with which we number are different, nor are they the images of these, and therefore they indeed are." Compare the argument about circles in his earlier *De immortalitate animae* of 387.

38. The formulation is Albertson's, *Mathematical Theologies*, 77, whose reading of *De genesi ad litteram* we are following here. Albertson's treatment of subsequent medieval mathematical theologies is illuminating, although we will not follow him into that future.

39. We will touch on a number of questions about the translation below. Here we simply note that in its Qur'anic context the word translated as "likeness" here (*kufuww*) is an equivalent in rank: a meaning essential to the (anti-Christian) polemical purpose of the sura, but muffled in our translation.

40. The exasperated monarch is Abū al-Walīd Ismāʿīl, who assumed the throne of the Nasrid kingdom of Granada in 1314. See Mohamad Ballan, "The Scribe of the Alhambra: Lisān Al-Dīn Ibn Al-Khaṭīb, Sovereignty and History in Nasrid Granada" (PhD Dissertation: University of Chicago, 2019), 73, 109.

41. "Came to be regarded" and "highest authority": Uri Rubin, "Al-Ṣamad and the High God: An Interpretation of Sūra CXII," *Der Islam* 61 (1984), 197–217. "Solid beaten metal" is from the early (ninth century?) Greek translations of the Qur'an, which used the words *sphyropēktos*, "beaten solid into a ball," and *holosphyros*, "of hammer-beaten metal." On the Greek translation's approach to our term, see Christos Simelidis, "The Byzantine Understanding of the Qur'anic Term al-Ṣamad and the Greek Translation of the Qur'an," *Speculum* 86 (2011), 887–913. On the Greek translation more generally, see Christian Høgel, "An Early Anonymous Greek Translation of the Qur'ān: The Fragments from Niketas Byzantios' Refutatio and the Anonymous Abjuratio," *Collectanea Christiana Orientalia* 7 (2010), 65–119. (Our thanks to Alexandre Roberts for this reference.) "Solid," etc.: for an early tenth-century example, al-Ṭabarī, *The Commentary on the Qur'ān by Abū Jaʿfar Muḥammad b. Jarīr al-Ṭabarī*, trans. J. Cooper, vol. 1 (Oxford: Oxford University Press, 1987), 215; cited in Rubin, "Al-Ṣamad," 213–14.

42. Nader el-Bizri, ed. and trans., *Epistles of the Brethren of Purity: On Arithmetic and Geometry; An Arabic Critical Edition and English Translation of Epistles 1 & 2* (Oxford: Oxford University Press, 2012), 65, 97, 99.

43. El-Bizri, 68, 71–73.

44. El-Bizri, 96–99.

45. Muslim thinkers could point to verses in the Qur'an that could justify the application of the laws of reason to the soul: "By the soul, and by what / [he who] gives her proportion and order," teaches sūra 91.7 (*As-Shams*, The Sun). Other verses could be used to justify contingency, such as Qur'an 65:1 (aṭ-Ṭalāq): "You never know; God may bring about some new event [*yuḥdithu . . . amran*]."

46. As is often the case in this book, beneath the tip of each of these examples lies a long history. The question of burning, for example, was raised (according to Diogenes Laertius) by the ancient Greek skeptics. See Richard Sorabji, "The Origins

of Occasionalism," in *Time, Creation and the Continuum. Theories in Antiquity and the Early Middle Ages* (London: Duckworth, 1983), 30.

47. Tracing the transmission of these questions across faiths and cultures is important if we wish to understand the long history of the relationship between science and theology. Early and important examples include the Jew (and for a time, convert to Christianity) Dawud al-Muqammas (early ninth century) and the tenth-century Syriac Christian Yaḥyā ibn 'Adi, whose "Treatise on the Unity" represents an attempt to defend Christian trinitarianism by integrating Aristotelian teachings with a theological one. See Dawud al-Muqammas, *Twenty Chapters*, ed. and trans. Sarah Stroumsa (Provo, UT: Brigham Young University Press, 2015), especially chap. 8; Yaḥyā b. 'Adī, *Maqālah fī'l-Tawḥīd*, ed Samīr Khalil (Beirut: Al-Maktaba al-Būlusiyya, 1980). On Yaḥya's polemical context, see A. Périer, "Un traité de Yaḥyâ ben 'Adî: Défense du dogme de la trinité contre les objections d'al-Kindî," *Revue de l'Orient chrétien*, 3rd ser., 2 (1920–1921), 3–21; Cornelia Schöck, "The Controversy between al-Kindī and Yaḥyā b. 'Adī on the Trinity, Part One: A Revival of the Controversy between Eunomius and the Cappadocian Fathers," *Oriens* 40 (2012), 1–50; "The Controversy between al-Kindī and Yaḥyā b. 'Adī on the Trinity, Part Two: Gregory of Nyssa's and Ibn 'Adī's Refutation of Eunomius' and al-Kindī's 'Error,'" *Oriens* 42 (2014): 220–53.

48. Those interested in the work of al-Kindī may turn to his *Rasâ'il al-Kindî al-Falsafîya*, ed. Abû Rîdah, M.'A.H., 2 vols. (Cairo: Dâr al-Fikr al-'Arabî, 1950–1953); and *Al-Kindî's Metaphysics*, trans. Alfred Ivry (Albany: State University of New York Press, 1974). Razi's *Opera philosophica* were edited by Paul Kraus (Cairo, 1939). For the translation of some relevant texts, see Arthur J. Arberry, *The Spiritual Physick of Rhazes* (London: John Murray, 1950), reprinted as *Razi's Traditional Psychology* (Damascus: Islamic Book Service, [2007?]).

49. "Necessary, universal premises": al-Fārābī, "*Fusūl al-Madanī* (Selected Aphorisms)," in *The Political Writings: Selected Aphorisms and Other Texts*, trans. C. E. Butterworth (Ithaca, NY: Cornell University Press, 2015), §34, 28–29. These premises follow those of Aristotle, *Metaphysics*, 5, 1013a18 (cf. *Physics*, 184a10–21, and *Nicomachean Ethics*, 1095a2–4). Compare the "common ideas" (κοιναὶ ἔννοιαι) provided by Euclid in *Elements*, bk. I: (1) Things which equal the same thing also equal one another. (2) If equals are added to equals, then the wholes are equals. (3) If equals are subtracted from equals, the remainders are equals. (4) Things which coincide with one another equal one another. (5) The whole is greater than the part.

50. Immortality and happiness: al-Fārābī, *Risalat fī'l-'aql (Letter on the Intellect)*, ed. M. Bouyges (Beirut: Imprimerie Catholique, 1938) 27, 31–32; "The Letter Concerning the Intellect," in *Philosophy in the Middle Ages. The Christian, Islamic, and Jewish Traditions*, trans. A. Hyman (Indianapolis, IN: Hackett, 1973), 219–20; "On the Intellect," in *Classical Arabic Philosophy. An Anthology of Sources*, trans. J. McGinnis and D. C. Reisman (Indianapolis, IN: Hackett, 2007), 75–76. In the *Enumeration of the Sciences*, al-Fārābī treats the sciences of number in chap. 3 on the propaedeutic sciences (propaedeutic to the study of the soul, that is), describing "the science of theoretical number" as a science of "the same [equal] and the different [nonequal]."

51. "First existent . . . matter": al-Fārābī, *Kitāb al-siyāsa al-madaniyya (also Known as the Treatise on the Principles of Beings)*, ed. F. M. Najjar (Beirut: Imprimerie Catholique, 1964), 31; "Political Regime," in *The Political Writings II: "Political Regime" and "Summary of Plato's Laws*," trans. C. E. Butterworth (Ithaca, NY: Cornell University Press, 2015), 29. "One, unique . . . itself": al-Fārābī, *Mabādi' ārā' ahl al-madīnah al-fāḍilah (al-Farabi on the Perfect State)*, trans. R. Walzer (Oxford: Clarendon Press, 1985), 56–89. On the neo-Platonic sources of al-Fārābī's cosmology see D. Janos "The Greek and Arabic Proclus and al-Fārābī's Theory of Celestial Intellection and Its Relation to Creation," *Documenti e Studi sulla tradizione filosofica medievale* 19 (2010): 19–44; D. Janos, *Method, Structure, and Development in al-Fārābī's Cosmology* (Leiden: Brill, 2012), 4–6, 11–37; C. D'Ancona, "Aux Origines du *Dator Formarum*: Plotin, l'*Épître sur la Science Divine*, et al-Fārābī," in *De l'antiquité tardive au Moyen Âge*, ed. E. Coda and C. Martini Bonadeo (Paris: Vrin, 2014), 381–413.

52. As al-Fārābī puts it in the *Risalat fī'l-'aql (Letter on the Intellect)*, the actual intellect becomes the actual intelligibles. They are one and the same: the intellect is not different from what is understood (trans. Bouyges, 19–21; "The Letter Concerning the Intellect," trans. Hyman, 216–17; "On the Intellect," trans. McGinnis and D. C. Reisman, 72–73).

53. Joel Kraemer, *Philosophy in the Renaissance of Islam: Abū Sulaymān Al-Sijistānī and His Circle* (Leiden: Brill, 1986), 179–80. Al-Sijistānī was a student of Yaḥyā b. 'Adī's. His list owes a good deal to the more detailed one of al-Muqammas.

54. Abū Ḥayyān al-Tawḥīdī, *Al-Muqābasāt*, ed. Ḥassan al-Sandūbī (Kuwait: Dār Su'ād al-Ṣabāḥ, 1992), 315–17. We are grateful to Daniel Watling for this observation and for a number of suggestions about our treatment of Islamic philosophy in the following paragraphs.

55. By way of comparison, contrast Al-Sijistānī's colloquy with Avicenna's chapter "On Discussing the One" and "On Ascertaining the One and the Many and Showing That Number Is an Accident," in *The Metaphysics of "The Healing,"* trans. Michael E. Marmura (Provo, UT: Brigham Young University Press, 2005), 3.2–3.3 (74–84).

56. Abū 'Alī al-Ḥusayn Ibn Sīnā, "Tafsīr sūra al-ikhlāṣ," in Avicenna, *Al-Tafsīr al-qur'ānī wa'l-lugha al-ṣūfiyya fī falsafa Ibn Sīnā*, ed. Ḥasan 'Āṣī (Beirut: al-Mu'assasa al-Jāmi'iyya lil-Dirāsāt wa'l-Nashr wa'l-tanzī', 1983), 104–13; *Tafsīr sūra al-ikhlāṣ*, edited by Abū'l-Qāsim Muḥammad (Delhi: 'Abd al-Raḥmān, 1893/1894).

57. Avicenna, *Metaphysics of* The Healing, 2.2–6 (74–102), 7.1–7.2 (236–49). We have slightly altered the translation of the chapter heading "On the Characteristics of Unity" (7.1), replacing Marmura's "Appendages" (for Arabic *lawāḥiq*) with "characteristics" (following Greek *sumbebēkota* or *huparchonta*) and "haecceity" with "identity" (the Arabic *huwiyya* means "it-ness," "identity," or "being").

58. The verse is cited by Abū Ḥāmid al-Ghazālī, *The Incoherence of the Philosophers*, trans. Michael E. Marmura (Provo, UT: Brigham Young University Press, 2000), 103. God's customary choices ('*ādat Allāh*) result in the habitual behavior of things (*al-'āda*). On the question of divine power and the possibility of other (even better) worlds, see Eric Linn Ormsby, *Theodicy in Islamic Thought: the Dispute over al-Ghazali's Best of All Possible Worlds* (Princeton, NJ: Princeton University Press,

1984), esp. 135–81. Dominik Perler and Ulrich Rudolph's *Occasionalismus: Theorien der Kausalität im arabisch-islamischen und im europäischen Denken* (Göttingen: Vandenhoek & Ruprecht, 2000), surveys Islamic "occasionalism" as well as its influence on medieval and early modern Christian philosophical thought in the West.

59. See here especially Frank Griffel, *Al-Ghazālī's Philosophical Theology* (New York: Oxford, 2009), 7, 97–122, 175–208, 275 ff. Put briefly, "even if God chooses always to connect the cause with its effect, the possibility of synchronic alternative to God's action means that this connection is not necessary. As far as practical human knowledge is concerned, however, . . . the connection is permanent, and there is no synchronic alternative" (279).

60. Averroes took Aristotle's characterization of the intellect in *On the Soul* 430a to be a "question [that] is very difficult and has the greatest ambiguity." He did not approve of al-Fārābī's solution and even less of Avicenna's, but he too insisted that the material intellect, the part of the mind that receives and registers apathic thoughts (i.e., logical and mathematical), must be one and the same for all human beings. See Averroes's *Long Commentary on the "De anima" of Aristotle*, trans. R. C. Taylor and T-A. Druart (New Haven, CT: Yale University Press, 2009), 317, 387–88. See also Richard Taylor, "The Agent Intellect as 'Form for Us' and Averroes's Critique of al-Fârâbî," *Proceedings of the Society for Medieval Logic and Metaphysics* 5 (2005): 18–32.

61. It should not be necessary to state, were it not for the polemics of our age, that we in no way mean to imply that Islamic philosophy concluded with Averroes or Ibn Ṭufayl any more than we meant to suggest that Christian philosophy ended with Augustine or Jewish philosophy with Philo. We call Ibn Ṭufayl Averroes's patron because according to tradition, it was at Ibn Ṭufayl's request that Averroes wrote the commentaries on Aristotle for the caliph Abū Yaʿqūb Yūsuf (r. 1163–84), because Ibn Ṭufayl was too busy to do so himself. The story is told by ʿAbd al-Wāḥid al-Marrākushī in his history of the Almohads, *Al-Muʿjib fī talkhīṣ akhbār al-Maghrib* (Beirut: al-Maktaba al-ʿAṣriyya, 2006). The historian names as his informant Ibn Rushd's student, Abū Bakr b. Bundūd b. Yaḥyā al-Qurṭubī.

62. We are utilizing the translation by Lenn Evan Goodman of Ibn Ṭufayl, *Ibn Ṭufayl's Hayy Ibn Yaqzan: A Philosophical Tale* (Chicago: University of Chicago Press, 2009). Page references are to that volume. Goodman's translation is based on and paginated to the edition by Léon Gauthier, *Hayy ben Yaqdhān: Roman philosophique d'Ibn Thofaïl*, 2nd ed. (Beirut: Imprimerie catholique, 1936). "Commander of the Faithful": 4.

63. Ibn Ṭufayl, *Hayy Ibn Yaqzān*. "All objects": 119. Necessary Existent: 135–42. "Will live on": 138.

64. Ibn Ṭufayl, *Hayy Ibn Yaqzān*, 150–51. For the Arabic *ajsām* we have chosen to translate "bodies" rather than Goodman's "physical things."

65. The bat is alluding to the primary intelligibles and common notions that many Hellenistic, Jewish, Christian, and Muslim (e.g., al-Fārābī) philosophers believed innate in humans. See note 49 above. The Arabic word translated here as "axiom" (*hukm/ahkām*) generally refers to legal statutes, especially to the five fundamental categories of statutes in Islamic law, from the forbidden to the obligatory.

66. In al-Ghazālī's *Mishkāt al-anwār*, the bat is used to demonstrate the limitations of discursive (Aristotelean) reason.

67. "What every sound mind": Ibn Ṭufayl, *Hayy Ibn Yaqzān*, 151. "Two wives": 154.

68. Simone Pétrement, *Simone Weil: A Life* (New York: Schocken Books, 1988), 272.

69. Letter of October 1942 to the French-Jewish philosopher Jean Wahl, who was then a refugee in the United States, in Simone Weil, *Œuvres* (Paris: Gallimard, 1999), 977–80. "Il y a quelques textes qui indiquent avec certitude que la géométrie grecque a son origine dans une pensée religieuse: et il semble bien qu'il s'agisse d'une pensée proche du christianisme presque jusqu'à l'identité." For Simone Weil's Pythagorean meditations, see her writings collected in *Œuvres*, especially 595–627, and in *Simone Weil* (Paris: Cahiers de l'Herne, 2014), esp. 15–26.

70. If a and b are numbers, the most popular way to find a number in between is to take their average or arithmetic mean (AM), equal to $(a + b)/2$. But there is also a geometric mean (GM). Assuming that a and b are not 0, we say that the number h is the GM of a and b if $h/a = b/h$, or equivalently, $h^2 = ab$. Finally, the harmonic mean (HM) h of a and b is the reciprocal of the arithmetic mean of their reciprocals: $h^{-1} = (a^{-1} + b^{-1})/2$. Numerical example: take $a = 1$ and $b = 2$; then AM = $\frac{3}{2}$ = 1.5, GM = $\sqrt{2}$ = 1.414 . . . , and HM = $\frac{4}{3}$ = 1.333. . . .

71. Weil, *Œuvres*, 599, emphasis added.

72. Weil, 979.

73. Weil, 602, "Il ne faut pas oublier qu'en grec *arithmos* et *logos* sont deux termes exactement synonymes."

74. Pindar, Olympian Ode 13, 45–46, trans. J. E. Sandys in *Pindar*, Loeb Classical Library (Cambridge, MA: Harvard University Press, 1968): ὡς μὰν σαφὲς οὐκ ἂν εἰδείην λέγειν ποντιᾶν ψάφων ἀριθμόν.

75. This is not merely a quibble about Greek. Weil calls her mathematical ideal love. Qua mathematical, love is apathic. Love for another cannot depend on whether one has ever met that other, on affinities between self and other, or on either's idiosyncrasies. It must be as impartial as sunlight. "God loves not as I love, but as an emerald is green. . . . And I too, if I were in the state of perfection, would love as an emerald is green. I would be an impersonal person." The ideal is inhuman, and we would not call it love. Simone Weil, *First and Last Notebooks*, trans. R. Rees (Oxford: Oxford University Press, 1970), 129. See also Robert C. Reed, "Decreation as Substitution: Reading Simone Weil through Levinas," *Journal of Religion* 93 (2013): 28. Emmanuel Levinas (1906–1995) provides a useful comparison. His ethics are based on a pathic human experience: the face-to-face encounter with the other.

76. Daniel J. Cohen, *Equations from God: Pure Mathematics and Victorian Faith* (Baltimore: Johns Hopkins University Press, 2007), 7–8. Cohen provides many additional Anglo-American examples.

77. Benjamin Pierce, "Address of Professor Benjamin Pierce, President of the American Association for the Year 1853," *Proceedings of the American Association for the Advancement of Science* 8 (1855), 6. On Pierce, see Cohen, *Equations from God*, 56.

78. Cicero, *De natura deorum* 2:5, trans. H. Rackham, Loeb Classical Library (Cambridge MA: Harvard University Press, 1933), 137–38.

79. "Die ganzen Zahlen hat der liebe Gott gemacht, alles andere ist Menschen-werk." According to the mathematician Heinrich Weber, who first printed the quote, Kronecker pronounced this at a paper he delivered in 1886 to the Berliner Naturforscher-Versammlung: see Heinrich Weber: "Leopold Kronecker," in *Jahres-berichte der Deutschen Mathematiker-Vereinigung* 2 (1891/1892): 19; reprinted in *Mathematische Annalen* 43 (1893): 1–25, here 15.

80. Roger Penrose, *Shadows of the Mind: A Search for the Missing Science of Conscious-ness* (Oxford: Oxford University Press, 1994), 413.

Chapter Five

1. Johann Gottfried von Herder, *Philosophical Writings*, ed. M. N. Forster (Cam-bridge: Cambridge University Press, 2002), 3.

2. Ludwig Wittgenstein, *Blue and Brown Books* (New York: Harper, 1965), 18.

3. "The authority, methodology, and nature of the claims of [Newton's] *Principia* be-came one of the most contested areas of eighteenth-century philosophy," as Eric Schliesser puts it in "Newton and Newtonianism," *The Routledge Companion to Eighteenth Century Philosophy*, ed. Aaron Garrett (New York: Routledge, 2014), 62. Sharp though the contest was, we will suggest that participants on multiple sides of the contest shared a tendency toward solutions of "sameness."

4. For Alexander Pope's own efforts in this regard, see his *Essay on Man*, composed in the 1730s.

5. *Phaedo*, 73c–78b, 85b–d. Socrates argues that two empirical objects—he gives the usual examples of pebbles or pieces of wood or parts of a geometric diagram—resemble the pure, abstract idea of sameness (αὐτὸ τὸ ἴσον, "sameness in itself") very imperfectly. We must therefore have this abstract idea of "sameness in itself" previously in our mind in order for us to be able to realize how poorly those objects represent it. For a similar but more explicitly mathematical argument, see Plato's *Meno*, 84a–86d, in which Socrates demonstrates the immortality of the soul by finding knowledge of geometry innate in an uneducated slave boy. On this last, see G. E. R. Lloyd, "The Meno and the Mysteries of Mathematics," *Phronesis* 37 (1992), 166–83.

6. *Phaedrus*, 265e–266c.

7. See David Lewis, "New Work for a Theory of Universals," *Australasian Journal of Philosophy* 61 (1984), 343–77. Lewis alludes explicitly to Plato's metaphor. The de-scriptor "most important" is from Brian Weatherson, "David Lewis," *Stanford En-cyclopedia of Philosophy*, https://plato.stanford.edu/entries/david-lewis/. The role of mathematical physics in Lewis's model is noteworthy.

8. Zhuang Zhou's story about the cook carving an ox for King Hui of Liang appears in chap. 3 of *Zhuangzi, The Complete Writings*, trans. Brook Ziporyn (Indianapo-lis, IN: Hackett, 2020), 29–30. Franz Kafka, *Die Erzählungen*, ed. Roger Hermes (Frankfurt am Main: Fischer, 1996), 307–9. Gottfried von Strassburg, *Tristan and Isolde*, trans. Francis G. Gentry (New York: Continuum, 2003), 46.

9. Descartes, Second Meditation, 7.24. In this sense (as in others), Descartes's proj-ect is quite Aristotelean. On the role of Aristotelean theories of cognition on Des-

cartes, see Gary Hatfield, "The Cognitive Faculties," in *The Cambridge History of Seventeenth-Century Philosophy*, ed. Daniel Garber and Michael Ayers (Cambridge: Cambridge University Press, 1998), 953–1002.

10. As is so often the case, Descartes's certainties would in time seem error to others. Numerous later natural philosophers of the scientific revolution thought of Descartes's subjectivist criteria for truth as "a primary pathology in the learning of their time." The quote is from Matthew L. Jones, *The Good Life in the Scientific Revolution: Descartes, Pascal, Leibniz, and the Cultivation of Virtue* (Chicago: University of Chicago Press, 2006), 57, and for an excellent discussion of those criteria, 58–86.

11. *Meditations on First Philosophy* in The *Philosophical Writings of Descartes*, trans. John Cottingham, Robert Stoothoff, and Dugald Murdoch (Cambridge: Cambridge University Press, 1985), Third Meditation, 2:35.

12. See Louis Loeb, "The Cartesian Circle," in *The Cambridge Companion to Descartes*, ed. J. Cottingham (Cambridge University Press, 1992), 200–235; "The Priority of Reason in Descartes," *Philosophical Review* 99 (1990), 3–43.

13. *Parmenides* 166c and *Sophist* 238b. Both statements are synonymous, since the one must be (or there must be units) for number to be, and reciprocally, if there is one, there is also two, and so on.

14. See, for example, Descartes, *Rules for the Direction of the Mind*, rule three: "Everyone can mentally intuit that he exists . . . that a triangle is bounded by just three lines, and a sphere by a single surface, and the like . . . many facts which are not self-evident are known with certainty, provided they are inferred from true and known principles through a continuous and uninterrupted movement of thought in which each individual proposition is clearly intuited." *The Philosophical Writings of Descartes*, trans. John Cottinghman, Robert Stoothoff, and Dugald Murdoch (Cambridge University Press, 1984–1985), 1:14–15. On 2 + 3 = 5, see vol. 2, *Meditations on First Philosophy*, First Meditation, 15, Third Meditation, 25. Al-Ghazālī approaches God and certainty in the same way in *Deliverance from Error* and elsewhere. All knowledge outside of his "intimacy" (*dhawq*; literally "taste" or "first-hand experience") with God is subject to doubt.

15. On Descartes's heuristic expectations for his geometry, see Jones, *Good Life*, 15–53.

16. There are those who have extended necessity into language by championing the idea of innate syntax and semantics. See most famously Noam Chomsky, *Syntactic Structures* (Paris: Mouton, 1957); *Knowledge of Language: Its Nature, Origin, and Use* (Westport, CT: Praeger, 1986); and Jerry Fodor, *The Language of Thought* (Cambridge, MA: MIT Press, 1975). Those views remain vulnerable to criticism that syntax is an emergent property of language (e.g., a consequence of semantic complexity), or that grammar is not innate. See, for example, P. Thomas Schoenemann, "Syntax as an Emergent Property of Semantic Complexity," *Minds and Machines* 9 (1999): 309–46.

17. For a famous statement of our point 2, see Saul Kripke, *Naming and Necessity* (Cambridge MA: Harvard University Press, 1980). We are of course aware that certain behaviors or instincts are understood to be "innate" in the sense of biological. There may even be some grammatical features common to all human natural

languages that are innate. But such innateness does not imply *necessity*: the former is a product of the evolution of life on earth, and we need not imagine that the same conditions must hold in any possible world. The exclusive presence of levogyre amino acids in proteins and of dextrogyre sugars in genetic material may be a general rule for life here on earth, but no one thinks it is a necessary universal law.

18. As with nearly all subjects treated in this book, the literature on Leibniz's argument with Locke is vast, some of it written by characters who have appeared elsewhere in our pages, such as John Dewey's *Leibniz's New Essays Concerning the Human Understanding: A Critical Exposition* (Chicago: S. C. Griggs, 1888). For a more recent example see Jonathan Francis Bennett, "Knowledge of Necessity," in *Learning from Six Philosophers: Descartes, Spinoza, Leibniz, Locke, Berkeley, Hume* (Oxford: Clarendon Press, 2001), 2:34–58.

19. John Locke, *An Essay Concerning Human Understanding*, ed. Peter H. Nidditch (Oxford: Clarendon Press, 1975).

20. Locke's attack on "Primary notions" takes aim at some of the attempts we have encountered in previous chapters to provide a foundation of universal human agreement about sameness and difference, such as those of Aristotle (*Metaphysics*, 5.1013a18), Euclid's "common ideas" (κοιναὶ ἔννοιαι, Elements 1), and al-Fārābī, "Fusūl al-Madanī (Selected Aphorisms)," in *The Political Writings: Selected Aphorisms and Other Texts*, trans. C. E. Butterworth (Ithaca, NY: Cornell University Press, 2015), §34, 28–29.

21. John Locke, *Of the Conduct of the Understanding* (Oxford: Clarendon, 1901), §7, 23. The work was originally published in 1706.

22. Locke explores what we can know, and what we can know with certainty, in book 4, which begins with his definition of knowledge (4.1.1). Note that for Locke we can know of God's existence only with the second highest degree of confidence, namely, that of demonstration. To our own existence the highest degree of certainty applies. Locke did hold our knowledge of mathematical and some moral truths to be certain because they are ideal models produced within our minds that things in the world must fit rather than types external to us that we can only grasp inadequately.

23. A fuller history would explore the similar (and earlier) claims of Thomas Hobbes in his *Leviathan*, bk. 1.

24. "Those who take this view of the soul are treating it as fundamentally corporeal." G. W. Leibniz, *Sämtliche Schriften und Briefe*, ed. Deutsche Akademie der Wissenschaften zu Berlin (Darmstadt: O. Reichl; Leipzig: Koehler und Amelang, 1938; Berlin: Akademie, 1923–), ser. 6, 6:110; *New Essays on Human Understanding*, trans. Peter Remnant and Jonathan Bennett (Cambridge: Cambridge University Press, 1981), 110.

25. Locke expressed trenchant opposition to "innate" or "engraven" truths in Essay 1.2.9 and 1.3.13. Leibniz, on the other hand, does not hide his allegiance to Plato on the origin of ideas: *Sämtliche Schriften*, ser. 6, 6:48; *New Essays*, 48.

26. Leibniz, *New Essays*, 80.

27. The passage continues, "All other truths are reduced to first truths with the aid of definitions or by the analysis of concepts; in this consists proof a priori, which is

independent of experience." The translation is from *Leibniz: Philosophical Papers and Letters*, 2nd ed., ed. and trans. Leroy E. Loemker (Dordrecht: D. Reidel, 1989), 267.

28. Years later in his fifth letter to Clarke (1716), Leibniz would criticize the sameness truths of mathematicians in terms we would endorse more generally as applying perhaps even to his own thought: "the Mind, not contented with an Agreement, looks for an Identity, for something that should be truly the same." Leibniz, *Die philosophischen Schriften*, ed. C. I. Gerhardt (Hildesheim: Olms, 1960–1961), 7:401.

29. G. W. Leibniz and Samuel Clarke, *Correspondence*, ed. Roger Ariew (Indianapolis, IN: Hackett, 2000), 7. The first edition: *A Collection of Papers Which Passed between the Late Learned Mr. Leibnitz, and Dr. Clarke, in the Years 1715 and 1716. Relating to the Principles of Natural Philosophy and Religion* (London: James Knapton, 1717). On the correspondence see Ezio Vailati, *Leibniz and Clarke: A Study of Their Correspondence* (New York: Oxford University Press, 1997).

30. Like Descartes and many others (but unlike Locke), Leibniz also develops arguments for the existence of God. See, for example, *That a Most Perfect Being Exists* (*Quod ens perfectissimum existit*) of 1676, many of whose arguments depend on sameness/difference assertions. Here Leibniz defines a "perfection" as a "simple quality which is positive and absolute, or which expresses without any limits whatever it does express," so that it cannot be divided or enclosed within limits and cannot therefore be in any way inconsistent with any other perfection. Leibniz then used modal logic to prove that if a "necessary being is possible, it follows that it exists actually"; therefore, God exists. *Sämtliche Schriften*, ser. 6, 3:578, 583; Leibniz, *De summa rerum: Metaphysical Papers, 1675–1676*, ed. and trans. G. H. R. Parkinson (New Haven, CT: Yale University Press, 1992), 101, 107. Leibniz offers another argument for God's existence in his *Monadology*, secs. 36–45.

31. Leibniz and Clarke, *Correspondence*, letter 4, 22. The second formulation (*solo numero*) is adapted from the negative formulation in *Sämtliche Schriften*, ser. 6, 4:1541; *G. W. Leibniz: Philosophical Essays*, ed. and trans. Roger Ariew and Dan Garber (Indianapolis, IN: Hackett, 1989), 42. For time and space, see the *New Essays*, 230: "time and place do not constitute the core of identity and diversity." See also "First Truths," 268 (citing and generalizing Thomas Aquinas). For an extensive treatment of the philosophical context and stakes, see R. Rodriguez-Pereyra, *Leibniz's Principle of Identity of Indiscernibles* (Oxford: Oxford University Press, 2014), and specifically 104–17 on the correspondence with Clarke.

32. *Discourse on Metaphysics*, sec. 9, in *Leibniz: Philosophical Papers and Letters*, 308. We should note that this is not Leibniz's only derivation of the Principle of the Identity of Indiscernibles. Leibniz's multiple formulations are quite diverse in justification, scope, and modal strength. See Anja Jauernig, "The Modal Strength of Leibniz's Principle of the Identity of Indiscernibles," *Oxford Studies in Early Modern Philosophy* 4 (2008), 191–225.

33. On the relationship between his mathematics and his metaphysics, see Christia Mercer, "Leibniz on Mathematics, Methodology, and the Good: A Reconsideration of the Place of Mathematics in Leibniz's Philosophy," *Early Science and Medicine* 11 (2006), 424–54.

34. Many of Leibniz's writings on the continuum have been conveniently gathered in *The Labyrinth of the Continuum: Writings on the Continuum Problem, 1672-1686*, ed. and trans. Richard T. W. Arthur (New Haven, CT: Yale University Press, 2002).

35. Leibniz, *Specimen Dynamicum*, in Leibniz, *Mathematische Schriften*, ed. C. I. Gerhardt (Hildesheim: Georg Olms, 1963), 6:243, translated in Leibniz, *Philosophical Essays*, 126.

36. The application of logic to the imagination of what any plurality of worlds must look like is as fine an example of philosophical hubris as any. For a prominent example of how these ideas serve foundational arguments in contemporary analytic philosophy, see David Lewis, *On the Plurality of Worlds* (Oxford: Blackwell, 1986), 90.

37. On Leibniz's metaphysics, Christia Mercer, *Leibniz's Metaphysics: Its Origins and Development* (Cambridge: Cambridge University Press, 2001), is a valuable guide. On Leibniz's Platonism, see pages 173-204. For Leibniz's Aristotelianism, see her "Leibniz, Aristotle, and Ethical Knowledge," in *The Impact of Aristotelianism on Modern Philosophy*, ed. Riccardo Pozzo (Washington, DC: Catholic University of America Press, 2004), 113-47.

38. David Hume, *A Treatise of Human Nature*, 2nd ed., ed. L. A. Selby-Bigge and rev. P. H. Nidditch (Oxford: Clarendon Press, 1975), 3 (1.1.1).

39. Hume, 4.

40. For the "Copy Principle" see Don Garrett, *Cognition and Commitment in Hume's Philosophy* (New York: Oxford University Press, 1997), 41-57. For a treatment of Hume's understanding of ideas in relation to Locke's (and Descartes's), see David Owen, *Hume's Reason* (Oxford: Oxford University Press, 1999), 62-82.

41. Hume would have rejected this choice of word. *Mechanical* for him was just as Descartes conceived it, that is, the interchange of momentum between parts *in contact* with one another, as in a clockwork. Newton, on the contrary, had based his revolution in astronomy on gravity, a force *acting at a distance* between two point masses whose intensity is proportional to both masses and to the inverse square of their distance from each other. Despite his efforts, Newton's system came to be widely characterized as "mechanical," just as Hume's did.

42. *The Principia: Mathematical Principles of Natural Philosophy; A New Translation*, trans. I. B. Cohen and Anne Whitman (Berkeley: University of California Press, 1999), 382.

43. Hume states his emulation of Newton in a number of places, including *An Enquiry Concerning Human Understanding*, 1.15, 14. See also his *An Enquiry Concerning the Principles of Morals*, 1.10, 173-74. Page references are to David Hume, *Enquiries Concerning Human Understanding and Concerning the Principles of Morals*, 3rd ed., ed. L. A. Selby-Bigge, rev. P. H. Nidditch (Oxford: Clarendon Press, 1975).

44. Hume's association was a basic force, like Newton's gravitation in the sense that one should not hypothesize about its causes ("hypotheses non fingo," in Newton's words), since these could not be explained except as "original qualities of human nature, which I pretend not to explain." *A Treatise of Human Nature*, 2nd ed., ed. L. A. Selby-Bigge, rev. P. H. Nidditch (Oxford: Clarendon Press, 1975), 1.1.4.6, 12-13. The forces of association he found to be of three types: resemblance, conti-

guity in time or place, and causation. Note that all of these are types of sameness. In the first two cases this is obvious. As for causation, it was for Hume an idea derived exclusively from the experience of the "constant conjunction" of two perceptions. *Treatise of Human Nature*, 1.3.3–6, 78–94, passim. Causation is what links our past and present experiences to our expectations of the future. For Hume, all reasoning about matters of fact is built from this relation of cause and effect: *An Enquiry Concerning Human Understanding*, 4.1.4, 26.

45. Hume's statement of the question comes in the introduction to the *Treatise of Human Nature*, xiii–xix, for example, at xvi. "Airy sciences": *An Enquiry Concerning Human Understanding* 1.12, 12. "A compleat system": *Treatise of Human Nature*, Introduction, xvi.

46. For an example of this type of argument, see *Treatise of Human Nature*, 1.2.2, 32, on the impossibility of infinite division of extension and the existence of points. Hume uses mental possibility to restrict knowledge within certain theoretical horizons, but he could just as well have used it to expand beyond those horizons. Compare Al-Ghazālī on creation ex nihilo: since we can *imagine* the possibility that something comes from nothing, the possibility is real. On Hume and infinity, see Dale Jacquette, *David Hume's Critique of Infinity* (Leiden: Brill, 2001), 34 ff. Though it would today be difficult to imagine a philosopher who rejects the possibility of infinity, the question of what "belief" in infinite sets might mean remains an open one. See, for example, George Boolos, "Must We Believe in Set Theory?," in *Logic, Logic, and Logic*, ed. Richard Jeffrey (Cambridge, MA: Harvard University Press, 1998), 120–32.

47. Samuel Taylor Coleridge, *Biographia Literaria*, in *The Complete Works*, ed. William Shedd (New York: Harper & Brothers, 1858), 222.

48. While this dependence may have inclined them toward certain positions we are criticizing (such as a foundational identity), it also may have disinclined them from other positions we would also criticize (such as determinism). Some philosophical movements have continued to reach back to the medieval Thomist tradition as an antidote to perceived philosophical excesses of logic and number. In the late twentieth century, for example, Peter Geach drew on scholastic discussions of the Trinity in order to develop a critique of strict identity. For an accessible and brief example of his approach, see Peter Geach, "Truth and God," *Proceedings of the Aristotelean Society*, Supplement 56 (1982): 83–97.

49. The words are from Leibniz's fifth letter to Clarke, §47, in *Die philosophischen Schriften*, 7:401. We quote from the translation of Matthew L. Jones, "Space, Evidence, and the Authority of Mathematics," *Routledge Companion to Eighteenth Century Philosophy*, ed. Aaron Garrett (London: Routledge, 2014), 211.

50. Immanuel Kant, *Critique of Pure Reason*, A727/B755, trans. Paul Guyer and Allen W. Wood (Cambridge: Cambridge University Press, 1998), 637 (all citations are to this translation). George Berkeley had made a similar point a half century earlier in *On Motion*, §72. On these debates, see especially Matthew L. Jones, "Space, Evidence, and the Authority of Mathematics," 203–31, from whence we have drawn the quote from Leibniz in the previous paragraph as well as the citations in this one.

51. It is for this reason that we here neglect the equally influential Benedict/Baruch

Spinoza, who though just as concerned with the logic and mathematics of his day took a distinctive approach to sameness and difference.

52. "Mere operation of thought": *Enquiry Concerning Human Understanding*, 4.1.1, 25. Triangles are an often-given example by many of these thinkers. See, for example, John Locke, *Essay*, 4.2.2, 4.4.6, 4.13.3; Leibniz, *New Essays*, 446–47.

53. Hume, *Treatise of Human Nature*, 1.3.1, 71. Note the important distinction Hume makes between arithmetic and geometry: many in his day would not have demoted geometry in this way. The "precise standard" he describes is called a one-to-one correspondence between two sets, and it is the definition of equality of number, or of cardinality. Dealing with it in his *Die Grundlagen der Arithmetik* (Breslau: Wilhelm Koebner, 1884), §73, Frege quotes Hume, and this definition has come to be called "the Hume Principle." See George Boolos, *Logic, Logic, and Logic*, ed. Richard Jeffrey (Cambridge, MA: Harvard University Press, 1998), 301–14. The "standard" or definition is, however, much older, and Galileo discusses it in his dialogue *Two New Sciences* (1638). For an acute exploration of the differences between the logical underpinnings of arithmetic and geometry, see two classic essays by Carl Hempel, "Geometry and Empirical Science," *American Mathematical Monthly* 52 (1945): 7–17; and "On the Nature of Mathematical Truth," *American Mathematical Monthly* 52 (1945), 543–56.

54. Note the difference between contradiction and falsehood or error. For Hume it is a contradiction to deny a priori truths dependent only on the "operations of thought," for example, to say that 2 + 2 is not equal to 4, or that the interior angles of a Euclidean triangle do not add up to 180°. It is a falsehood or error to deny truths in the world, for example, to say that the Mediterranean is larger than the Atlantic.

55. Bertrand Russell, *The Problems of Philosophy* (London: Williams & Norgate, 1912), 85, quoted in Ian Hacking, "What Mathematics Has Done to Some and Only Some Philosophers," in *Mathematics and Necessity: Essays in the History of Philosophy*, ed. Timothy Smiley (Oxford: Oxford University Press on behalf of the British Academy, 2000), 84. Hacking's essay provides, among other things, a philosophical lexicon of how a selection of philosophers used the terms of art central to the question we are exploring here: terms like *a priori, necessary, analytic, inconceivable, certainty, apodictic*.

56. Immanuel Kant, *Prolegomena to Any Future Metaphysics*, trans. James W. Ellington (Indianapolis, IN: Hackett, 1977), 5.

57. Much of the mathematics that Kant considered a priori (e.g., Galileo and Newton's rational mechanics) philosophers of mathematics today might consider empirical: see Hacking, "What Mathematics Has Done," 88. We should not forget that Kant's early work was itself intended to contribute to the natural sciences of his day: among his first books was a *Universal Natural History and Theory of the Heavens* (1755).

58. Kant, *Critique of Pure Reason*, B11, 107. Compare Ayn Rand, whose hero declares that "the noblest act you have ever performed is the act of your mind in the process of grasping that two and two make four": *Atlas Shrugged: 35th Anniversary Edition* (New York: Signet, 1992), 969.

59. No further mark: "Kein noch näheres Merkmal der Wahrheit," *Critique of Pure Reason*, A 260, B 316, 366.

60. For a negative judgment on Kant's grasp on the mathematics of his time, see Paul Rusnock, "Was Kant's Philosophy of Mathematics Right for Its Time?" *Kant-Studien* 95 (2004): 426–42. For a much more positive judgment, see Michael Friedman, *Kant and the Exact Sciences* (Cambridge, MA: Harvard University Press, 1992), 95.

61. Definition of "pure" at *Critique of Pure Reason*, B24, 132. Three questions at *Critique of Pure Reason*, B 20, 146, posed as well in his *Prolegomena to any Future Metaphysics* of 1783. On the "reliable" and therefore presumably extendable example of mathematics, see *Critique of Pure Reason*, B8, 128.

62. *Critique of Pure Reason*, B10, 107. "*Mathematics* and *physics* are the two theoretical sciences which have to determine their objects *a priori*. The former is purely *a priori*, the latter is partially so, but is also dependent on other sources of knowledge other than reason." Kant's goal in asking these questions is not only to provide foundations to metaphysics but perhaps even to then ask whether there can be a true natural science that is not grounded in metaphysics. For a version of that question, see the excerpt from his *Physical Monadology* of 1756, cited by Friedman, *Kant and the Exact Sciences*, 2.

63. Aristotle, *De anima*, iii.3, 4, *On the Soul*, trans. Joe Sachs (Santa Fe, NM: Green Lion Press, 2004), 430a1–2, 141–42: "it [the intellect] is in potency in the same way a tablet is, when nothing written is present in it actively—this is exactly what happens to the intellect." Diogenes Laertius, 7, 46: "stamped and imprinted," *enapesphragismenēn kaì enapomemagmemēn*. Note the formidable finality of the perfect tenses of the passive participles with the double prefix, *en* and *apo*. The passage is cited in R. J. Hankinson, "Stoic Epistemology," in *The Cambridge Companion to the Stoics*, ed. Brad Inwood (Cambridge: Cambridge University Press, 2003), 60. For medieval models of the soul as wax and the divine as seal, see among the works of Brigitte Bedos-Rezak, her "Semiotic Anthropology. The Twelfth Century Experiment," in *European Transformations: The Long Twelfth Century*, ed. Thomas F. X. Noble and John Van Engen (Notre Dame, IN: University of Notre Dame Press, 2012), 426–67. On the powerful influence of new mediatic technology such as movies, cameras (both still and video), flashbulbs, and so forth, on how modern Americans think about memory, see Alison Winter, *Memory: Fragments of a Modern history* (Chicago: University of Chicago Press, 2012).

64. Kant credits Thales with discovering the apriorism of mathematics that is so important to Kant's own arguments, a discovery he ranks higher than those of Magellan's voyages: *Critique of Pure Reason*, B11–12, 107–8. First Copernicus and now Magellan. Kant clearly liked to compare himself favorably to the revolutionaries of the natural sciences.

65. Pascal, *Pensées*, ed. Léon Brunschvicg (Paris: Hachette, 1897), nos. 72, 425, and 434.

66. *Critique of Pure Reason*, B 40, 175. On the importance of the distinction between concepts and intuitions in Kant's understanding of the mathematical sciences, see

Friedman, *Kant and the Exact Sciences*, 96–135. Again, compare Rusnock, "Was Kant's Philosophy of Mathematics Right for Its Time?," for example at 428.

67. Kant, *Critique of Pure Reason*, A 38–39, B 55–6, 183. Kant died a half century before the publication of Riemann's non-Euclidean geometries. On their reception by philosophers like Frege (who compared them to alchemy) see Hans Freudenthal, "The Main Trends in the Foundations of Geometry in the 19th Century," *Studies in Logic and the Foundations of Mathematics* 44 (1966): 613–21.

68. For an excellent example of such a later philosopher see W. v. O. Quine, "Two Dogmas of Empiricism." *Philosophical Review* 60 (1951): 20–43.

69. As he put it in the paragraph before this one (A37, B54), if someone, myself or another being (*ein ander Wesen*), could look into myself without the space-time conditioning of our sensibility (*Sinnlichkeit*), he or she or it would see *something* timeless and unchanging.

70. *Critique of Pure Reason*, A263, B319, 368. Emphasis in original, but we have translated the Latin terms. The German for identity and difference is *Einerleiheit und Verschiedenheit*.

71. In suggesting that Kant put pebbles of sameness at the foundations of the mind, we don't mean to imply that he advocated for a mathematical psychology. On Kant's stance that psychology could not be mathematicised (or rather, that geometry could not be applied to it) and the relationship of his view to the chemistry of his time, see Thomas Sturm, "Kant on Empirical Psychology," in *Kant and the Sciences*, ed. Eric Watkins (Oxford: Oxford University Press, 2001), 163–80. See also Philip Kitcher, *Kant's Transcendental Psychology* (New York: Oxford 1990), 11. Kitcher discusses Kant's position, in the *Metaphysical Foundations of Natural Science*, that psychology can be neither an exact science (because the phenomena it studies cannot be expressed mathematically) nor an experimental one (because thoughts cannot be isolated from one another in order to be counted).

72. Connectedness as necessary for unity is related to the Principle of Locality in physics, whereby any interaction between two points must be mediated by *something* (e.g., a field like gravitation) that travels through the space between the two points. The consensus today is that quantum entanglement violates the Principle of Locality.

73. Of course Kant might object that musical notes are not the kinds of objects he had in mind here, that physics has nothing to do with art, and that critiques of reason and of aesthetics are entirely separate things. We would reply that, as with his boundaries between philosophy and mathematics, Kant's boundaries between reason and aesthetics are also full of assumptions and commitments that need not be universally shared.

74. W. Kaufmann, *Discovering the Mind: Goethe, Kant, and Hegel* (New York: McGraw Hill, 1980), 83.

75. See, for example, Heidegger's critique of Kant in Martin Heidegger, *Basic Problems of Phenomenology*, trans. Albert Hofstadter (Indianapolis: Indiana University Press, 1982), 77 ff.

76. H. Poincaré, *La Science et l'hypothèse* (Paris: Flammarion, 1917), 24. Poincaré was

characteristically prudent about any extension of this principle: "L'induction, appliquée aux sciences physiques, est toujours incertaine, parce qu'elle repose sur la croyance à un ordre général de l'Univers, ordre qui est en dehors de nous. L'induction mathématique, c'est-à-dire la démonstration par récurrence, s'impose au contraire nécessairement, parce qu'elle n'est que l'affirmation d'une propriété de l'esprit lui-même." Compare a little later: "On est donc forcé conclure que cette notion a été créée de toutes pièces par l'esprit, mais que c'est l'expérience qui lui en a fourni l'occasion." (35) The principle was first used, without explicitly stating it, by Blaise Pascal in his *Traité du triangle arithmétique* (Paris: Guillaume Desprez, 1665).

77. See George Boolos, "Introductory Note to Kurt Gödel's 'Some Basic Theorems on the Foundations of Mathematics and their Implications," in *Logic, Logic, and Logic*, ed. Richard Jeffrey (Cambridge, MA: Harvard University Press, 1998), 105–19.

78. Stephen Hawking, "Unified Theory—Stephen Hawking at European Zeitgeist 2011," posted by Google Zeitgeist, May 18, 2011, video, 26:04, https://www.you tube.com/watch?v=r4TO1iLZmcw. Hawking's argument is that philosophers were previously justified in considering significant questions about the nature of reality, but now philosophy is dead because it has not assimilated recent discoveries made in the field of physics.

79. *Critique of Pure Reason*, B14–16, 144. In this section, perhaps following Aristotle, Kant treats the Principle of Non-Contradiction as basic to logic and mathematics and the Principle of Identity as derived from it. Given our focus on sameness and difference, in the next chapter we will stress the place of the Principle of Identity in meeting the (non a priori) needs of number.

Chapter Six

1. "The Metaphysician" in H. L. Mencken, *A Mencken Chrestomathy* (New York: Vintage, 1982), 13–14.

2. And not only human. Some theoretical biologists describe recognition of same/ different as a most basic biological function, the basis of the "autopoeisis" through which the organism takes shape. See Humberto R. Maturana and Francisco J. Varela, *Autopoesis and Cognition: The Realization of the Living* (Dordrecht: D. Reidel, 1980).

3. Nader el-Bizri, ed. and trans., *Epistles of the Brethren of Purity: On Arithmetic and Geometry; An Arabic Critical Edition and English Translation of Epistles 1 & 2* (Oxford: Oxford University Press, 2012), 73.

4. Johann Wigand, *De neutralibus et mediis libellus* (1562), cited in F. P. Wilson, *The Oxford Dictionary of English Proverbs*, 3rd ed. (Oxford: Clarendon Press, 1970), 849. Alfred North Whitehead, *An Introduction to Mathematics* (New York: Henry Holt, 1911), 9.

5. Nor will we cheat by invoking other number structures differently defined, like integers modulo three, where two plus two is equal to one.

6. In this we are following the example of countless predecessors, including all of those mentioned thus far in this chapter, as well as Kant, whose example of

"7 + 5 = 12" concluded our previous chapter. Our method and conclusions, however, are different.

7. Compare the famous discussion about "distinguishing the one, the two, and the three," that is, number and counting, in Plato's *Republic*, 522c–526c.

8. T is here following Gottlob Frege, *Die Grundlagen der Arithmetik* (Breslau: Wilhelm Koebner, 1884), 5–8 (§§5–6); *The Foundations of Arithmetic*, trans. J. L. Austin (Evanston, IL: Northwestern University Press, 1978), 5–8 (§§5–6).

9. See, for example, in chap. 1, §2, of H. Poincaré's *La Science et l'hypothèse* (Paris: Flammarion, 1917), a proof that 2 + 2 = 4 in which it is assumed that our *italic* and roman numerical expressions are the same, and then the general Associative Property of the Sum is proved by induction, or recurrence. Mathematical induction is, according to Poincaré, the only statement to which the Kantian category "synthetic a priori" may be applied (Poincaré, *La Science et l'hypothèse*, 12–13).

10. The Zermelo-Fraenkel axioms are today a standard form of axiomatic set theory, and as such the most common foundation of mathematics. For a list of the axioms and an introduction to them, any textbook of set theory can serve. See, for example, Kenneth Kunen, *Set Theory: An Introduction to Independence Proofs* (Amsterdam: North-Holland, 1980); or Olivier Daiser, *Einführung in die Mengenlehre*, 2nd ed. (Springer: Berlin, 2000).

11. Our teacher might be mortified to realize that she has just quoted William Wordsworth, *The Prelude, or Growth of a Poet's Mind* (Oxford: Oxford University Press, 1933), 90 (lines 186–87). "In verity, an independent world / Created out of pure Intelligence."

12. Such as, for example, inaccessible cardinals. See, inter alia, Akihiro Kanamori, *The Higher Infinite* (Berlin: Springer, 1994), introduction.

13. A phenomenon common in many languages called in English by many names, including *enantiosemy, antilogy, auto-antonym, contronym, Janus words, addad* (from the Arabic), as well as others.

14. We have not yet interrogated the commonsense idea of "things" as mutually exclusive or asked "what is a thing?" But we have at least described some of the conditions of possibility for applying number to things.

15. Martin Stone, "Interpretation and Pluralism in Law and Literature: Some Grammatical Remarks," in *Reason and Reasonableness*, ed. Riccardo Dottori (Münster: Lit Verlag Münster, 2005), 147.

16. Friedrich Nietzsche, *The Gay Science*, trans. Walter Kaufmann (New York: Vintage, 1974), §121, 177.

17. The transition from an Aristotelian-Linnaean approach based on "sameness" to a Darwinian one based on "difference" is a chapter in the scientific history of sameness and difference that Nietzsche attributed to Hegel's idea of development (*Gay Science* §357, 305) and that both Joel E. Cohen and Lorraine Daston have suggested to us we should more fully explore.

18. Given the speed with which a politics is nowadays ascribed to authors, we should point out that we are not associating "freedom" here with any politics but rather with an attitude toward thought: namely, the attitude that we are free to choose

our approach to sameness or difference and are not entirely constrained by the essential necessity of any one foundational sameness. This is not, we should add, a metaphysical claim about free will.

19. To our examples in the previous chapter we might add Thomas Hobbes, *De Corpore* (1655-1656), chap. 1, where ratiocination is defined as computation, and the latter by the two operations of collecting together and disaggregating. "*Ratiocination*, therefore, is the same with *addition* and *subtraction*." See *The Metaphysical System of Hobbes: In Twelve Chapters from "Elements of Philosophy Concerning Body," Together with Briefer Extracts from "Human Nature" and "Leviathan,"* ed. Mary Whiton Calkins (Chicago: Open Court, 1913), 7. See also *Leviathan*, chap. 5.

20. Gottlob Frege, "Über Sinn und Bedeutung," *Zeitschrift für Philosophie und philosophische Kritik*, n.s., 100 (1892): 25-50.

21. David Lewis, *Parts of Classes* (Oxford: Blackwell, 1991), 6.

22. The answer to this last can be found in Lewis's late essay, in which he attempts to derive all that is from the laws of physics: "How Many Lives Has Schrödinger's Cat?" *Australasian Journal of Philosophy* 82 (2004): 3-22.

23. The so-called *Poena cullei*, from Latin "penalty of the sack." The ancient Roman practice seems to have preferred snakes, dogs, and even monkeys to cats. For the Saxon practice, see the thirteenth-century compilation of law by Eike von Repgow and Johann von Buch's fourteenth-century commentary on it: Eike von Repgow, *Der Sachsenspiegel*, ed. August Lüppen and Friedrich Kurt von Alten (Amsterdam: Rodopi, 1970). Johann von Buch, *Glossen zum Sachenspiegel*, ed. Frank-Michael Kaufmann (Hannover: Hahn, 2002).

24. At the Courant Institute of New York University in 1963, Ricardo took Prof. Jack Schwartz's course on nonlinear functional analysis. Three doctoral students from Argentina — Héctor Fattorini, Horacio Porta, and Ricardo — were assigned the task of preparing the notes, which a few years later were published as a book. One day after class they met Gian-Carlo Rota, who had been Schwartz's doctoral student at Yale and was then at the Rockefeller Institute. The four repaired to McSorley's Tavern, where amid tobacco smoke and ale foam, Rota advised Ricardo to read Ortega y Gasset's *En torno a Galileo*. That book from 1933 preceded by two or three years Husserl's *The Crisis of European Sciences and Transcendental Phenomenology*, and both (Ricardo discovered later) were at the core of Rota's philosophical thought. Horacio Porta has written about these experiences in "In Memory of Jacob Schwartz," *Notices of the AMS* 62, no. 5 (2015): 488.

25. Gian-Carlo Rota, "*Fundierung* as a Logical Concept," *Monist* 72 (1989): 70-77. Husserl defines *Fundierung* in his *Logische Untersuchungen*, pt. 3, "On the Theory of Wholes and Parts." Husserl, *Logical Investigations*, trans. by J. N. Findlay from the 2nd German ed. (1913) (London: Routledge, 1970).

26. English: Honoré de Balzac, *Séraphîta* (1834) in *The Novels of Honoré de Balzac*, vol. 42 (New York: G. D. Sproul, 1895-1900), 290. French: Honoré de Balzac, *Séraphîta* (Paris: L'Harmattan, 1995), 119.

27. Carl A. H. Burckhardt, *Goethes Unterhaltungen mit dem Kanzler Friedrich von Müller* (Stuttgart: Cotta'sche, 1898), 187 (§213, Sunday, June 18, 1826).

28. The commonality does not, however, detract from the resultant truths. Here we

disagree with yet another writer, T. S. Eliot: "In a sense, I think, *all* significant truths are private truths; they must be made mine before they can be true for me. As they become common, they become either inarticulate (in gaining in density) or they become insignificant. 2 + 2 = 4. But is this a *truth* at all?" T. S. Eliot, "The Validity of Artificial Distinctions," *Times Literary Supplement*, May 30, 2014, 14 (a hitherto unpublished essay).

29. Friedrich Nietzsche, *Human, All Too Human: A Book for Free Spirits* trans. R.J. Hollingdale (Cambridge: Cambridge University Press, 1996), 1.19, 22.

30. Kurt Gödel, "What Is Cantor's Continuum Problem," in *Collected Works*, vol. 2, *Publications 1938–1974*, ed. Solomon Feferman (Oxford: Oxford University Press, 1990), 268.

Chapter Seven

1. Werner Heisenberg's report of his early 1926 Berlin meeting with Einstein is in "Die Quantenmechanik und ein Gespräch mit Einstein (1925–1926)," published as chap. 5 in Heisenberg's *Der Teil und das Ganze: Gespräche im Umkreis der Atomphysik* (Munich: R. Piper, 1969), 85–100. The phrase attributed to Einstein, "Erst die Theorie entscheidet darüber, was man beobachten kann," is at 92.

2. John Dewey, "Matthew Arnold and Robert Browning," *Andover Review*, August 1891, and collected with the title "Poetry and Philosophy" in *Characters and Events: Popular Essays in Social and Political Philosophy by John Dewey*, ed. Joseph Ratner (New York: Henry Holt, 1929), 1:3–17.

3. We disagree with the philosopher Gaston Bachelard's claim that the Identity Principle is "tranquilly fundamental for common sense" (du principe d'identité, si tranquillement fondamental pour la connaissance commune): G. Bachelard, *Le Rationalisme appliqué* (Paris: Presses Universitaires de France, 1949), 101–2. To the contrary, the Identity Principle should be a rock of scandal for everyone but the mathematician.

4. Francis Bacon, *Novum organum scientiarum*, bk. 1, aphorism 51, in *Works*, ed. James Spedding, Robert Leslie Ellis, and Douglas Denon Heath (Boston: Taggard and Thompson, 1864), vol. 1.

5. Émile Meyerson, *Identity and Reality*, trans. Kate Loewenberg (London: George Allen & Unwin, 1930), 282. The first French edition is *Identité et réalité* (Paris: F. Alcan, 1908). Meyerson himself came to understand his philosophy of science as based on an awareness of the choice between "the same" and "the different," an awareness he attributed to his reading of Plato's *Timaeus*. See his letter to Félicien Challaye from 1924 in Émile Meyerson, *Lettres françaises*, ed. Bernadette Bensaude-Vincent and Eva Telkes-Klein (Paris: CNRS, 2009), 109.

6. Rudolf Carnap, *The Logical Structure of the World*, trans. R. A. George (Chicago: Open Court, 2003), 9.

7. "Great concert" is from *Poesías de Fray Luis de León*, ed. A. C. Vega (Madrid: Saeta 1955), 473. To the poet's contrast between the eternal freedom of heaven and the dark prison of earth, compare Plotinus, *Enneads* 4.8.1.

8. Jerónimo Muñoz, *Libro del nuevo cometa* (Valencia: Pedro de Huete, 1573). For a modern edition bringing together several of Muñoz's writings on the subject, see

Libro del nuevo cometa, Valencia, Pedro de Huete, 1573; Littera ad Bartholomaeum Reisacherum, 1574; Summa del prognostico del cometa: Valencia, [. . .] *1578* (Valencia: Hispaniae Scientia, 1981).

9. In his prologue addressed to King Phillip II, Muñoz specifically indicts Plato's *Timaeus* as well as Aristotelian cosmology. He cleverly (given the theological stakes) implies that Moses and scripture spoke more truly than the philosophers when Genesis describes the initial space of creation as *tohu va bohu* (which he translates as chaos): fols. 2r–5r.

10. The literature on Galileo's conflicts with church authorities is vast. A brief starting point may be found in David Lindberg's "Galileo, the Church, and the Cosmos," in David Lindberg and Ronald Numbers, *When Science and Christianity Meet* (Chicago: University of Chicago Press, 2003). Disputes over the nature of comets loomed large in Galileo's conflicts as well. The texts on that controversy are conveniently gathered in Drake and O'Malley, trans., *The Controversy on the Comets of 1618: Galileo Galilei, Horatio Grassi, Mario Guiducci, Johann Kepler* (Philadelphia: University of Pennsylvania Press, 1960).

11. "Nam cum scientiae omnes nihil aliud sint quam humana sapientia, quae semper vna & eadem manet, quantumvis differentibus subjectis applicata, nec majorem ab illis distinctionem mutuatur, quam Solis lumen a rerum, quas illustrat, varietate." *Œuvres de Descartes*, ed. Charles Adam and Paul Tannery (Paris: Cerf, 1897), 10: 360. Translation from "Rules for the Direction of the Mind" in *The Philosophical Writings of Descartes*, trans. John Cottingham, Robert Soothoff, and Dugald Murdoch (Cambridge: Cambridge University Press, 1984), 1:9. The book was not published during his lifetime.

12. Whatever the truths of this formulation, it omits the possibility of alternate ways of understanding both the initial thing and everything else. We quote from Leibniz's letter to Des Bosses of 1710: "Quin imo qui unam partem materiæ comprehenderet, idem comprehenderet totum universum ob eandem περιχώρησιν quam dixi. Mea principia talia sunt, ut vix a se invicem develli possint. Qui unum bene novit, omnia novit." See *The Leibniz-Des Bosses Correspondence*, ed. and trans. Donald Rutherford and Brandon Look (New Haven, CT: Yale University Press, 2007), 188 (our translation). The Leibnizean term περιχώρησιν, here rendered by "mutual union and communion," is the object of scholarly debate: see Antonio Lamarra, "Leibniz e la περιχώρησις," *Lexicon Philosophicum* 1 (1985), Daphnet Digital Library, http://scholarlysource.daphnet.org/index.php/DDL/article/view/21/10.

13. The echo with Keats's negative capability is intended: John Keats, *The Complete Poetical Works and Letters of John Keats, Cambridge Edition* (London: Houghton, Mifflin, 1899), 277.

14. Plato, *Timaeus*, 49a5–6, 52a8, d3: space (χώρα) is the "receptacle (ὑποδοχή) of all becoming." Hence the pot is identical to itself.

15. This is what Husserl was doing in chap. 1 when he spoke of points as "spatio-temporal idealities."

16. For a succinct history of Euclid's definition of point, see Thomas L. Heath's comments in his edition of Euclid, *The Thirteen Books of Euclid's "Elements"* (Mineola, NY: Dover, 1956), 1:155–58.

17. Already Schopenhauer observed that we can appeal to the Principle of Sufficient Reason to argue that there can be no two individuals exactly alike, because there would otherwise be no sufficient reason why one of the individuals was in one place while the other individual was in another. *On the Fourfold Root of the Principle of Sufficient Reason and On the Will in Nature*, trans. Karl Hillebrand (London: George Bell and Sons, 1903).

18. David Hilbert, *Foundations of Geometry* (Chicago: Open Court, 1971), 3. Hilbert added the axioms of completeness ensuring, for example, that the two circles in Euclid's Proposition One do meet at two points.

19. On this subject generally see the essays collected in Jeremy Gray, ed., *The Symbolic Universe: Geometry and Physics 1890-1930* (Oxford: Oxford University Press, 1999). Jonathan Swift, *The Benefit of Farting Explain'd* (1722) (Exeter: Old Abbey Press, 1996), 9. Paul Valéry, "La Crise de l'esprit, deuxième lettre" (1919) in *Œuvres* (Paris: Gallimard, 1957), 1:1412-15: "La Grèce a fondé la géométrie. C'était une entreprise insensée : nous *disputons* encore sur la *possibilité* de cette folie." H. Poincaré, "L'Espace et la géométrie," *Revue de métaphysique et de morale* 3 (1895): 638. The physicist Erwin Schrödinger made a similar point in the third of his William James Lectures of 1954, treating this extrapolation of habit and experience into science as an aspect of human nature he called "forming invariants." See Erwin Schrödinger, *The Interpretation of Quantum Mechanics: Dublin Seminars (1949-1955) and Other Unpublished Essays*, ed. Michel Bitbol (Woodbridge, CT: Oxbow Press, 1995), 146.

20. Isaac Newton, The *Principia: Mathematical Principles of Natural Philosophy; A New Translation*, trans. I. B. Cohen and Ann Whitman (Berkeley: University of California Press, 1999), law 3, 417. By the motion of a body Newton means the product of its mass times its velocity, its "momentum," as we say.

21. Ludwig Boltzmann, *Theoretical Physics and Philosophical Problems* (Dordrecht: D. Reidel, 1974), 228-31. The citation is from 228, emphasis added. A similar point is made by Leibniz in *New Essays on Human Understanding*, trans. Peter Remnant and Jonathan Bennett (Cambridge: Cambridge University Press, 1981), preface, 57-58.

22. "No variation" is from Maxwell, *Theory of Heat* (London: Longmans, Green, 1872), 330-32. It is worth pointing out that Maxwell is here, among other things, marking distance from Charles Darwin's *On the Origin of Species* (1859). He is emphasizing a distinction between physics, in which there is no difference within species, and Darwin's biology, in which difference and variability within species drive evolution. On the discoveries of spectroscopy in Maxwell's day, see W. McGucken, *Nineteenth Century Spectroscopy: Development of the Understanding of Spectra, 1802-1897* (Baltimore: Johns Hopkins University Press, 1969). See also Peter Pešić, *Seeing Double: Shared Identities in Physics, Philosophy, and Literature* (Cambridge, MA: MIT Press, 2002), 82-84.

23. More recent philosophers have tried to solve the problem by attacking only one of those "laws of thought," Leibniz's "Identity of Indiscernibles." See, for example, Max Black, "The Identity of Indiscernibles," *Mind* 61 (1952): 153-64, reprinted in Max Black, *Problems of Analysis* (Ithaca, NY: Cornell University Press, 1954), 204-16, and in J. Kim and E. Sosa, eds., *Metaphysics: An Anthology* (London: Blackwell,

1999), 66–71. We have many quibbles with Black's approach, which seems to us in any event irrelevant to our problem, since Black grants that his model requires the absence of any observer. By excluding the possibility of an observer Black gives up his purchase on our problem: if one wants to say something about discernibility one must assume the presence, or at least the possibility, of some discerning.

24. Boltzmann, *Theoretical Physics*, 230–31. Leibniz, *New Essays*, preface, 56.

25. Meyerson, *Identity and Reality*, 384.

26. Einstein's changing sense of the relationship between theory and experience is addressed in Erhard Scheibe, "Albert Einstein: Theorie, Erfahrung, Wirklichkeit," *Heidelberger Jahrbücher* 36 (1992): 121–38, here 125. The question of whether or not Leibniz's Principle of the Identity of Indiscernibles applies at the quantum level remains a pressing one in the philosophy of physics. For a review of shifting positions over the past decade (from a consensus that it does not to proposals for how it could), see Adam Caulton's "Discerning 'Indistinguishable' Quantum Systems," *Philosophy of Science* 80 (2013), 49–72; and "Issues of Identity and Individuality in Quantum Mechanics" (PhD diss., Cambridge University, 2015). See also Adam Caulton and Jeremy Butterfield, "On Kinds of Indiscernibility in Logic and Metaphysics," *British Journal for the Philosophy of Science* 63 (2012): 27–84.

27. Pešić explains that importance in *Seeing Double*, 64–77. Though we are using Gibbs to make a different (albeit related) point in this section, Pešić treats the problem we are about to take up much more briefly in this chapter: the challenges posed to classical ideas of individuality, distinguishability, and identity at the quantum level. See also his more extended treatment of Gibbs in Peter Pešić, "The Principle of Identicality and the Foundations of Quantum Theory I. The Gibbs Paradox," *American Journal of Physics* 59 (1991): 971–74.

28. Gibbs's report can be found in his *Elementary Principles in Statistical Mechanics* (Mineola, NY: Dover, 1960), 187–207. Readers interested in the relationship of Boltzmann's and Gibbs's statistical mechanics may turn to Carlo Cercignani, *Ludwig Boltzmann: The Man Who Trusted Atoms* (Oxford: Oxford University Press, 1998/2006), chap. 7.

29. Edwin T. Jaynes, "The Gibbs Paradox," in *Maximum Entropy and Bayesian Methods*, ed. C. R. Smith, G. J. Erickson, P. O. Neudorfer (Dordrecht: Kluwer Academic, 1992), 1–22.

30. In this sense, we can never achieve the ancient view of objectivity so starkly stated in Thomas Aquinas's principle *veritas est adaequatio rei et intellectus*. *De veritate*, Cb.1, *respondeo*. (Aquinas attributed this statement to the Andalusī Jewish philosopher Isaac Israeli's dialogue *Fons vitae*.)

31. E. T. Jaynes, "Gibbs vs Boltzmann Entropies," *American Journal of Physics* 33 (1965): 398. This earlier article contains much of the general framework for the 1992 piece we have been discussing.

32. Wigner seems not to have asked himself whether there are *any* concepts that are *not* anthropomorphic at some historical era of *anthropos* taken in both its mental as well as its physical sense.

33. Max Planck, *Acht Vorlesungen über Theoretische Physik, gehalten an der Columbia*

University, NY (Leipzig: S. Hirzel, 1910); translation in: *Eight Lectures on Theoretical Physics*, trans. A. P. Wilson (New York: Columbia University Press, 1915), 3; reprinted (Mineola, NY: Dover, 1998). These lectures also make clear that Planck did not think much of the difficulties we have been outlining in the previous pages: "Mechanics requires for its foundation in principle nothing more than the concepts of space, of time, and of that which is moving, whether one considers this as a substance or a state" (Die Mechanik bedarf zu ihrer Begründung prinzipiell nur der Begriffe des Raumes, der Zeit, und dessen was sich bewegt, mag man es nun als Substanz oder als Zustand bezeichnen" (9).

34. Eugene P. Wigner, "The Unreasonable Effectiveness of Mathematics in the Natural Sciences," *Communications on Pure and Applied Mathematics* 13 (1960): 14. On the question of consciousness, see his *Philosophical Reflections and Syntheses*, ed. Jagdish Mehra and Arthur S. Wightman (Berlin: Springer, 1995), 14. Wigner's European background may have affected his epistemological interests. During a visit to the United States in 1929, Werner Heisenberg noted that American physicists seemed untroubled by the aspects of quantum theory (wave-particle duality, purely statistical character, etc.) that prompted philosophical debate among Europeans. Nancy Cartwright has suggested that the lack of "epistemological discussion" among US physicists was due to the influence of American pragmatism and operationalism. See Nancy Cartwright, "Philosophical Problems of Quantum Theory: The Response of American Physicists," in *The Probabilistic Revolution*, vol. 2, *Ideas in the Sciences*, ed. Lorenz Krüger, Gerd Gigerenzer, and Mary S. Morgan (Cambridge, MA: MIT Press, 1987), 417–35, Heisenberg's observation quoted at 418. For an example of a more recent manifesto about the epistemological and ontological challenges posed by quantum mechanics from adherents of one particular school, see C. A. Fuchs and Asher Peres, "Quantum Theory Needs No 'Interpretation,'" *Physics Today* 53 (2000): 70.

35. Many basic explanations of the two-slit experiment are available. See, for example, Richard Feynman, *Lectures on Physics* (Boston: Addison-Wesley, 1963), 3:1-1-1-11.

36. In terms of trigonometry: if I_1 and I_2 are two waves, the intensity I_{12} of the combined wave is given by $I_{12} = I_1 + I_2 + 2\sqrt{(I_1 I_2)} \cos \delta$, where δ is the phase difference between the two waves. This is a simple consequence of the "cosine theorem," which is itself a generalization of Pythagoras Theorem: in any triangle with sides a, b, c, we have $c^2 = a^2 + b^2 - 2ab \cos \delta$, where δ is the angle between a and b. For the application of complex numbers to this problem, see Feynman, *Lectures on Physics*, vol. 1, chap. 29, §5.

37. C. F. von Weizsäcker, "Zur Deutung der Quantenmechanik," *Zeitschrift für Physik* 118 (1941): 489–509, cited in X.-S. Ma, J. Kofler, and A. Zellinger, "Delayed-Choice Gedanken Experiments and Their Realization," *Reviews of Modern Physics* 88 (2016), article 015005, https://doi.org/10.1103/RevModPhys.88.015005.

38. "An unavoidable effect" is from J. A. Wheeler, "Law Without Law," in *Quantum Theory and Measurement* (Princeton, NJ: Princeton University Press, 1983). See also Wheeler, "The 'Past' and the 'Delayed-Choice' Double-Slit Experiment," in *Mathematical Foundations of Quantum Theory*, ed. A. R. Marlow (Cambridge, MA:

Academic Press, 1978). An accessible description of these experiments and their implications is given by Philip Ball in his introduction to the epistemological challenges of quantum physics, *Beyond Weird: Why Everything You Thought You Knew about Quantum Physics Is Different* (Chicago: University of Chicago Press, 2018), 92–95.

39. We say "as if" in order to avoid taking any position on the reasons for or ontological meaning of these results. For an early and important example of that disagreement, compare Albert Einstein, B. Podolsky, and N. Rosen, "Can Quantum-Mechanical Description of Physical Reality Be Considered Complete?," *Physical Review* 47 (1935), 777–80; with Niels Bohr's response by the same title in *Physical Review* 48 (1935): 696–702. Evolving technology continues to make possible new kinds of double-slit experiments with which to test competing theories of quantum behavior. For a recent one, see Lothar Schmidt, J. Lower, T. Jahnke, S. Schößler, M. S. Schöffler, A. Menssen, C. Lévêque, N. Sisourat, R. Taïeb, H. Schmidt-Böcking, and R. Dörner, "Momentum Transfer to a Free-Floating Double Slit: Realization of a Thought Experiment from the Einstein-Bohr Debates," *Physical Review Letters* 111 (2013). Though the experimental results are consistent, physicists continue to disagree sharply on their ontological implications. Compare, for example, the "realist" approach to delayed-choice experiments in Eduardo V. Flores, "A Model of Quantum Reality," unpublished, https://ui.adsabs.harvard.edu/abs/2013arXiv1305.621 9F/abstract, with the antirealist view of X. S. Ma et al., "Quantum Erasure with Causally Disconnected Choice," *Proceedings of the National Academy of Sciences* 110 (2013), 1221–26.

40. At time of writing, one of us (David) served on the board of governors of the Argonne National Laboratory, one of the leading sites of US investment in research in quantum computing and other "entanglement related" fields.

41. P. A. M. Dirac, *The Principles of Quantum Mechanics*, 2nd ed. (Oxford: Clarendon Press, 1947), 12.

42. Erwin Schrödinger, "July 1952 Colloquium," in *Interpretation of Quantum Mechanics*, 32. See also "Notes for the 1955 Seminar" on measuring velocities by the police or race course method, 113–14. The "one electron" thesis was proposed by John Wheeler to Richard Feynman in 1940. See also Peter Pešić, "The Principle of Identicality and the Foundations of Quantum Theory II: The Role of Identicality in the Formation of Quantum Theory," *American Journal of Physics* 59 (1991): 975–78.

43. Y. Aharonov, F. Colombo, S. Popescu, I. Sabadini, D. C. Strouppa, and J. Tollaksen, "Quantum Violation of the Pigeonhole Principle and the Nature of Quantum Correlations," *Proceedings of the National Academy of Sciences* 113 (2016): 532–35.

44. John von Neumann, *Mathematische Grundlagen der Quantenmechanik* (Berlin: Springer, 1932); *Mathematical Foundations of Quantum Mechanics*, trans. Robert T. Beyer (Princeton, NJ: Princeton University Press, 1955).

45. Compare, for example, the positions taken by Steven French and Décio Krause, *Identity in Physics: A Historical, Philosophical, and Formal Analysis* (Oxford: Oxford University Press, 2006) (on quasi sets see chap. 7, 272–319); Simon Saunders, "Indistinguishability," in *Oxford Handbook in Philosophy of Physics*, ed. R. Batter-

man (Oxford University Press, 2013), 340–80; and Adam Caulton, "Issues of Identity and Individuality in Quantum Mechanics" (PhD Thesis, University of Cambridge, 2015), and "Discerning 'Indistinguishable' Quantum Systems." See most recently A. S. Sant' Anna, "Individuality, Quasi-Sets and the Double-Slit Experiment," *Quantum Studies: Mathematical Foundations* (2019), https://link.springer.com/article/10.1007/s40509-019-00209-2. On quasi sets, see L. Dalla Chiara and G. Toraldo di Francia, "Individuals, Kinds and Names in Physics," in *Bridging the Gap: Philosophy, Mathematics, Physics*, ed. Giovanna Corsi, Maria Luisa Dalla Chiara, and Gian Carlo Ghirardi (Dordrecht: Kluwer Academic, 1993), 261–83; and "Identity Questions from Quantum Theory," in *Physics, Philosophy and the Scientific Community*, ed Kostas Gavroglu, John Stachel, and Marx W. Wartofsky (Dordrecht: Kluwer Academic, 1995), 39–46.

46. Lucretius, *On the Nature of Things*, trans. Martin Ferguson Smith (Indianapolis, IN: Hackett, 2001), 2.256, 41.

47. Lucretius's impact on the Renaissance has recently come to broad attention through Stephen Greenblatt's *The Swerve: How the World Became Modern* (New York: W. W. Norton, 2011). Ada Palmer provides a reconstruction of the Renaissance reception in *Reading Lucretius in the Renaissance* (Cambridge, MA: Harvard University Press, 2014). Michel Serres, *The Birth of Physics* (Manchester: Clinamen Press, 2000), 191. It will shortly become clear why we cannot subscribe to Serres's plea for liquidity, sympathetic though we are to his critique.

48. Carl G. Jung and Wolfgang Pauli, *The Interpretation of Nature and the Psyche* (Princeton, NJ: Princeton University Press, 1969). More interesting to our taste is the correspondence, published as *Atom and Archetype: The Pauli/Jung Letters (1932–1958)*, ed. C. A. Meier (Princeton, NJ: Princeton University Press, 2001). See pages 160–69 for Pauli's two dreams in the context of the nonconservation of parity experiments. Jung treats the dreams as examples of synchronicity.

49. "The Arithmetical Paradox: The Oneness of Mind," in Erwin Schrödinger, *Mind and Matter* (Cambridge: Cambridge University Press, 1958), 52–68, here 62. Compare from his slightly later *My View of the World* (Cambridge: Cambridge University Press, 1964), 17: "Shared thoughts, with several people really thinking the same thing . . . really are thoughts in common, and they are *single* occurrences; any numerical statement about how many of them there are, based on a count of the number of individuals engaged in thinking, is quite without meaning in respect of what is being thought."

50. "The Principle of Objectivation," in Schrödinger, *Mind and Matter*, 36–51. The continuation of the passage is significant: "The relatively new science of psychology imperatively demands living space, it makes it unavoidable to reconsider the initial gambit. This is a hard task, we shall not settle it here and now, we must be content at having pointed it out." A historian of philosophy should object that Schrödinger's use of the words *subject* and *object* is anachronistic. In Greek philosophy these terms were grammatical and were not used to represent "the relationship of the observer to the observed" before the medieval Latin philosophers. "The song seraphically free": George Meredith, "The Lark Ascending," lines 93–94.

51. David Bohm, *Wholeness and the Implicate Order* (London: Routledge, 2002), 12. Notice that Bohm associates the division with an atomist view of the universe as built out of solid bodies, a view that, after relativity, can no longer be defended.

52. *Bohm*, 147. Compare Bohm's dictum with our citation from Valéry's "Mauvaises pensées" in chap. 2 of this book.

53. Bohm, *Wholeness and the Implicate Order*, 20-21.

54. On Rheomode, see Bohm, *Wholeness and the Implicate Order*, 34-60, here 37.

55. Aristotle, *Physics*, 185b. Compare Nietzsche, *On the Genealogy of Morals*, trans. Walter Kaufmann and R. J. Hollindale (New York: Vintage, 1967), 1.13.

56. Hermann Weyl, *Gruppentheorie und Quantenmechanik* (Leipzig: S. Hirzel, 1928); *The Theory of Groups and Quantum Mechanics*, trans. H. P. Robertson (Mineola, NY: Dover, 1950).

57. Collected in Hermann Weyl, *Mind and Nature* (Princeton, NJ: Princeton University Press, 2009), 194-203.

58. Weyl, 203.

59. H. Weyl, *Philosophy of Mathematics and Natural Science* (New York: Atheneum, 1963), 113. Cf. our chap. 1 on Husserl's opposite views about synthetic versus analytic geometry and our comments on synthetic geometry in this chapter, "The Celestial Abode of Sameness."

60. Weyl, *Mind and Nature*, 199.

61. Weyl, 195.

62. Weyl, 202. Distinctions like this one were especially important in the influential vision of Hermann von Helmholtz (1821-1894) for the disciplinary organization of the German academy and in the writings of philosophers like Wilhelm Dilthey (1833-1911).

63. Weyl, *Mind and Nature*, 202-3.

64. W. H. Auden, "The Shield of Achilles," in *Selected Poems* (New York: Vintage, 1979), 198-200, lines 1-15.

65. Dewey, "Matthew Arnold and Robert Browning," *The Andover Review*, August 1891, under the title "Poetry and Philosophy," collected in *Characters and Events*, 3-17.

66. Richard Rorty, *Contingency, Irony, and Solidarity* (Cambridge: Cambridge University Press, 1989), 7-8.

67. Richard Rorty, *Philosophy as Poetry* (Charlottesville: University of Virginia Press, 2016), 3-4; "Pragmatism and Romanticism," in *Philosophy as Cultural Politics* (Cambridge: Cambridge University Press, 2007), 4:107.

68. Yves Bonnefoy, "Paul Valéry," in *L'Improbable et autres essais* (Paris: Gallimard, 1983), 99-105. (The essay first appeared in 1963.) Translated in *The Act and the Place of Poetry: Selected Essays by Yves Bonnefoy*, ed. and trans. John T. Naughton (Chicago: University of Chicago Press, 1989), 96-100. On Bonnefoy as poet of the ephemeral, see Richard Stamelman, *Lost Beyond Telling: Representations of Death and Absence in Modern French Poetry* (Ithaca, NY: Cornell University Press, 1990), chaps. 5-6. Czesław Miłosz, "Shestov, or the Purity of Despair," in his *Emperor of the Earth: Modes of Eccentric Vision* (Berkeley: University of California Press, 1977), 102, emphasis added.

69. Martin Heidegger, "Wozu Dichter" ("What Are Poets For?") was written in 1946 and published in *Holzwege* (Frankfurt am Main: Vittorio Klostermann,1950), translated by Albert Hofstadter in Martin Heidegger, *Poetry, Language, Thought* (New York: Harper Collins, 1971), 87–139, here 92. We don't know what Heidegger himself thought of Valéry's poetry. Heidegger did cite Valéry's 1919 letters on the crisis of Europe ("La crise de l'esprit," discussed in chap. 1 of this book) in his November 1945 "de-Nazification" petition to be admitted to teach as an emeritus professor at Freiburg, invoking Valéry as a French (non-Nazi) parallel to his own efforts to address "the crisis of the Western spirit" in those years. He alludes to the same work again in his 1959 lectures on "Hölderlins Erde und Himmel." See Martin Heidegger, *Gesamtausgabe*, vol. 16, *Reden und andere Zeugnisse eines Lebensweges*, ed. H. Heidegger (Frankfurt am Main: Vittorio Klostermann, 2000), 398; and vol. 4, *Erläuterung zu Hölderlins Dichtung*, ed. F.-W. von Herrmann (Frankfurt am Main: Vittorio Klostermann, 1996), 176. Both are cited by Morkore Stigel Hansen, "The Spirit of Europe: Heidegger and Valéry on the 'End of Spirit'" (PhD diss., London School of Economics and Political Science, 2017), 20.

70. For Musil's eulogy, see Robert Musil, *Tagebücher, Aphorismen, Essays und Reden*, ed. Adolf Frisé (Hamburg: Rowohlt, 1955), 885–98. Heidegger would not have approved of Musil's statement. For him, "no poet of this world era could overtake" Hölderlin, the "precursor of poets in a destitute time." Rilke represented, for Heidegger, the greatest poetry possible in this world era for a poet who remained within classical (i.e., non-Heideggerean) metaphysics. See "What Are Poets For?," in Heidegger, *Poetry, Language, Thought*, 126, 139.

71. Monique Saint-Hélier, *À Rilke pour Noël* (Berne: Éditions du Chandelier, 1927). On Rilke's translations of Valéry in this period, see Karin Wais, *Studien zu Rilkes Valéry-Übertragungen* (Tübingen: Max Niemayer, 1967). Rilke wrote to many people about his encounter with Valéry's work. On April 28, for example, he wrote to André Gide, and on November 26 to Gertrud Ouckama Knoop (to whose deceased daughter the *Sonnets to Orpheus* are dedicated): see Rilke, *Andé Gide: Correspondances 1909–1926* (Paris: Corrá, 1952), 151; and Rilke, *Briefe aus Muzot, 1921–1926* (Leipzig: Insel, 1936), 49–50. "Boundless storm" is from a letter dated February 11, 1922. Rilke to Andreas-Salomé, *Rainer Maria Rilke, Lou Andreas-Salomé: Briefwechsel* (Leipzig: Insel, 1952), 464. Compare "a nameless storm, a hurricane in the spirit," Rilke to Marie von Thurn und Taxis, February 11, 1922, *Letters of Rainer Maria Rilke*, vol. 2, trans. Jane Bannard Greene and M. D. Herter Norton (New Haven, CT: Yale University Press, 1947), 290.

72. For a physicist's brief criticism of poets, see Feynman, *Lectures on Physics*, vol. 1, lecture 3, 6n: "What men are poets who can speak of Jupiter if he were a man, but if he is an immense spinning sphere of methane and ammonia must be silent?"

73. We do worry that the word *truly* (*wahrhaft*) may hint at commitments similar to those we have criticized in Heidegger and others. On figures, see the 1938 study by Erich Auerbach, "Figura," included in *Scenes from the Drama of European Literature*, trans. Ralph Manheim (Minneapolis: University of Minnesota Press, 1984), 11–76.

74. In Hannah Vandegrift Eldridge, *Lyric Orientation: Hölderlin, Rilke, and the Poetics of Community* (Ithaca, NY: Cornell University Press, 2015), 164.

75. The German word *ergänzen* (to complete, add, supplement) is also a mathematical term, as in *quadratisch ergänzen* (complete the square). Characteristically, Heidegger discusses Rilke's fondness for spheres and centers only in order to dismiss the geometric content of the metaphor: "What Are Poets For?," in Heidegger, *Poetry, Language, Thought*, 120–21.

76. We are reminded of "Number there in love was slain," from Shakespeare's *The Phoenix and the Turtle*, a poem no less mysterious than this one of Rilke's. The only rendering we know of along these lines is the French translation by J.-F. Angelloz in Rainer Maria Rilke, *Les Elégies de Duino, Sonnets à Orphée* (Paris: Aubier, 1943), which has "anéantis le nombre."

77. Heidegger: "What Are Poets For?" in Heidegger, *Poetry, Language, Thought*, 124–26, and on atomic physics specifically, 109–10. Musil: "eine klare Stille in einer niemals anhaltenden Bewegung," is at *Tagebücher, Aphorismen, Essays und Reden*, 892. Valéry, "Mauvaises pensées," in *Œuvres* (Paris: Pléiade, 1960), 2:794.

Chapter Eight

1. John Stuart Mill, *The Collected Works of John Stuart Mill*, vol. 1, *Autobiography and Literary Essays*, ed. John M. Robson and Jack Stillinger with an introduction by Lord Robbins (Toronto: University of Toronto Press, 1981), 225–26.

2. F. A. Hayek, "The Use of Knowledge in Society," *American Economic Review* 35 (1945): 520.

3. Jean-Jacques Rousseau, *Œuvres complètes* (Paris: Gallimard, 1964), 3:419–20 (the citation is from *The Social Contract*). Bentham's "fundamental axiom" appears in the anonymously published *A Fragment on Government* (1776); modern edition: *A Comment on the Commentaries and A Fragment on Government*, ed. J. H. Burns and H. L. A. Hart (Oxford:Clarendon Press, 1977), 3. From this perspective, modern econometric institutions like the National Bureau of Economic Research in the United States can be considered Rousseau's children. On the aspirations of the NBER (but *sans* Rousseau), see Robert W. Fogel, Mark Guglielmo, and Nathanial Grotte, *Political Arithmetic: Simon Kuznets and the Empirical Tradition in Economics* (Chicago: University of Chicago Press, 2013).

4. Daniel Halévy, *Essai sur l'accélération de l'histoire* (Paris: Éditions Self/Les Îles d'Or, 1948), 97: "Les Encyclopédistes avaient rêvé d'une politique déduite à la manière des théorèmes d'Euclide, et menant l'homme, de conséquence en conséquence, à cet état d'équilibre que nous appelons, en termes de sentiment, bonheur. C'est cette idée, adoptée par les foules, qui va déterminer la combustion d'un peuple." For Condorcet, Keith Michael Baker provides a good starting point: *Condorcet: From Natural Philosophy to Social Mathematics* (Chicago: University of Chicago Press, 1975). On probability, reason, and the Enlightenment see especially Lorraine Daston, *Classical Probability in the Enlightenment* (Princeton, NJ: Princeton University Press, 1988).

5. On the difference between eighteenth- and nineteenth-century hopes for the application of probability to society, see Lorraine J. Daston, "Rational Individuals versus Laws of Society: From Probability to Statistics," in *The Probabilistic Revolution*,

vol. 1, *Ideas in History*, ed. Lorenz Krüger, Lorraine J. Daston, and Michael Heidelberger (Cambridge, MA: MIT Press, 1987), 295–304. "Avalanche": Ian Hacking, "Biopower and the Avalanche of Printed Numbers," *Humanities in Society* 5 (1982): 279. Joshua Cole provides a case study in *The Power of Large Numbers: Population, Politics, and Gender in Nineteenth-Century France* (Ithaca, NY: Cornell University Press, 2000).

6. Michel Foucault, "*Omnes et singulatim*: Towards a Criticism of 'Political Reason,'" in *The Tanner Lectures on Human Values*, ed. S. McMurrin and trans. P. E. Dauzat (Salt Lake City: University of Utah Press, 1981), 2:223–54. Karl Marx declares his goal "to lay bare the economic law of motion of modern society" in the preface to *Capital: A Critique of Political Economy* (New York: International, 2003), 1:10. See John P. Burkett, "Marx's Concept of an Economic Law of Motion," *History of Political Economy* 32 (2000): 381–94.

7. Our reformulation also corrects a mathematical "misunderstanding" in Bentham's aphorism pointed out by John von Neumann and Oskar Morgenstern, *A Theory of Games and Economic Behavior* (Princeton, NJ: Princeton University Press, 1944). Here and throughout we cite from the sixtieth anniversary edition (Princeton, NJ: Princeton University Press, 2004), 11: "A guiding principle cannot be formulated by the requirement of maximizing two (or more) functions at once."

8. For instance, Leonhard Euler, *Methodus Inveniendi Lineas Curvas Maximi Minimive Proprietate Gaudentes* (1744; reprinted in *Leonhardi Euleri opera omnia*, ser. 1, vol. 24, ed. C. Carathéodory, Zurich: Orell Füssli, 1952).

9. Pierre Louis Morceau de Maupertuis, "Les Loix de mouvement et du repos, déduites d'un principe de métaphysique," in *Histoire de l'Académie Royale des Sciences et des Belles Lettres* (Paris: Académie Royale des Sciences et des Belles Lettres, 1746), 267–94. For the history of the principle and its changes during the eighteenth century and beyond, see Herman H. Goldstine, *A History of the Calculus of Variations from the 17th through the 19th Century* (New York: Springer, 1980). Incidentally, Goldstine was a collaborator of von Neumann in the EDVAC computer project and the von Neumann computer architecture.

10. Joseph-Louis Lagrange, *Mécanique analytique* (1788–89), in *Œuvres*, vols. 11–12 (Paris: Gauthier-Villars et Fils, 1867–1892).

11. "Pleasure or pain": we borrow the expression from William Stanley Jevons, one of the authors of the marginal revolution in economics, "Brief Account of a General Mathematical Theory of Political Economy," *Journal of the Royal Statistical Society* 29 (1866): 282–87. For an example from aesthetics, see François Hemsterhuis's definition of the Beautiful in his *Lettre sur la sculpture* (Amsterdam: Marc Michel Rey, 1769), 5.

12. The original titles: *Sur l'homme et le développement de ses facultés, ou Essai de physique sociale; Sur la statistique morale et les principes qui doivent en former la base.* Quetelet is extensively discussed in Theodore M. Porter's *The Rise of Statistical Thinking, 1820–1900* (Princeton, NJ: Princeton University Press, 1986), which we have drawn on extensively in this and the following paragraphs. On the influence and reception of Quetelet's methods in England and France, see Bernard-

Pierre Lécuyer, "Probability in Vital and Social Statistics: Quetelet, Farr, and the Bertillons," in *The Probabilistic Revolution*, vol. 1, *Ideas in History*, ed. Lorenz Krüger, Lorraine J. Daston, and Michael Heidelberger (Cambridge, MA: MIT Press, 1987), 317-35.

13. Joseph Fourier, "Extrait d'un mémoire sur la théorie analytique des assurances," *Annales de chimie et de physique*, 2nd ser., 10 (1819): 188-89, cited and translated in Porter, *Rise of Statistical Thinking*, 99.

14. We suspect that philosophy has not yet caught up with the ontological implications for mereology of the statistical revolution, but we will not attempt to do so here. Already in the early twentieth century Karl Pearson—in his lectures as the first holder of the Galton Chair in Eugenics (later renamed Genetics) at the University of London—called attention to some of the theological stakes. See his *The History of Statistics in the 17th and 18th Centuries against the Changing Background of Intellectual, Scientific, and Religious Thought*, ed. E. S. Pearson (London: Charles Griffin, 1978). On this fascinating figure, see Theodore M. Porter, *Karl Pearson: The Scientific Life in a Statistical Age* (Princeton, NJ: Princeton University Press, 2004).

15. Henry Thomas Buckle, *History of Civilization in England*, 2 vols. (1857-1861; reprint, New York: D. Appleton, 1913), as cited by Porter, *Rise of Statistical Thinking*, 61.

16. Friedrich Nietzsche, *Untimely Meditations*, trans. R. J. Hollingdale (Cambridge: Cambridge University Press, 1983), 113.

17. For Quetelet's influence on Maxwell see Porter, *Rise of Statistical Thinking*, 118. "Go to!": The poem is cited by Porter in his chap. 7, "Time's Arrow and Statistical Uncertainty in Physics and Philosophy," here 196, from Maxwell, *Theory of Heat* (London: Longmans, Green, 1872), 635-36.

18. Friedrich Nietzsche, *Beyond Good and Evil*, trans. Walter Kaufmann (New York: Vintage, 1966) 15-16. Those extremists still exist today, albeit in evolving forms. For a typical manifesto, see Daniel Dennett, *Darwin's Dangerous Idea: Evolution and the Meaning of Life* (New York: Simon and Schuster, 1995).

19. Ludwig von Mises, *Epistemological Problems of Economics* (Auburn, AL: Ludwig von Mises Institute, 1960, originally published in German in 1933), 13-18 (I, §6). Von Mises's goal is to overthrow the Dilthean distinction between the history-based *Geistwissenschaften* and the mathematical *Naturwissenschaften*, traditional in German universities, by showing that economics is a mathematical natural science.

20. Oskar Morgenstern, *Wirtschaftsprognose* (Vienna: Springer, 1928).

21. John von Neumann, "Zur Theorie der Gesellschaftsspiele," *Mathematische Annalen* 100 (1928): 295-320. Oskar Morgenstern, "Logistik und Sozialwissenschaften," *Zeitschrift für Nationalökonomie* 7 (1936): 1-24 . "To read": von Neumann and Morgenstern, *Theory of Games*, 714.

22. Von Neumann and Morgenstern, *Theory of Games*, 2.

23. Von Neumann and Morgenstern, 3-4.

24. Von Neumann and Morgenstern, 4.

25. Their point here is similar to that of Morgenstern's teacher von Mises cited above: it should be presumed that mathematical notions and the science of human action

are the same. What would be startling—"a major revolution"—is if the two diverged.

26. Von Neumann and Morgenstern, *Theory of Games*, 6–7.

27. Von Neumann and Morgenstern, 8–9.

28. John von Neumann "The Mathematician," in *Works of the Mind*, ed. Robert B. Heywood (Chicago: University of Chicago Press, 1947), 181.

29. Herbert A. Simon, "Review of Theory of Games and Economic Behavior," *American Journal of Sociology* 50, 6 (1945): 558–60. See here Joel Isaac, "Tool Shock: Technique and Epistemology in the Postwar Social Sciences," *History of Political Economy* 42, suppl. 1 (2010), 133–64.

30. W. H. Auden, *Selected Poems* (New York: Vintage, 1979), 178–83.

31. Readers looking for a history of this development may turn to Paul Erickson, *The World the Game Theorists Made* (Chicago: University of Chicago Press, 2015).

32. On the general failure of the predictive pretensions of economics, see Alexander Rosenberg, *Economics: Mathematical Politics, or Science of Diminishing Returns?* (Chicago: University of Chicago Press, 1992).

33. Von Neumann and Morgenstern, *Theory of Games*, 16: But just as in their thermodynamics the quantification of temperature, for example, is ultimately based on "the immediate sensation" of one body feeling warmer than another—a sensation "which possibly cannot and certainly need not be analyzed any further"—so, in economics, in the case of utilities, quantification must be ultimately based on "the immediate sensation of preference—of one object or aggregate of objects as against another." The analogy is misleading in that "immediate sensation" here represents a spurious attempt to simplify the problem and remove subjectivity. It would have been impossible to quantify temperature if it had been just up to our immediate sensations as living beings and if it weren't also the case that matter suffers measurable changes that enable us to build thermometers: metals contract or dilate, some metal couples exhibit the thermoelectric effect, and so forth. Familiar experiences, for example, of having one hand warmer than the other, so that the same object feels warmer when touched with one hand than the other, argue for the opposite of what von Neumann and Morgenstern wished to show.

34. For an extended critique see Philip Mirowski, *More Heat Than Light: Economics as Social Physics, Physics as Nature's Economics* (Cambridge: Cambridge University Press 1989). See also Stephen Toulmin, "Economics, or the Physics That Never Was," in his *Return to Reason* (Cambridge: Harvard University Press, 2001).

35. A contemporary economist would not embrace such a unilateral causality. But even if we treat the micro as a function of the macro or treat both micro and macro as always applicable, we will not escape the difficulties with sameness assumptions that we will demonstrate below.

36. Von Neumann and Morgenstern, *Theory of Games*, 10.

37. Von Neumann and Morgenstern, 8. Their formulation seems influenced here as elsewhere by their set-theoretical predilections: the identity of a set A is completely determined by the (independent) identity of its elements. If two sets A and B have the *same* elements, then they are the *same* set (this is one of the Zermelo-Fraenkel axioms, called the Axiom of Extension).

38. Von Neumann and Morgenstern, *Theory of Games*, 8, 10.

39. Fyodor Dostoyevsky, *Notes from Underground*, trans. Constance Garnett (Mineola, NY: Dover, 1992), 8.

40. Von Neumann and Morgenstern, *Theory of Games*, 17, emphasis added.

41. Von Neumann and Morgenstern, 30.

42. Von Neumann and Morgenstern, 27. Von Neumann and Morgenstern are willing to concede "that one may doubt whether a person" is really capable of such clarity of decision. "But whatever the merits of this doubt . . . the completeness of the system (of individual preferences) . . . must be assumed" (28–29). Richard Robb provides a clarifying explanation of the problem in *Willful: How We Choose What We Do* (New Haven, CT: Yale University Press, 2019), 158–59.

43. Von Neumann and Morgenstern, *Theory of Games*, 45.

44. Von Neumann and Morgenstern, 74.

45. Von Neumann and Morgenstern, 77.

46. C. S. Peirce, "The logic of mathematics in relation to education," in *Collected Papers*, 3.559, 348–50, cited in Subroto Roy, *Philosophy of Economics: On the Scope of Reason in Economic Inquiry* (London: Routledge, 1989), 163.

47. Karl Marx made a similar point: "Since Robinson Crusoe's experiences are a favorite theme with political economists, let us take a look at him on his island." Marx, *Capital*, 1:81. (We thank Richard Robb for the reference.)

48. Ibn Ṭufayl, *Philosophus autodidacticus*, trans. Edward Pococke (Oxford, 1671).

49. Daniel Defoe, *Robinson Crusoe* (New York: Penguin, 2001), 6–7.

50. Defoe, 10–12.

51. Defoe, 91.

52. Pensée 97, in *Pensées*, ed. Léon Brunschvicg (Paris: Hachette, 1897), 26. "The most important thing in our life is the choice of career: chance decides it." Blaise Pascal, *Pensées and Other Writings*, trans. Honor Levi, Oxford World's Classics (Oxford: Oxford University Press, 1999), 125. A recent economics paper disagrees with Pascal. Lars J. Kirkeboen, Edwin Leuven, and Magne Mogstad, "Field of Study, Earnings, and Self-Selection," *Quarterly Journal of Economics* 131 (2016), 1057–111, finds that different fields of study have substantially different labor market payoffs; that the effect on earnings of field choice is greater than that from attending a more selective institution; and that the estimated payoffs are consistent with individuals choosing fields in which they have a comparative advantage.

53. Jonathan Swift, "Abstract of Collins's 'Discourse on Free Thinking,'" in *The Prose Works of Jonathan Swift*, ed. Temple Scott (London: George Bell & Sons, 1898), 3:182. For an early statement of the findings of this new "behavioral" school, see Amos Tversky and Daniel Kahneman, "Rational Choice and the Framing of Decisions," in *Rational Choice: The Contrast between Economics and Psychology*, ed. R. Hogarth and M. Reder (Chicago: University of Chicago Press, 1986), 67–94. For retrospective (and accessible) summary of developments in the field that pays deep attention to issues of resemblance, sameness, and difference in the psychology of economic behavior, see Daniel Kahneman, *Thinking, Fast and Slow* (New York: Farrar, Straus and Giroux, 2011). Foundational, in our view, is Amos Tversky's cri-

tique of symmetrical or transitive models of similarity in his "Features of Similarity," *Psychological Review* 84 (1977): 327-52.

54. Samuel Johnson, *The History of Rasselas, Prince of Abyssinia*, Oxford World's Classics (Oxford: Oxford University Press, 2009), 43.

55. Gary Becker, *The Economic Approach to Human Behavior* (Chicago: University of Chicago Press, 1976), 14.

56. John Ruskin, *Unto This Last, and Other Writings* (New York: Penguin, 1986), 167. Ruskin's preference for a chemical rather than a mathematical or mechanistic model may have been influenced by Hegel's "chemism" in his *Science of Logic*, or Kant's in the *Critique of Pure Reason*. The move, however, does not eliminate assumptions of apathic sameness; it only pushes them one level down, so to speak.

57. See Daniel Hausman, "John Stuart Mill's Philosophy of Economics," *Philosophy of Science* 48 (1981): 363-85.

58. Alfred Marshall's *Principles of Economics*, were first published in 1880. The quote is from the ninth variorum edition (London: Macmillan, 1961), 36, as cited by James Heckman, "Causal Parameters and Policy Analysis in Economics: A Twentieth Century Retrospective," *Quarterly Journal of Economics* 115, no. 1 (2000): 46.

59. Daniel Hausman, *The Separate and Inexact Science of Economics* (Cambridge: Cambridge University Press, 1991), 136-37. Hausman specifically uses the example of transitivity: "Thus to believe that, ceteris paribus, everybody's preferences are transitive is to believe that anything that satisfies the ceteris paribus condition and is a human being has transitive preferences. One need not be disturbed by intransitive preferences caused by, for example, changes in taste, because such counterexamples to the unqualified generalization lie outside C." The passage is also cited by Rosenberg (*Economics*, 114-24) in his useful discussion of our difficulty.

60. The comparison is all the more telling because it was invited by economists themselves. Milton Friedman famously argued that even if the psychological assumptions of the economists were incorrect, they would nevertheless be justified by proving predictive. See Milton Friedman, "The Methodology of Positive Economics," in *Essays in Positive Economics* (Chicago: University of Chicago Press, 1953), 3-43, reprinted in *The Methodology of Positive Economics: Reflections on the Milton Friedman Legacy*, ed. Uskali Mäki (Cambridge: Cambridge University Press, 2009), 3-42.

61. This paragraph is derived from Heckman's magisterial retrospective, "Causal Parameters and Policy Analysis." The quote is from 49. The structural parameters approach was mathematically founded by Leonid Hurwicz, "On the Structural Form of Interdependent Systems," *Logic, Methodology, and Philosophy of Science: Proceedings of the 1960 International Congress*, edited by Ernest Nagel, Patrick Suppes, and Alfred Tarski (Stanford, CA: Stanford University Press, 1962), 232-39. For an additional retrospective on that approach, see Richard Blundell, "What Have We Learned from Structural Models?," *American Economic Review* 107, no. 5 (2017): 287-92.

62. On randomization and its limits, see James A. Heckman, "Randomization and Social Policy Evaluation," in *Evaluating Welfare and Training Programs*, ed. Charles F.

Manski and Irwin Garfinkel (Cambridge: Harvard University Press, 1992), 201–30. Heckman outlines the kinds of "sameness" assumptions that advocates of randomization often only implicitly make, including "that randomization does not alter the program being studied." (203) On the burgeoning field of experimental gaming, see J. H. Kagel and A. E. Roth, eds. *Handbook of Experimental Economics* (Princeton, NJ: Princeton University Press, 1995), chaps. 1–4.

63. Steven D. Levitt and John A. List, "What Do Laboratory Experiments Measuring Social Preferences Reveal about the Real World?" *Journal of Economic Perspectives*, 21, no. 2 (2007), 153, 154.

64. For recent mathematical approaches to the problem of contextual interference, for example, see Emir Kamenica, "Contextual Inference in Markets: On the Informational Content of Product Lines," *American Economic Review* 98 (2008), 2127–2149; Paulo Natenzon, "Random Choice and Learning," *Journal of Political Economy* 127 (2019), 419–57. "Do not transfer comfortably" is from Andrew M. Colman, "Cooperation, Psychological Game Theory, and Limitations of Rationality in Social Interaction," *Behavioral and Brain Sciences* 26 (2003): 152.

65. On this general topic we recommend once again Richard Robb, *Willful.*

66. Gary Becker and Kevin Murphy, "A Theory of Rational Addiction," *Journal of Political Economy* 96 (1988): 675: "we claim that addictions, even strong ones, are usually rational in the sense of involving forward-looking maximization with stable preferences. Our claim is even stronger: a rational framework permits new insights into addictive behavior."

67. Plato, *Protagoras* 358b–d. Franz Kafka, *Aphorisms*, trans. W. and E. Muir and M. Hofmann (New York: Schocken Books, 2015), no. 81: 80. Onomastics is capable of irony. Gregory S. Kavka, apparently unaware of the writings of his predecessor, published a thought experiment (in exchange for a million dollars, can you fully intend to drink a painful toxin, knowing that you are free to change your mind after having fully and truly intended it?) that took the question of whether we are capable of willing our own harm to rationalist extremes of paradox: Gregory S. Kavka, "The Toxin Puzzle," *Analysis* 43 (1983): 33–36.

68. *Phaedrus* 245c–257b.

69. Donald H. Davidson, "How Is Weakness of the Will Possible?" originally published 1970, republished in Davidson, *Essays on Actions and Events* (Oxford: Clarendon Press, 1980), 21–42; see also Davidson's 1978 essay "Intending," 83–102.

70. The epigraph is taken from Donald McIntosh, *The Foundations of Human Society* (Chicago: University of Chicago Press, 1969): "The idea of self-control is paradoxical unless it is assumed that the psyche contains more than one energy system, and that these energy systems have some degree of independence from each other."

71. Richard Thaler and H. M. Shefrin, "An Economic Theory of Self Control," *Journal of Political Economy* 89 (1981): 394. Thomas Schelling, "Egonomics, or the Art of Self-Management." *American Economic Review* 68 (1978): 290–94.

72. Thaler and Shefrin, "Economic Theory of Self Control," 394 and 404.

73. George Ainslie: *Picoeconomics: The Strategic Interaction of Successive Motivational States Within the Person* (Cambridge: Cambridge University Press, 1992); *Breakdown of Will* (Cambridge: Cambridge University Press, 2005). The citation is from

Ainslie, "Précis of *Breakdown of Will*," in *Behavioral and Brain Sciences* 28 (2005): 635–73.

74. Sigmund Freud, *New Introductory Lectures on Psycho-analysis* (1933), vol. 22 of *The Standard Edition of the Complete Psychological Works of Sigmund Freud*, trans. and ed. James Strachey with Anna Freud (New York: Vintage, 1999) (henceforth *SE*), 73.

75. Sigmund Freud, *An Outline of Psycho-analysis* (1940), *SE*, 23:159.

76. Freud, "The Unconscious," (1915), *SE*, 14:186. For an attempt to mathematicise the unconscious despite the nonapplicability of these laws of thought, see Ignacio Matte Blanco, *The Unconscious as Infinite Sets: an Essay in Bi-Logic* (London: Duckworth, 1975).

77. Such a view is not acceptable to a rational theorist like Gary Becker, who refused to consider any attempt to explain a rise or fall in death rates as expressing changes in people's desire for death.

78. Freud, *Beyond the Pleasure Principle* (1920), *SE*, 18:59.

79. Ludwig Wittgenstein, *Philosophical Investigations*, 3rd ed. (Oxford: Blackwell, 1958), 232e. This paragraph was placed at the end by the editors, G. E. M. Anscombe and R. Rhees.

80. Those seeking a more formal exposition may turn to Gerard Debreu, *Theory of Value: An Axiomatic Analysis of Economic Equilibrium* (New Haven, CT: Yale University Press, 1959). Debreu's goal is to treat economics "with the standard of rigor of the contemporary formalist school of mathematics" (x). He presents all the Zermelo-Fraenkel axioms plus the fixed point theorems needed for Nash equilibrium.

Chapter Nine

1. "Ma mission est de tuer le temps et la sienne de me tuer à son tour. On est tout à fait à l'aise entre assassins. " From "Écartèlement," in Emil Cioran, *Œuvres* (Paris: Gallimard, 1995), 1465.

2. "Moreno, voy a decir / Sigún mi saber alcanza: / El tiempo sólo es tardanza / De lo que está por venir." José Hernández, *El gaucho Martín Fierro y La vuelta de Martín Fierro* (Bueno Aires: El Ateneo, 1950), 374, vv. 4350–53.

3. Peter Damian, Letter 119, De divina omnipotentia, here 611D–612B. We cite from Peter Damian, *Pierre Damien, Lettre sur la toute-puissance divine*, ed. and trans. A. Cantin, Sources chrétiennes 191 (Paris: Cerf, 1972). (Cantin reproduces the column numbers from Migne's early nineteenth-century edition in the Patrologia Latina.) For an English translation see Peter Damian, *Letters 91–120*, trans. O. J. Blum, The Fathers of the Church: Mediaeval Continuation vol. 5 (Washington, DC: Catholic University of America Press, 1998), 344–86.

4. Plato, *Parmenides*, 156c–e. On the ἐξαίφνης, see Niko Strobach, *The Moment of Change: A Systematic History in the Philosophy of Space and Time* (Dordrecht: Springer, 1998).

5. Pierre Hadot, *The Inner Citadel: The Meditations of Marcus Aurelius* (Cambridge, MA: Harvard University Press, 1998), 119. Cf. Marcus Aurelius, *Meditations*, 2.14

6. Condolence letter from Einstein to the family of Michele Besso, dated March

21, 1955, in Albert Einstein and Michele Besso, *Correspondance 1903-1955*, trans. Pierre Speziali (Paris: Hermann, 1979), cited by Jimena Canales, *The Physicist and the Philosopher: Einstein, Bergson, and the Debate That Changed Our Understanding of Time* (Princeton, NJ: Princeton University Press, 2015), 339.

7. Aristotle, *Physics*, 4.11.219b.10-15; 4.11.220a.5-10, 4.11.220a.15-20. See Ursula Coope, *Time for Aristotle: Physics 4.10-14* (Oxford: Oxford University Press, 2009); David Bostock, *Space, Time, Matter, and Form: Essays on Aristotle's Physics* (Oxford: Oxford University Press, 2006). For an excellent introduction to these topics, see Richard Sorabji, *Time, Creation, and the Continuum: Theories in Antiquity and the Early Middle Ages* (Chicago: University of Chicago Press, 1983).

8. Rudolph Carnap, "Intellectual Biography," in *The Philosophy of Rudolf Carnap*, ed. P. A. Schilpp (La Salle, IL: Open Court, 1963), here 37-38.

9. Albert Einstein, "Zur Elekrodynamik bewegter Körper," *Annalen der Physik* 17 (1905): 891-921; Hermann Minkowski, "Raum und Zeit," *Jahresbericht der Deutschen Mathematiker-Vereinigung* 18 (1909): 75-88. For English translations of these papers, see Albert Einstein, "On the Electrodynamics of Moving Bodies," in *The Principle of Relativity*, trans. W. Perrett and G. B. Jeffery (New York: Dover, 1952), 35-65; H. Minkowski, "Space and Time," in *Principle of Relativity*, 73-91. The reality of Newtonian space-time was also the focus of logical critique in these same years, as in the essay by John M. E. McTaggart, "The Unreality of Time," *Mind*, n.s., 68 (1908): 457-84.

10. Roger Penrose, *The Emperor's New Mind: Concerning Computers, Minds, and Laws of Physics* (New York: Oxford University Press, 1989), 303-4, building on C. W. Rietdijk, "A Rigorous Proof of Determinism Derived from the Special Theory of Relativity," *Philosophy of Science*, 33 (1966), 341-44; "Special Relativity and Determinism," *Philosophy of Science* 43 (1976): 598-609; and Hilary Putnam, "Time and Physical Geometry," *Journal of Philosophy* 64 (1967): 240-47.

11. Aristotle, *Physics*, 6.1.231b.6-10. We cite this passage from the translation by Richard Sorabji, *The Philosophy of the Commentators, 200-600 AD*, vol. 2, *Physics* (Ithaca, NY: Cornell University Press, 2005), 190.

12. Remember the "Pythagorean" primal scene: a straight line, a unit-length PQ on it, and a unit-length QR perpendicular to PQ. The triangle PQR is the site of the crime, for PR has length whose square is 2. We put the needle of the compass on point P, the pen on point R, and draw the whole circle. This circle does not intersect our original line — no matter how much it might seem to on the paper or sand we are drawing on — since it would do so at the points $\sqrt{2}$ and $-\sqrt{2}$.

13. An ordered pair is not ipso facto a set, but an *ordered* pair (a, b), [*not* equal to (b, a)], can be defined by means of a set, for example as $\{\{a\}, a, b\}$.

14. Richard Dedekind, *Stetigkeit und Irrationale Zahlen* (Braunschweig: Friedrich Biewig und Sohn, 1872), translated as "Continuity and Irrational Numbers," in Richard Dedekind, *Essays on the Theory of Numbers* (Mineola. NY: Dover, 1963).

15. Mathematicians today refer to that property by saying that R is *complete*. R can be constructed alternatively, by taking all Cauchy sequences of rational numbers instead of Dedekind cuts; this was done by Georg Cantor, a friend of Dedekind, in that same year, 1872. What follows, mutatis mutandis, applies to both methods.

16. The spaces of interest to physicists are not always Euclidean, but they are practically always *locally* Euclidean. For example, a circle (one dimensional) is not *globally* Euclidean (since it is closed and not open), but any open arc in it is homeomorphic (i.e., topologically equivalent) to R. (If you are wondering how an open segment, say $(-1, 1)$, can be continuously transformed into the whole of R, just take the trig function $y = \tan 1/2\, \pi x$.)

17. Some philosophers think that physics could dispense with these vast constructions involving number. For an example, see Hartry Field, *Science without Numbers: A Defense of Nominalism*, 2nd ed. (Oxford: Oxford University, 2016). We are not convinced by Field's approach, among other reasons because we do not see how it can deal with the violation of the Principle of the Identity of Indiscernibles we discussed in chap. 7, which led Boltzmann to require the presence of coordinates (that is, of real numbers). Though we don't know of any physicists who argue against the use of "number" in their science, there certainly are physicists willing to consider the possibility of discrete, granular, or loopy spaces and times.

18. For an earlier effort by Cantor in this direction, see his 1873 letters to Richard Dedekind in *Briefwechsel Cantor-Dedekind*, ed. E. Noether und J. Cavaillès (Paris: Hermann, 1937), 12–18. For the diagonal procedure, see Georg Cantor, "Über eine elementare Frage der Mannigfaltigkeitslehre," *Jahresbericht der Deutschen Mathematiker-Vereinigung* 1 (1891): 75–78. English translation in *From Kant to Hilbert: A Source Book in the Foundations of Mathematics*, ed. William B. Ewald (Oxford: Oxford University Press, 1996), 2:920–22. On Cantor's mathematics and theology, see R. Thiele, "Georg Cantor, 1845–1918," in *Mathematics and the Divine: A Historical Study*, ed. Teun Koetsier and Luc Bergmans (Amsterdam: Elsevier, 2004), 523–48.

19. For simplicity, we have not taken into account the cases where a sequence has a tail end of only zeroes or only ones. For example, 1010111 (then ones forever), corresponds to the *same* real number as 1011000 (then zeroes forever). A slight modification of Cantor's change rule takes care of those cases.

20. Bergson, *Time and Free Will: An Essay on the Immediate Data of Consciousness*, trans. F. L. Pogson (London: Macmillan, 1910), 8–9. As an example of the fashion Bergson was reacting against, see Hyppolite Taine's *De l'intelligence* (Paris: L'Harmattan, 2005), which presented math as the perfection of all human knowledge, *because* it is based only on the Identity Principle.

21. James and Dewey: see Richard A. Cohen, "Philo, Spinoza, Bergson: the Ride of an Ecological Age," in *The New Bergson*, ed. John Mullarkey (Manchester: Manchester University Press, 1999), 18.

22. Carl Hoefer, "Causal Determinism," in *Stanford Encyclopedia of Philosophy*, https://plato.stanford.edu/entries/determinism-causal/. Compare the definition given by Moritz Schlick in chap. 1 of this book.

23. Bergson, *Time and Free Will*: "true duration," 92–93; "wholly dynamic," 9.

24. We use scare quotes here because the distinction is philosophically and anthropologically naive.

25. Bergson, *Time and Free Will*, 9.

26. In his treatment of space as susceptible to "clear and clean cut distinctions," and

in his dualistic distinction between extension and mind, Bergson is following Descartes. We point this out because Bergson presented his own philosophy as an overcoming of Cartesian dualism and even used the word *Cartesian* as a term of philosophical opprobrium to characterize the systems of his rivals (as he will do to Einstein). We want to emphasize, instead, the common ground beneath Descartes's and Bergson's feet.

27. Bergson, *Time and Free Will*, 107–8.

28. H. Poincaré, *Science and Method*, trans. Francis Maitland (New York: Dover: 1952), pt. 2, chap. 1, "The Relativity of Space," 1908. From Poincaré's passage we can see that space shares with time this fundamental property: we are unable to say or think anything about a spot either in space or in time without referring to some perception/memory associated with that spot. For time, it may be the position of the hands of a clock, for space, the Place du Panthéon. The poet W. B. Yeats, who found this perspective gratifying, put it this way: "The mathematician Poincaré considers time and space the work of our ancestors." N. Mann, M. Gibson, and C. Nally, eds., *W. B. Yeats's* A Vision: *Explications and Contexts* (Oxford: Oxford University Press, 2012), 172. For Poincaré's opinions on Bergsonian time, see his *Dernières pensées* (Paris: Flammarion, 1917), 42–43.

29. G. Santayana, *Winds of Doctrine* (New York: Scribner's, 1913), 58. For a detailed account of Bergson's immense prestige in this era, see François Azouvi, *La Gloire de Bergson: Essai sur le magistère philosophique* (Paris: Gallimard, 2007).

30. Letter to Bergson, June 25, 1934, in Henri Bergson, *Mélanges*, ed. André Robinet (Paris: Presses universitaires de France, 1972), 1511–12.

31. The interview, with Jonathan Cott, took place in 1978. Jonathan Cott, *Susan Sontag: The Complete Rolling Stone Interview*, by (New Haven, CT: Yale University Press, 2013), 65–66.

32. For the debate, see Einstein et al., "La Théorie de la relativité: Séance du 6 avril 1922," *Bulletin de la société française de la philosophie* 22, no. 3 (1922): 91–113. The transcripts can be downloaded at http://www.sofrphilo.fr/activites-scientifiques -de-la-sfp/conferences/grandes-conferences-en-telechargement/. We borrow the term *insurgent* from an essay written a year or two after the debate: Morris R. Cohen, "The insurgence against reason," *Journal of Philosophy* 22 (1925): 113–26. The Bergson-Einstein debate has been reexamined, most recently by Canales, *The Physicist and the Philosopher*.

33. In his *La Déduction relativiste* of 1925, Meyerson applied the philosophy of science that he had developed in *Identity and Reality* to Einstein's theory of relativity. Einstein himself wrote a glowing review of the book, which Meyerson treasured for the rest of his life as validation that his philosophy of science applied as much to the sciences of the present as those of the past. See Albert Einstein, "Á propos de *La Déduction relativiste* de M. Émile Meyerson," *Revue philosophique de la France et de l'étranger* 105 (1928): 161–66.

34. "On peut toujours choisir telle représentation qu'on veut si l'on croit qu'elle est plus commode qu'une autre pour le travail qu'on se propose; mais cela n'a pas de sens objectif." Canales, *The Physicist and the Philosopher*, 76, sees in the last clause

a pointed rejection of Poincaré's conventionalism (*commodisme*). That seems to us too strong a reading of a qualifier that remains quite cryptic.

35. Henri Bergson, *Durée et simultanéité: À propos de la théorie d'Einstein* (Paris: Felix Alcan, 1922). For an early and intelligent reception in English (quite sympathetic to Bergson) see the review by H. Wildon Carr, *Nature* 110 (1922): 503-5.

36. Albert Einstein, *The Travel Diaries of Albert Einstein: The Far East, Palestine & Spain, 1922-23*, ed. Z. Rosenkranz, (Princeton, NJ: Princeton University Press, 2018), 85.

37. Canales, *The Physicist and the Philosopher*, is largely a study of this expansion. On Maurice Merleau-Ponty, see her pages 48-50.

38. Henri Bergson, *Duration and Simultaneity: With Reference to Einstein's Theory*, trans. Leon Jacobson, with and introduction by Herbert Dingle (Indianapolis, ID: Bobbs-Merrill, 1965). See page xlii for the most apocalyptic claims.

39. Albert Einstein, "Die Grundlage der allgemeinen Relativitätstheorie," *Annalen der Physik* 49 (1916), 777. On the public controversies surrounding Einstein's work in 1920s Germany, see Milena Wazeck, *Einsteins Gegner: Die öffentliche Kontroverse um die Relativitätstheorie in den 1920er Jahren* (Frankfurt: Campus, 2009). Canales, *The Physicist and the Philosopher*, 85, proposes that these hardened Einstein's stance in Paris.

40. Flammarion was much inspired by the work of his friend Hippolyte Léon Denizard Rivail, a.k.a. Allan Kardec, the founder of Spiritism.

41. The notion of vital energy, a *tertium quid* between body and soul, should not be passed over without comment. Flammarion had published, in *Dieu dans la nature* (1869), a defense of vitalism, a doctrine taken over by him and by others (e.g., the homeopaths) from the German chemist Georg Ernst Stahl (1659-1734), who was also the sponsor of phlogiston. A variant of Flammarion's "vital energy" would reappear in Bergson's *Creative Evolution* of 1907 as *élan vital*. *Lumen*'s narrator insists in his first dialogue that his description, unlike that of Plato or of others, is neither metaphysics nor theology but pure science.

42. Camille Flammarion, *Lumen*, trans. A. A. M. and R. M. (New York: Dodd, Mead, 1897), 41-42.

43. Poincaré delivered reflections on Flammarion on the occasion of the author's jubilee: "Jubilé scientifique de Camille Flammarion," *L'Astronomie* 26 (1912): 97-153. See also F. Dutry, "Le Centenaire de Camille Flammarion," *Ciel et terre* 58 (1942), 166.

44. Jimena Canales, "Albert Einstein's Sci-Fi Stories," *New Yorker*, November 20, 2015.

45. Canales, *The Physicist and the Philosopher*, 54, for Langevin's claim at Bologna and Einstein's interest in it. See also 11 for Bergson's put-down, which continued, "we shall have to find another way of not aging."

46. Alain Connes, Danye Chéreay, Jacques Dixmier, *Le Théâtre quantique* (Paris: Odile Jacob, 2013).

47. We are aware that other subgenres of science fiction explore what may be lost in these fantastic futures. For one example, see *La Invención de Morel* (Buenos Aires: Editorial Losada, 1940) (*The Invention of Morel*, 1964), by the Argentine writer

Adolfo Bioy Casares. Borges may also have had Lumen's deficits in mind when he penned his story "El otro" ("The Other") in 1972. In it, old Borges encounters his much younger self sitting on a bench in Cambridge, Massachusetts. Young Borges is not a mere image. Borges junior talks, he corrects senior's mistakes, he may smell of eau de cologne, his shoulder can be touched, his hand shaken: a very different psychic experience from that of *Lumen*. On Borges's reading of Flammarion, see Adolfo Bioy Casares, *Borges*, ed. D. Martino (Barcelona: Destino, 2006), 1512.

48. Wyndham Lewis's *Time and Western Man* (London: Chatto & Windus, 1927), for example, is dedicated to attacking Proust and the "time-books" and "time-cult" that he saw Proust as having founded, a cult that included many "sub-Prousts," from James Joyce to Virginia Woolf and others. See Aaron Matz, "The Years of Hating Proust," *Comparative Literature* 60 (2008): 356. Lewis saw Proust as a shallow Bergsonian, an acolyte of *durée*. This is, as we shall see, a misprision.

49. Marcel Proust, *Du côté de chez Swann* (Paris: Bibliothèque de la Pléiade, 1966), 1:44-45.

50. Marcel Proust, *Le Temps retrouvé* (Paris: Bibliothèque de la Pléiade, 1966), 3:866-67.

51. A similar point about Proust was made by the literary scholar Georges Poulet (also an admirer of Bergson), who understood the work as simultaneously retrospective and prospective. See Georges Poulet, *Études sur le temps humain*, vol. 2, *De l'instant éphémère à l'instant éternelle* (Paris: Librairie Plon, 2017), 608-38, here especially 612-14, 620-21. We are not aware of any scholarly comparisons of Kierkegaard's and Proust's repetition.

52. Collected in *Otras inquisiciones* (Buenos Aires: Sur, 1952), 211-13. We cite from the translation "New Refutation of Time" by Anthony Kerrigan in Jorge Luis Borges, *A Personal Anthology*, ed. Anthony Kerrigan (New York: Grove Press, 1967), 44-64.

53. Borges seems to be thinking of time as a line, with P and Q two apparently different moments or points on the line that then prove to be the same point, and he takes this as proof that there is nothing, no moments, between P and Q. But it can be differently interpreted as proof that the time between P and Q is circular (that, for instance, the timeline curls on itself at point $P = Q$ and then goes on) or any number of curls and topologically different forms.

54. We are reminded of Isaac Bashevis Singer's comment that the Kantian categories disappear at dusk.

55. Plato, *Parmenides* 130a-e. Borges does not mention Plato in the story. He does, however, call out Heraclitus for deceiving us with his river.

56. J. W. Dunne, *An Experiment with Time*, 3rd ed. (London: Faber and Faber, 1934). J. L. Borges, "El tiempo y J. W. Dunne," first published in *Sur* 72 (September 1940) and collected in *Otras inquisiciones*. Borges rejected Dunne's multiplication of times, accusing Dunne, among other things, of incurring in the error Bergson had warned against, of confusing time with spatial dimension.

57. Vladimir Nabokov, *Insomniac Dreams. Experiments with Time*, ed. Gennady Barabtarlo (Princeton, NJ: Princeton University Press, 2018), 6-7, 19-20, 25-26. For Bergson's influence on Nabokov, see Michael Glynn, *Vladimir Nabokov: Bergson-*

ian and Russian Formalist Influences in His Novels (New York: Palgrave Macmillan, 2007), chap. 3.

58. See Richard A. Muller, *Now: The Physics of Time* (New York: W. W. Norton, 2016), 119 ff.

59. On the role of the olfactory bulb in the brain functions of rats see Leslie M. Kay, "Two Species of Gamma Oscillations in the Olfactory Bulb: Dependence on Behavioral State and Synaptic Interactions," *Journal of Integrative Neuroscience* 2 (2003): 31-44; Leslie M. Kay, "Two Minds about Odors," *Proceedings of the National Academy of Sciences* 101 (2004): 17569-70; Leslie M. Kay, M. Krysiak, L. Barlas, and G. B. Edgerton, "Grading Odor Similarities in a Go/No-Go Task," *Physiology and Behavior* 88 (2006): 339-46. Leslie M. Kay, and S. M. Sherman, "Argument for an Olfactory Thalamus," *Trends in Neurosciences* 30 (2006): 47-53. Kay has demonstrated that the cellular structures of a rat's olfactory bulb determine what odorant molecules the rat's brain can recognize as "the same" and that when previously encountered molecules are encountered in combinations (as they so often are in nature), rats have the ability to "choose" between recognizing the separate constituent parts of the odor individually as things that are "the same" as something previously known or recognizing the compound as something new and "different." However, that choice is not independent of their individual life histories. The electrophysiological activity in the olfactory bulb interacts with other regions of the brain in order to mediate not just odor perception but also many other critical functions of learning, memory and behavior. For rats, as for humans, the "sameness" necessary for counting smells and tastes (not to mention interpreting them) is not a fixed and stable natural given but contingent on experience, context, attention, and (of course not necessarily conscious) decision. On field of view, see Yang Zhou and David J. Freedman, "Posterior parietal cortex plays a causal role in perceptual and categorical decisions," *Science* 365, no. 6449 (July 12, 2019): 180-85, https://science.sciencemag.org/content/365/6449/180.full.

Chapter Ten

1. Søren Kierkegaard, *Concluding Unscientific Postscript to "Philosophical Fragments,"* ed. and trans. H. V. Hong and E. H. Hong (Princeton, NJ: Princeton University Press, 1992), 1:151.

2. Johann Wolfgang von Goethe, *Maximen und Reflexionen*, ed. Max Hecker (Weimar: Goethe-Gesellschaft, 1907), 132.

3. Einstein interviewed by G. S. Viereck, October 26, 1929, in Viereck's *Glimpses of the Great* (New York: Macaulay, 1930), 452. Viereck is an untrustworthy source, an admirer of Mussolini and Hitler with not a few anti-Jewish bees in his buffoon cap. Yet determinism is not foreign to Einstein's own writings. Be guided or be dragged is a dictum attributed by Seneca to Cleanthes the Stoic in Ep. 107, 11: "Ducunt volentem fata, nolentem trahunt" (The fates guide the willing and drag the unwilling).

4. Śāntarakṣita, *Madhyamakālaṃkāra*, verse 92. The translation from Tibetan is by Dan Arnold, who brought the text to our attention and generously provided us with his forthcoming translation and commentary of the text as edited by Masa-

michi Ichigō, *Madhyamakālaṃkāra of Śāntarakṣita with His Own Commentary or Vṛtti, and with the Subcommentary or Pañjikā of Kamalaśīla*, 2 vols. (Kyoto: Buneido, 1985).

5. See especially Brook Ziporyn, *Ironies of Oneness and Difference: Coherence in Early Chinese Thought; Prolegomena to the Study of Li* (Albany: State University of New York Press, 2013), and *Beyond Oneness and Difference: Li and Coherence in Chinese Buddhist Thought and Its Antecedents* (Albany: State University of New York Press, 2013). Fascinating work has been done on the differences and similarities between Buddhist, Hindu, and Greek philosophy and on the historical encounters between them. See, for example, Matthew Kapstein, "Mereological Considerations in Vasunandhu's 'Proof of Idealism,'" in *Reason's Traces: Identity and Interpretation in Indian and Tibetan Buddhist Thought* (Boston: Wisdom Publications, 2001), 181–204.

6. Kierkegaard, *Concluding Unscientific Postscript*, 1:151–52.

7. Olga Tokarczuk, *Flight*, trans. Jennifer Croft (London: Fitzcarraldo, 2017), 11.

8. Richard Dehmel, *Weib und Welt* (Berlin: Schuster und Loeffler, 1896).

9. The quote is from Johann Gottlieb Fichte's 1794 Jena lecture, *The Vocation of the Scholar*, trans. William Smith (Dumfries: Anodos Books, 2017), 10. In his *Foundations of the Science of Knowledge* (*Wissenschaftslehre*) published in the same year, Fichte also claimed that at its origins, the "I" posits its own being: "Das Ich setze ursprünglich schlechthin sein eigenes Sein." Johann Gottlieb Fichte, *Sämmtliche Werke*, ed. J. H. Fichte (Berlin: De Gruyter, 1845/1965), 1:98.

10. "Practical identity": Christine Korsgaard presents this practical identity and its integrity as a prerequisite for being a reflective ethical agent. See Christine Korsgaard, *The Sources of Normativity*, ed. Onora O'Neill (Cambridge: Cambridge University Press, 1996), 100–102.

11. Aristotle, *Politics* 8.1339a–8.1340b: *paidías*, or children's games, we translate as amusement, *anapaúseōs*—a pause from work, as relaxation.

12. Giambattista Vico, *On the Most Ancient Wisdom of the Italians*, trans. L. M. Palmer (Ithaca, NY: Cornell University Press, 1988), 46–47.

13. That bitter wine, as well as the mysterious lines of the first tercet about "der Kreuzweg deiner Sinne" (the crossroads of your senses), seem to us temptingly close to Plato's *Theaetetus* 159e, but we abstain from going there.

14. Kierkegaard, *Concluding Unscientific Postscript*, 1:130, emphasis added.

15. Obviously we do not agree with Heidegger's claim that Kierkegaard had nothing to say about the decisive problem of Being. See Clare Carlisle, "Kierkegaard and Heidegger," in *The Oxford Handbook of Kierkegaard*, ed. J. Lippitt and G. Pattison (Oxford: Oxford University Press, 2013), 422, quoting Heidegger's *Was heißt Denken?* of 1952.

16. We understand the question of why Kierkegaard adopted pseudonymous authorship in the first stage of his development as related to his view that the self is, at best, a work in progress. For the author to appear under a single name when his I is not one would be misleading. The books of his so-called second authorship (after *Concluding Unscientific Postscript*) were published under his own name because

those books come after his leap from the realm of the ethical to that of (Christian) faith.

17. We will not take on the topic of the so-called replication crisis in the laboratory psychological sciences of our own day, but we do think that our reading of Kierkegaard's *Repetition* is relevant to it.

18. Søren Kierkegaard, *Fear and Trembling/Repetition*, ed. and trans. H. V. Hong and E. H. Hong (Princeton, NJ: Princeton University Press, 1983), 131 ff. Despite the ironic treatment of Constantius, we don't doubt that for Kierkegaard himself, repetition was indeed a fundamental philosophical category. A long footnote in *The Concept of Anxiety* (1844) by its pseudonymous author Vigilius Haufniensis restates that importance: "To my knowledge, he [CC] is the first to have a lively understanding of 'repetition' and to have allowed the pregnancy of the concept to be seen in the explanation of the relation of the ethical and the Christian, by directing attention to the invisible point and to the *discrimen rerum* [turning point] where one science breaks against another until a new science comes to light. But what he has discovered he has concealed again by arraying the concept in the jest of an analogous conception." Søren Kierkegaard, *The Concept of Anxiety: A simple Psychologically Orienting Deliberation on the Dogmatic Issue of Hereditary Sin*, ed. and trans. Reidar Thomte and Albert B. Anderson (Princeton, NJ: Princeton University Press, 1980), 17–19.

19. Psychologists continue to devise theories about the pleasures (or lack thereof) of repetition and experiments with which to test them. See, for example, E. O'Brien, "Enjoy It Again: Repeat Experiences Are Less Repetitive Than People Think," *Journal of Personality and Social Psychology* 116, no. 4 (2019), 519–40.

20. Constantius himself repeatedly asserts the opposite. He claims, for example, that the paladin with the "courage to will repetition" will wear love as an "indestructible garment." Kierkegaard, *Fear and Trembling/Repetition*, 131–32.

21. Kierkegaard, 212–13.

22. Kierkegaard, 9.

23. Joakim Garff, *Søren Kierkegaard: A Biography*, trans. Bruce H. Kirmmse (Princeton, NJ: Princeton University Press, 2005), 199–213. A translation of Schelling's lectures appears in F. W. J. Schelling, *The Grounding of Positive Philosophy: The Berlin Lectures*, trans. Bruce Matthews (Albany: State University of New York Press, 2007). Kierkegaard's notes on the lectures are included as an appendix in his *The Concept of Irony*, ed. and trans. H. V. Hong and E. H. Hong (Princeton, NJ: Princeton University Press, 1989), 335–412. They are relevant to our topic, but pursuing that relevance would take us too far afield.

24. Garff, *Kierkegaard*, 211.

25. Garff, 231.

26. Concerning Abraham's descendants see Gen. 22:16–18. On Kierkegaard's use of this biblical motif, see Jon D. Levenson, *The Death and Resurrection of the Beloved Son: The Transformation of Child Sacrifice in Judaism and Christianity* (New Haven, CT: Yale University Press, 1993), 125–42. Horace, *Odes* 3.16, lines 21–22: *Quanto quisque sibi plura negaverit / ab dis plura feret.*

27. In characterizing Weyl as a philosopher as well as a mathematician, we are only slightly abusing his own characterization of himself (in 1947) as "a philosophically-minded mathematician." Hermann Weyl, *Philosophy of Mathematics and Natural Science* (New York: Atheneum, 1963), vi.

28. On these letters, see Saul Friedländer, *Franz Kafka: The Poet of Shame and Guilt* (New Haven, CT: Yale University Press, 2013), 5. In 1913 Kafka had read Kierkegaard's *Buch des Richters*, a selection, in German, from Kierkegaard's diaries of 1833-1855.

29. Franz Kafka, *Das Urteil und andere Erzählungen* (Berlin: Fischer Taschenbücher, 1952), 81-82; included in *Der Prozess* (Berlin: Die Schmiede, 1925). In his diary entry for December 13, 1914, Kafka wrote that "the Legend," that is, "Before the Law," inspires in him satisfaction and happiness, a rare event in Kafka's life as a writer.

30. Søren Kierkegaard, *Either/Or Part II*, ed. and trans. H. V. Hong and E. H. Hong (Princeton, NJ: Princeton University Press, 1987), 168.

31. Kafka, *Der Prozess*. We prefer our suggestion to that of Martin Buber, who maintained that Joseph K. was guilty but that the guilt was existential, related to original sin. See Martin Buber, *Schuld und Schuldgefühle* (Heidelberg: L. Schneider, 1958), 55.

32. Nathalie Sarraute, *L'Ère du soupçon* (Paris: Gallimard, 1956); *The Age of Suspicion*, trans. M. Jolas (New York: G. Braziller, 1963), 11 ff.

33. Sarraute, *The Age of Suspicion*, 33.

34. Stendhal, *Souvenirs d'égotisme* (Paris: Le Divan, 1927), 6: "Le génie poétique est mort, mais le génie du soupçon est venu au monde."

35. Edward Young, *Night Thoughts* (London: C. Whittingham, 1798), 32, Second Night, lines 485-95.

36. Borges, with his love of Old Norse, presumably knew that in Kierkegaard's Danish giving back is *Gjen-tagelse*, which is also the word for repetition. On the presence of Kierkegaard in Borges's library and readings, see L. Rosato and G. Álvarez, eds., *Borges, libros y lecturas: Catálogo de la colección Jorge Luis Borges en la Biblioteca Nacional* (Buenos Aires: Biblioteca Nacional, 2010), 215. See also Eduardo Fernández Villar, "Jorge Luis Borges: The Fear without Trembling," in *Kierkegaard's Influence on Literature, Criticism and Art*, ed. Jon Stewart (London: Routledge, 2013), 21-32.

37. Jorge Luis Borges, *Poesía Completa* (Barcelona: Random House Mondadori, 2011), 240. The word "everness" is not recorded in the Oxford English Dictionary. Borges thought it had been coined by Bishop Wilkins. See Adolfo Bioy Casares, *Borges*, ed. D. Martino (Barcelona: Destino, 2006), 971.

38. Rilke expressed a remarkably similar thought in a letter to Nora Purtscher-Wydenbruck (August 11, 1924): "Since my earliest youth I have entertained the conjecture (and have too, as far as I was able to, lived accordingly) that, at some deeper level of this pyramid of consciousness, simple Being could become a big event for us, an unbreakable being here, simultaneously, of all that which, at the upper 'normal' apex of self-consciousness, is vouchsafed us to experience only *seriatim*." *Rainer Maria Rilke Briefe*, ed. Ruth Sieber-Rilke and Karl Altheim (Ber-

lin, 1950), 2:450–56. For Kierkegaard's influence on Rilke, see Leonardo F. Lisi, "Rainer Maria Rilke: Unsatisfied Love and the Poetry of Living," in *Kierkegaard's Influence on Literature, Criticism and Art*, ed. Jon Stewart, Kierkegaard Research, vol. 12 (Abingdon: Routledge, 2016).

39. Borges, *Poesía Completa*, 241.
40. One might venture an alternative hypothesis that Borges is referring to God, who blessed the clay of which He molded us. But if that were so, instead of the Spanish form *lo que* = that which, one should expect *El que* or *Al que* = He who or Him who. In any event, Borges was more Spinozistic than biblical.
41. Including by Kierkegaard: *The Sickness Unto Death: A Christian Psychological Exposition for Upbuilding and Awakening*, ed. and trans. H. V Hong and E. H. Hong (Princeton, NJ: Princeton University Press, 1980), 13, 174n2.
42. Lao Tzu, *Tao Te Ching, A Book about the Way and the Power of the Way*, trans. Ursula K. Le Guin (Boston: Shambhala, 1997), 3.
43. Kierkegaard, *Concluding Unscientific Postscript*, 93 and 117.
44. Manilius, *Astronomica*, 4.392, trans. G. P. Goold, Loeb Classical Library (Cambridge, MA: Harvard University Press, 1977), 232–33: "The object of your quest is God; you are seeking to scale the skies and though born beneath the rule of fate, to gain knowledge of that fate; you are seeking to pass beyond your human limitations and make yourself master of the universe. The toil involved matches the reward to be won, *nor are such high attainments secured without a price.*" We are far from the first to be concerned about the price. For an environmentally oriented expression of that concern, see Hans Jonas, *The Imperative of Responsibility: In Search of an Ethics for the Technological Age* (Chicago: University of Chicago Press, 1984), 2.
45. Anthony Hecht, "The Transparent Man," in *The Transparent Man: Poems* (New York: Knopf, 1990), 72.

BIBLIOGRAPHY

The bibliography contains only cited works. Following convention, we do not list ancient authors unless we have cited a specific edition or translation. In all other cases our citations of ancient authors and their works can be found through their entries in the index.

Adelard of Bath. *Conversations with His Nephew: "On the Same and the Different," "Questions on Natural Science," and "On Birds."* Edited and translated by Charles Burnett. Cambridge: Cambridge University Press, 1998.

Adorno, Theodor. *Gesammelte Schriften in 20 Banden.* Vol. 10, bk. 1, *Kulturkritik und Gesellschaft 1: Prismen. Ohne Leitbild.* Frankfurt: Suhrkamp, 1977.

Adorno, Theodor. *Prisms.* Translated by Samuel and Shierry Weber. Cambridge, MA: MIT Press, 1981.

Aharonov, Y., F. Colombo, S. Popescu, I. Sabadini, D. C. Strouppa, and J. Tollaksen. "Quantum Violation of the Pigeonhole Principle and the Nature of Quantum Correlations." *Proceedings of the National Academy of Sciences* 113 (2016): 532–35.

Ahmed, Shahab. *What Is Islam? The Importance of Being Islamic.* Princeton, NJ: Princeton University Press, 2016.

Ainslie, George. *Breakdown of Will.* Cambridge: Cambridge University Press, 2005.

Ainslie, George. *Picoeconomics: The Strategic Interaction of Successive Motivational States Within the Person.* Cambridge: Cambridge University Press, 1992.

Ainslie, George. "Précis of *Breakdown of Will.*" *Behavioral and Brain Sciences* 28 (2005): 635–73.

Albertson, David. *Mathematical Theologies: Nicholas of Cusa and the Legacy of Thierry of Chartres.* Oxford: Oxford University Press, 2015.

Alexander, Samuel. *Space, Time and Deity.* 2 vols. London: MacMillan, 1927. Reprint: New York: Dover, 1966.

Annas, Julia. *Aristotle's Metaphysics Books M and N: English Translation and Commentary.* 2nd ed. Oxford: Oxford University Press, 1988.

Anscombe, G. E. M. "Parmenides, Mystery and Contradiction." *Proceedings of the Aristotelian Society* 69 (1968): 125–32.

Arberry, Arthur J. *The Spiritual Physick of Rhazes.* London: John Murray, 1950. Reprinted as *Razi's Traditional Psychology*, Damascus: Islamic Book Service, [2007?].

Aristotle. *The Complete Works of Aristotle.* 2 vols. Edited by Jonathan Barnes. Princeton, NJ: Princeton University Press, 1984.

Aristotle. *Metaphysics.* 2 vols. Translated by H. Tredennick. Loeb Classical Library. Cambridge, MA: Harvard University Press, 1947.

Aristotle. *The Nicomachean Ethics.* Loeb Classical Library. Cambridge, MA: Harvard University Press, 2003.

Aristotle. *On Sophistical Refutations, On Coming-to-Be and Passing-Away, On the Cosmos.* Translated by E. S. Forster and D. J. Furley. Loeb Classical Library. Cambridge, MA: Harvard University Press, 1955.

Aristotle. *On the Soul.* Translated by Joe Sachs. Santa Fe, NM: Green Lion Press, 2004.

Aristotle. *On the Soul, Parva Naturalia, On Breath.* Translated by W. S. Hett. Loeb Classical Library. Cambridge, MA: Harvard University Press, 1957.

Arnason, Johann P., S. N. Eisenstadt, and Björn Wittrock, eds. *Axial Civilizations and World History.* Leiden: Brill, 2005.

Assmann, Jan. *The Search for God in Ancient Egypt.* Ithaca, NY: Cornell University Press, 1984.

Aubrey, John. *Aubrey on Education: A Hitherto Unpublished Manuscript by the Author of Brief Lives.* Edited by J. E. Stephens. London: Routledge, 2012. First published 1972.

Auden, W. H. *Selected Poems.* New York: Vintage, 1979.

Auerbach, Erich. *Scenes from the Drama of European Literature.* Translated by Ralph Manheim. Minneapolis: University of Minnesota Press, 1984.

Augustine. *De doctrina christiana.* Edited and translated by R. P. H. Green. Oxford: Oxford University Press, 1996.

Augustine. *On the Free Choice of the Will, On Grace and Free Choice, and Other Writings.* Edited and translated by Peter King. Cambridge: Cambridge University Press, 2010.

Averroes. *Long Commentary on the "De anima" of Aristotle.* Translated by R. C. Taylor and T-A. Druart. New Haven, CT: Yale University Press, 2009.

Avicenna. *The Metaphysics of* The Healing. Translated by Michael E. Marmura. Provo, UT: Brigham Young University Press, 2005.

Avicenna. *Tafsīr al-qur'ānī wa'l-lugha al-ṣūfiyya fī falsafa Ibn Sīnā.* Edited by Ḥasan 'Āṣī. Beirut: al-Mu'assasa al-Jāmi'iyya lil-Dirāsāt wa'l-Nashr wa'l-tanzī', 1983.

Avicenna. *Tafsīr sūra al-ikhlāṣ.* Edited by Abū'l-Qāsim Muḥammad. Delhi: 'Abd al-Raḥmān, 1893/1894.

Azouvi, François. *La Gloire de Bergson: Essai sur le magistère philosophique.* Paris: Gallimard, 2007.

Bachelard, G. *Le Rationalisme appliqué.* Paris: Presses Universitaires de France, 1949.

Bacon, Francis. *Novum organum scientiarum.* In *Works,* edited by James Spedding, Robert Leslie Ellis, and Douglas Denon Heath. 15 vols. Boston: Taggard and Thompson, 1864.

Baker, Keith Michael. *Condorcet: From Natural Philosophy to Social Mathematics.* Chicago: University of Chicago Press 1975.

Ball, Philip. *Beyond Weird: Why Everything You Thought You Knew about Quantum Physics Is Different*. Chicago: University of Chicago Press, 2018.

Ballan, Mohamad. "The Scribe of the Alhambra: Lisān Al-Dīn Ibn Al-Khaṭīb, Sovereignty and History in Nasrid Granada." PhD diss., University of Chicago, 2019.

Balzac, Honoré de. *Séraphîta*. In *The Novels of Honoré de Balzac*. Vol. 42, *Séraphîta*. New York: G. D. Sproul, 1895–1900.

Balzac, Honoré de. *Séraphîta*. Paris: L'Harmattan, 1995.

Barford, Robert. "The Context of the Third Man Argument in Plato's *Parmenides*." *Journal of the History of Philosophy* 16 (1978): 1–11.

Barnes, Jonathan. "Aristotle's Theory of Demonstration." *Phronesis* 14 (1969): 123–52.

Barnes, Jonathan. *The Presocratic Philosophers*. London: Routledge, 1982.

Barns, J. W. B., G. M. Browne, and J. C. Shelton, eds. *Nag Hammadi Codices: Greek and Coptic Papyri from the Cartonnage of the Covers*. Leiden: Brill, 1981.

Becker, Gary. *The Economic Approach to Human Behavior*. Chicago: University of Chicago Press, 1976.

Becker, Gary, and Kevin Murphy. "A Theory of Rational Addiction." *Journal of Political Economy* 96 (1988): 675–700.

Bedos-Rezak, Brigitte. "Semiotic Anthropology: The Twelfth Century Experiment." In *European Transformations: The Long Twelfth Century*, edited by Thomas F. X. Noble and John Van Engen, 426–67. Notre Dame, IN: University of Notre Dame Press, 2012.

Beierwaltes, Werner. "*Aequalitas numerosa*: Zu Augustins Begriff des Schönen." *Weisheit und Wissenschaft* 38 (1975): 140–57.

Beierwaltes, Werner. *Denken des Einen: Studien zur Neuplatonischen Philosophie und ihrer Wirkungsgeschichte*. Frankfurt am Main: Vittorio Klostermann, 1985.

Benacerraf, Paul. "What Numbers Could Not Be." *Philosophical Review* 74 (1965): 47–73.

Benacerraf, Paul, and H. Putnam, eds. *Philosophy of Mathematics: Selected Readings*. 2nd ed. Cambridge: Cambridge University Press, 1983.

Bennett, Jonathan Francis. "Knowledge of Necessity." In *Learning from Six Philosophers: Descartes, Spinoza, Leibniz, Locke, Berkeley, Hume*, 2:34–58. Oxford: Clarendon Press, 2001.

Bentham, Jeremy. *A Comment on the Commentaries and A Fragment on Government*. Edited by J. H. Burns and H. L. A. Hart. Oxford: Clarendon Press, 1977.

Bergson, Henri. *Durée et simultanéité: À propos de la théorie d'Einstein*. Paris: Felix Alcan, 1922.

Bergson, Henri. *Mélanges*. Edited by André Robinet. Paris: Presses universitaires de France, 1972.

Bergson, Henri. *Time and Free Will: An Essay on the Immediate Data of Consciousness*. Translated by F. L. Pogson. London: Macmillan, 1910.

Bergson, Henri. *Duration and Simultaneity: With Reference to Einstein's Theory*. Translated by Leon Jacobson and with an introduction by Herbert Dingle. Indianapolis, ID: Bobbs-Merrill, 1965.

Bernoulli, Daniel, and Johann Bernoulli. *Hydrodynamics* and *Hydraulics*. Translated by Thomas Carmody and Helmut Kobus. Mineola, NY: Dover, 1968.

Betegh, G. "What Makes a Myth *Eikôs*? Remarks Inspired by Myles Burnyeat's 'Eikôs

Mythos.'" In *One Book, The Whole Universe: Plato's "Timaeus" Today*, edited by R. Mohr, K. Sanders, and B. Sattler, 213–24. Las Vegas, NV: Parmenides, 2009.

Bioy Casares, Adolfo. *Borges*. Edited by D. Martino. Barcelona: Destino, 2006.

Bioy Casares, Adolfo. *La Invención de Morel*. Buenos Aires: Editorial Losada, 1940.

el-Bizri, Nader, ed. and trans. *Epistles of the Brethren of Purity: On Arithmetic and Geometry; An Arabic Critical Edition and English Translation of Epistles 1 & 2*. Oxford: Oxford University Press, 2012.

Black, Max. "The Identity of Indiscernibles." *Mind* 61 (1952): 153–64.

Black, Max. *Problems of Analysis*. Ithaca, NY: Cornell University Press, 1954.

Blumenberg, Hans. *The Laughter of the Thracian Woman: A Protohistory of Theory*. Translated by S. Hawkins. London: Bloomsbury, 2015.

Blundell, Richard. "What Have We Learned from Structural Models?" *American Economic Review* 107, no. 5 (2017): 287–92.

Bohm, David. *Wholeness and the Implicate Order*. London: Routledge, 1980. Reprinted 2002.

Bohr, Niels. "Can Quantum-Mechanical Description of Physical Reality Be Considered Complete?" *Physical Review* 48 (1935): 696–702.

Boltzmann, Ludwig. *Theoretical Physics and Philosophical Problems*. Dordrecht: D. Reidel, 1974.

Bonnefoy, Yves. *The Act and the Place of Poetry: Selected Essays by Yves Bonnefoy*. Edited and translated by John T. Naughton. Chicago: University of Chicago Press, 1989.

Bonnefoy, Yves. "Paul Valéry." In *L'Improbable et autres essais*, 99–105. Paris: Gallimard, 1983.

Boolos, George. "Introductory Note to Kurt Gödel's 'Some Basic Theorems on the Foundations of Mathematics and Their Implications.'" In *Logic, Logic, and Logic*, edited by Richard Jeffrey, 105–119. Cambridge, MA: Harvard University Press, 1998.

Boolos, George. "Must We Believe in Set Theory?" In *Logic, Logic, and Logic*, edited by Richard Jeffrey, 120–32. Cambridge, MA: Harvard University Press, 1998.

Borges, Jorge Luis. *Cuentos Completos*. Barcelona: Random House Mondadori, 2011.

Borges, Jorge Luis. *A Personal Anthology*. Edited by Anthony Kerrigan. New York: Grove Press, 1967.

Borges, Jorge Luis. *Otras inquisiciones*. Buenos Aires: Sur, 1952.

Borges, Jorge Luis. *Poesía completa*. Barcelona: Random House Mondadori, 2011.

Bostock, David. *Space, Time, Matter, and Form: Essays on Aristotle's Physics*. Oxford: Oxford University Press, 2006.

Brading, Katherine, and Elena Castellani, eds. *Symmetries in Physics: Philosophical Reflections*. Cambridge: Cambridge University Press, 2003.

Brewster, Sir David. *Memoirs of Sir Isaac Newton*. 2 vols. 1855. Reprint, New York: Johnson Reprint, 1965.

Broch, Hermann. *Geist and Zeitgeist: The Spirit in an Unspiritual Age*. Edited and translated by John Hargraves. New York: Counterpoint, 2002.

Brown, Peter. *Augustine of Hippo: A Biography*. Berkeley: University of California Press, 2000.

Buber, Martin. *Schuld und Schuldgefühle*. Heidelberg: L. Schneider, 1958.

Buch, Johann von. *Glossen zum Sachsenspiegel*. Edited by Frank-Michael Kaufmann. Hannover: Hahn, 2002.

Buckle, Henry Thomas. *History of Civilization in England*, 2 vols. 1857–1861. Reprint, New York: D. Appleton, 1913.

Burckhardt, Carl A. H. *Goethes Unterhaltungen mit dem Kanzler Friedrich von Müller*. Stuttgart: Cotta'sche, 1898.

Burkett, John P. "Marx's Concept of an Economic Law of Motion." *History of Political Economy* 32 (2000): 381–94.

Burkert, Walter. *Lore and Science in Ancient Pythagoreanism*. Cambridge, MA: Harvard University Press, 1972.

Burnyeat, M. F. "Eikôs Mythos." *Rizai* 2, no. 2 (2005): 143–65.

Burnyeat, M. F. "Enthymeme: Aristotle on the Logic of Persuasion." In *Aristotle's Rhetoric: Philosophical Essays*, edited by D. J. Furley and A. Nehemas, 3–55. Princeton, NJ: Princeton University Press, 1994.

Burnyeat, M. F. "Platonism and Mathematics: A Prelude to Discussion." In *Mathematics and Metaphysics in Aristotle*, edited by A. Graeser, 213–40. Berne: P. Haupt, 1987.

Burnyeat, M. F. "Platonism in the Bible." In *Metaphysics, Soul, and Ethics in Ancient Thought: Themes from the Work of Richard Sorabji*, edited by R. Salles, 143–69. Oxford: Oxford University Press, 2005.

Burnyeat, M. F. "Plato on Why Mathematics Is Good for the Soul." In *Mathematics and Necessity: Essays in the History of Philosophy*, Proceedings of the British Academy 103, edited by T. Smiley, 1–81. Oxford: Oxford University Press, 2000.

Canales, Jimena. "Albert Einstein's Sci-Fi Stories." *New Yorker*, November 20, 2015.

Canales, Jimena. *The Physicist and the Philosopher: Einstein, Bergson, and the Debate That Changed Our Understanding of Time*. Princeton, NJ: Princeton University Press, 2015.

Cantor, Georg. *Briefwechsel Cantor-Dedekind*. Edited by E. Noether and J. Cavaillès. Paris: Hermann, 1937.

Cantor, Georg. "Über eine elementare Frage der Mannigfaltigkeitslehre." *Jahresbericht der Deutschen Mathematiker-Vereinigung* 1 (1891): 75–78. English translation in *From Kant to Hilbert: A Source Book in the Foundations of Mathematics*, edited by William Bragg Ewald, 2:920–22. Oxford: Oxford University Press, 1996.

Carlisle, Clare. "Kierkegaard and Heidegger." In *The Oxford Handbook of Kierkegaard*. Edited by J. Lippitt and G. Pattison, 421–39. Oxford: Oxford University Press, 2013.

Carnap, Rudolf. *Der logische Aufbau der Welt*. Berlin: F. Meiner, 1928.

Carnap, Rudolph. "Intellectual Biography." In *The Philosophy of Rudolf Carnap*, edited by P. A. Schilpp, 3–84. La Salle, IL: Open Court, 1963.

Carnap, Rudolf. *The Logical Structure of the World*. Translated by R. A. George. Chicago: Open Court, 2003.

Carr, H. Wildon. Review of *Durée et simultanéité*, by Henri Bergson. *Nature* 110 (1922): 503–5.

Carraud, Vincent. *Causa sive ratio: La raison de la cause, de Suarez à Leibniz*. Paris: Presses universitaires de France, 2002.

Cartwright, Nancy. "Philosophical Problems of Quantum Theory: The Response of

American Physicists." In *The Probabilistic Revolution*, vol. 2, *Ideas in the Sciences*, edited by Lorenz Krüger, Gerd Gigerenzer, and Mary S. Morgan, 417-35. Cambridge, MA: MIT Press, 1987.

Cassirer, Ernst. "Der Begriff der Symbolischen Form im Aufbau der Geisteswissenchaften." In *Vorträge der Bibliothek Warburg*, edited by Fritz Saxl, 1:11-39. Leipzig: B. G. Teubner, 1921

Cassirer, Ernst. *Substance and Function and Einstein's Theory of Relativity*. Translated by William Curtis Swabey and Marie Collins Swabey. Chicago: Open Court, 1923.

Caulton, Adam. "Discerning 'Indistinguishable' Quantum Systems." *Philosophy of Science* 80 (2013): 49-72.

Caulton, Adam. "Issues of Identity and Individuality in Quantum Mechanics." PhD diss., Cambridge University, 2015.

Caulton, Adam, and Jeremy Butterfield. "On Kinds of Indiscernibility in Logic and Metaphysics." *British Journal for the Philosophy of Science* 63 (2012): 27-84.

Cercignani, Carlo. *Ludwig Boltzmann: The Man Who Trusted Atoms*. Oxford: Oxford University Press, 1998.

Cherniss, H. F. "The Characteristics and Effects of Presocratic Philosophy." *Studies in Presocratic Philosophy*, edited by D. J. Furley and R. E. Allen, 1-28. New York: Humanities Press, 1970.

Chiara, L. Dalla, and G. Toraldo di Francia. "Identity Questions from Quantum Theory." In *Physics, Philosophy and the Scientific Community*, edited by Kostas Gavroglu, John Stachel, and Marx W. Wartofsky, 39-46. Dordrecht: Kluwer Academic, 1995.

Chiara, L. Dalla, and G. Toraldo di Francia. "Individuals, Kinds and Names in Physics." In *Bridging the Gap: Philosophy, Mathematics, Physics*, edited by Giovanna Corsi, Maria Luisa Dalla Chiara, and Gian Carlo Ghirardi, 261-83. Dordrecht: Kluwer Academic, 1993.

Chomsky, Noam. *Knowledge of Language: Its Nature, Origin, and Use*. Westport, CT: Praeger, 1986.

Chomsky, Noam. *Syntactic Structures*. Paris: Mouton, 1957.

Cicero. *De natura deorum*. Translated by H. Rackham. Loeb Classical Library. Cambridge MA: Harvard University Press, 1933.

Cioran, Emil. *Œuvres*. Paris: Gallimard, 1995.

Claudel, Paul. *Œuvres en prose*. Paris: Pléiade, 1965.

Clement of Alexandria. *Clemens Alexandrinus Stromata*. 3rd ed. Vols. 1-2. Edited by O. Stählin, L. Früchtel and U. Treu. Berlin: Akademie, 1960-1985.

Cohen, Daniel J. *Equations from God: Pure Mathematics and Victorian Faith*. Baltimore: Johns Hopkins University Press, 2007.

Cohen, Gustave, and Paul Valéry. *Essai d'explication du "Cimetière marin": Précédé d'un avant-propos de Paul Valéry au sujet du "Cimetière marin."* Paris: Gallimard, 1933.

Cohen, Morris R. "The Insurgence against Reason." *Journal of Philosophy* 22 (1925): 113-26.

Cohen, Richard A. "Philo, Spinoza, Bergson: the Ride of an Ecological Age." In *The New Bergson*, edited by John Mullarkey, 18-31. Manchester: Manchester University Press, 1999.

Cole, Joshua. *The Power of Large Numbers: Population, Politics, and Gender in Nineteenth-Century France.* Ithaca, NY: Cornell University Press, 2000.

Coleridge, Samuel Taylor. *The Complete Works.* Edited by William Shedd. New York: Harper & Brothers, 1858.

Colman, Andrew M. "Cooperation, Psychological Game Theory, and Limitations of Rationality in Social Interaction." *Behavioral and Brain Sciences* 26 (2003): 139–98.

Connes, Alain, Danye Chéreay, and Jacques Dixmier. *Le Théâtre quantique.* Paris: Odile Jacob, 2013.

Coope, Ursula. *Time for Aristotle: Physics 4.10–14.* Oxford: Oxford University Press, 2009.

Cornelli, G., R. McKirahan, and C. Macris, eds. *On Pythagoreanism.* Berlin: De Gruyter, 2013.

Cott, Jonathan. *Susan Sontag: The Complete Rolling Stone Interview.* New Haven, CT: Yale University Press, 2013.

Couprie, Dirk L. "The Discovery of Space: Anaximander's Astronomy." In *Anaximander in Context*, edited by D. Couprie, R. Hahn, and G. Nadaff, 167–240. Albany: State University of New York Press, 2003.

Curd, Patricia. "'Parmenides' 131c–132b: Unity and Participation." *History of Philosophy Quarterly* 3 (1986): 125–36.

Daiser, Olivier. *Einführung in die Mengenlehre.* 2nd ed. Springer: Berlin, 2000.

Damian, Peter. *Letters 91–120.* Translated by O. J. Blum. The Fathers of the Church: Mediaeval Continuation, vol. 5, Washington, DC: Catholic University of America Press, 1998.

Damian, Peter. *Pierre Damien, Lettre sur la toute-puissance divine.* Edited and translated by A. Cantin. Sources chrétiennes 191. Paris: Cerf, 1972.

D'Ancona, C. "Aux Origines du *Dator Formarum*: Plotin, l'*Épître sur la Science Divine*, et al-Fārābī." In *De l'antiquité tardive au Moyen Âge*, edited by E. Coda and C. Martini Bonadeo, 381–413. Paris: Vrin, 2014.

Daston, Lorraine. *Classical Probability in the Enlightenment.* Princeton, NJ: Princeton University Press, 1988.

Daston, Lorraine. "Rational Individuals versus Laws of Society: From Probability to Statistics." In *The Probabilistic Revolution.* Vol. 1, *Ideas in History*, edited by Lorenz Krüger, Lorraine J. Daston, and Michael Heidelberger, 295–304. Cambridge, MA: MIT Press, 1987.

Davidson, Donald H. "How Is Weakness of the Will Possible?" In *Essays on Actions and Events*, 21–42. Oxford: Clarendon Press, 1980.

Davidson, Donald H. "Intending." In *Essays on Actions and Events*, 21–42. Oxford: Clarendon Press, 1980.

Davis, Michael. *The Soul of the Greeks: An Inquiry.* Chicago: University of Chicago Press, 2011.

Debreu, Gerard. *Theory of Value: An Axiomatic Analysis of Economic Equilibrium.* New Haven, CT: Yale University Press, 1959.

Dedekind, Richard. *Essays on the Theory of Numbers.* Mineola, NY: Dover, 1963.

Dedekind, Richard. *Stetigkeit und Irrationale Zahlen.* Braunschweig: Friedrich Biewig und Sohn, 1872.

Defoe, Daniel. *Robinson Crusoe*. New York: Penguin, 2001.

Defoort, Carine. *The Pheasant Cap Master (He guan zi): A Rhetorical Reading*. Albany: State University of New York Press, 1997.

Dehmel, Richard. *Weib und Welt*. Berlin: Schuster und Loeffler, 1896.

Dehn, Max. *Über die geistige Eigenart des Mathematikers: Rede anlässlich der Gründungsfeier des Deutschen Reiches am 18. Januar 1928*. Frankfurter Universitätsreden 28. Frankfurt am Main: Werner und Winter, 1928.

Dennett, Daniel. *Darwin's Dangerous Idea: Evolution and the Meaning of Life*. New York: Simon and Schuster, 1995.

Descartes, René. *Œuvres de Descartes*. 13 vols. Edited by Charles Adam and Paul Tannery. Paris: Cerf, 1897.

Descartes, René. *The Philosophical Writings of Descartes*. Translated by John Cottingham, Robert Stoothoff, and Dugald Murdoch. 3 vols. Cambridge: Cambridge University Press, 1984–1985.

Detienne, Marcel. *The Masters of Truth in Archaic Greece*. Translated by J. Lloyd. Boston: Zone Books, 1996.

Dewey, John. *Characters and Events: Popular Essays in Social and Political Philosophy by John Dewey*. Edited Joseph Ratner. 2 vols. New York: Henry Holt, 1929.

Dewey, John. *Leibniz's New Essays Concerning the Human Understanding: A Critical Exposition*. Chicago: S. C. Griggs, 1888.

Dewey, John. "Matthew Arnold and Robert Browning." *Andover Review*, August 1891.

Dewey, John. *The Quest for Certainty: A Study of the Relation of Knowledge and Action*. New York: Minton, Balch, 1929.

Dilcher, Roman. "How Not to Conceive Heraclitean Harmony." In *Doctrine and Doxography: Studies on Heraclitus and Pythagoras*, edited by David Sider and Dirk Obbink, 263–80. Berlin: De Gruyter, 2013.

Diogenes Laertius. *Lives of Eminent Philosophers*. Translated by R. D. Hicks. 2 vols. Loeb Classical Library. Cambridge, MA: Harvard University Press, 1925.

Dirac, P. A. M. *The Principles of Quantum Mechanics*. 2nd ed. Oxford: Clarendon Press, 1947.

Dodds, E. R. "The *Parmenides* of Plato and the Origin of the Neoplatonic One." *Classical Quarterly* 22 (1928): 129–41.

Dostoyevsky, Fyodor. *Notes from Underground*. Translated by Constance Garnett. Mineola, NY: Dover, 1992.

Drake, Stillman, and C. D. O'Malley, trans. *The Controversy on the Comets of 1618: Galileo Galilei, Horatio Grassi, Mario Guiducci, Johann Kepler*. Philadelphia: University of Pennsylvania Press, 1960.

Dreyfus, Hubert L. *What Computers Still Can't Do: A Critique of Artificial Reason*. Cambridge, MA: MIT Press, 1992.

Dunne, J. W. *An Experiment with Time*. 3rd ed. London: Faber and Faber, 1934.

Dutry, F. "Le Centenaire de Camille Flammarion." *Ciel et terre* 58 (1942): 166.

Ehrenfels, Christian von. *Das Primzahlengesetz*. Leipzig: O. R. Reisland, 1922.

Einstein, Albert. "Á propos de *La Déduction relativiste* de M. Émile Meyerson." *Revue philosophique de la France et de l'étranger* 105 (1928): 161–66.

Einstein, Albert. "Die Grundlage der allgemeinen Relativitätstheorie." *Annalen der Physik* 49 (1916): 769-822.

Einstein, Albert. "On the Electrodynamics of Moving Bodies." In *The Principle of Relativity*, translated by W. Perrett and G. B. Jeffery, 33-65. New York: Dover, 1952.

Einstein, Albert. *The Travel Diaries of Albert Einstein: The Far East, Palestine & Spain, 1922-23*. Edited by Z. Rosenkranz. Princeton, NJ: Princeton University Press, 2018.

Einstein, Albert. "Zur Elekrodynamik bewegter Körper." *Annalen der Physik* 17 (1905): 891-921.

Einstein, Albert, et al. "La Théorie de la relativité: Séance du 6 avril 1922." *Bulletin de la société française de la philosophie* 22, no. 3 (1922): 91-113.

Einstein, Albert, and Michele Besso. *Correspondance 1903-1955*. Translated by Pierre Speziali. Paris: Hermann, 1979.

Einstein, Albert, Hedwig Born, and Max Born. *Briefwechsel 1916-1955*. Edited by M. Born. Munich: Nymphenburger, 1969.

Einstein, Albert, B. Podolsky, and N. Rosen. "Can Quantum-Mechanical Description of Physical Reality Be Considered Complete?" *Physical Review* 47 (1935): 777-80.

Eisenstadt, Shmuel. "The Axial Age in World History." In *The Cultural Values of Europe*, edited by Hans Joas and Klaus Wiegandt, 22-42. Liverpool: Liverpool University Press, 2008.

Eldridge, Hannah Vandegrift. *Lyric Orientation: Hölderlin, Rilke, and the Poetics of Community*. Ithaca, NY: Cornell University Press, 2015.

Eliot, T. S. "A Commentary." *Monthly Criterion* 6, no. 4 (October 1927).

Eliot, T. S. "The Validity of Artificial Distinctions." *Times Literary Supplement*, May 30, 2014.

Erickson, Paul. *The World the Game Theorists Made*. Chicago: University of Chicago Press, 2015.

Erickson, Paul, et al. *How Reason Almost Lost Its Mind: The Strange Career of Cold War Rationality*. Chicago: University of Chicago Press, 2013.

Euclid. *The Thirteen Books of Euclid's "Elements."* Translated by Thomas L. Heath. 3 vols. Mineola, NY: Dover, 1956.

Euler, Leonhard. *Letters of Euler on Different Subjects in Physics and Philosophy Addressed to a German Princess*. Translated by H. Hunter. 2nd ed. London: Murray and Highley, 1802.

Euler, Leonhard. *Lettres à une Princesse d'Allemagne sur divers sujets de Physique & de Philosophie*. Saint Petersbourg: De l'Imprimerie de l'Académie impériale des sciences, 1768.

Euler, Leonhard. *Lettres de M. Euler à une princesse d'Allemagne sur différentes questions de physique et de philosophie. Nouvelle Édition, Avec des Additions, par MM. le Marquis de Condorcet et De La Croix, Tome Premier*. Paris, 1787.

Euler, Leonhard. *Methodus Inveniendi Lineas Curvas Maximi Minimive Proprietate Gaudentes*. 1744. Reprinted in *Leonhardi Euleri opera omnia*, ser. 1, vol. 24, edited by C. Carathéodory, Zurich: Orell Füssli, 1952.

al-Fārābī, Abū Naṣr. *Kitāb al-siyāsa al-madaniyya (also Known as the Treatise on the Principles of Beings)*. Edited by F. M. Najjar. Beirut: Imprimerie Catholique, 1964.

al-Fārābī, Abū Naṣr. "The Letter Concerning the Intellect." In *Philosophy in the Middle Ages: The Christian, Islamic, and Jewish Traditions*, translated by A. Hyman, 215–21. Indianapolis, IN: Hackett, 1973.

al-Fārābī, Abū Naṣr. *Mabādi' ārā' ahl al-madīnah al-fāḍilah (al-Farabi on the Perfect State)*. Translated by R. Walzer. Oxford: Clarendon Press, 1985.

al-Fārābī, Abū Naṣr. "On the Intellect." In *Classical Arabic Philosophy: An Anthology of Sources*, translated by J. McGinnis and D. C. Reisman, 68–78. Indianapolis, IN: Hackett, 2007.

al-Fārābī, Abū Naṣr. "Political Regime." In *The Political Writings II: "Political Regime" and "Summary of Plato's Laws,"* translated by C. E. Butterworth, 27–94. Ithaca, NY: Cornell University Press, 2015.

al-Fārābī, Abū Naṣr. *The Political Writings: Selected Aphorisms and Other Texts*. Edited and translated by C. E. Butterworth. Ithaca, NY: Cornell University Press, 2001.

al-Fārābī, Abū Naṣr. *Risalat fī'l-'aql*. Edited by M. Bouyges. Beirut: Imprimerie Catholique, 1938.

Feke, Jacqueline. "Mathematizing the Soul: The Development of Ptolemy's Psychological Theory from *On the Kritērion* and *Hēgemonikon* to the *Harmonics*." *Studies in History and Philosophy of Science* 43 (2012): 585–94.

Fernández Villar, Eduardo. "Jorge Luis Borges: The Fear without Trembling." In *Kierkegaard's Influence on Literature, Criticism and Art*, edited by Jon Stewart, 5:21–32. London: Routledge, 2013.

Ferreirós, José. "The Crisis in the Foundations of Mathematics." In *Princeton Companion to Mathematics*, edited by I. Leader, J. Barrow-Green, T. Gowers, 142–56. Princeton, NJ: Princeton University Press, 2010.

Feynman, Richard. *Lectures on Physics*. 3 vols. Boston: Addison-Wesley, 1961–1963.

Fichte, Johann Gottlieb. *Sämmtliche Werke*. 8 vols. Edited by J. H. Fichte. Berlin: De Gruyter, 1965. First published 1845–1846 by Veit (Berlin).

Fichte, Johann Gottlieb. *The Vocation of the Scholar*. Translated by William Smith. Dumfries: Anodos Books, 2017.

Field, Hartry. *Science without Numbers: A Defense of Nominalism*. 2nd ed. Oxford: Oxford University Press, 2016.

Flammarion, Camille. *Lumen*. Translated by A. A. M. and R. M. New York: Dodd, Mead, 1897.

Flores, Eduardo V. "A Model of Quantum Reality." Unpublished. https://ui.adsabs.harvard.edu/abs/2013arXiv1305.6219F/abstract.

Fodor, Jerry. *The Language of Thought*. Cambridge, MA: MIT Press, 1975.

Fogel, Robert W., Mark Guglielmo, and Nathanial Grotte. *Political Arithmetic: Simon Kuznets and the Empirical Tradition in Economics*. Chicago: University of Chicago Press, 2013.

Forman, Paul. "*Kausalität, Anschaulichkeit*, and *Individualität*, or How Cultural Values Prescribed the Character and Lessons Ascribed to Quantum Mechanics." In *Society and Knowledge*, edited by Nico Stehr and Volker Meja, 333–47. New Brunswick, NJ: Transaction Books, 1984.

Forman, Paul. *Weimar Culture, Causality, and Quantum Theory: Adaptation by German*

Physicists and Mathematicians to a Hostile Environment. Historical Studies in the Physical Sciences 3. Philadelphia: University of Pennsylvania Press, 1971.

Foucault, Michel. *"Omnes et singulatim:* Towards a Criticism of 'Political Reason.'" In *The Tanner Lectures on Human Values,* edited by S. McMurrin and translated by P. E. Dauzat, 2:223–54. Salt Lake City: University of Utah Press, 1981.

Fourier, Joseph. "Extrait d'un mémoire sur la théorie analytique des assurances." *Annales de chimie et de physique,* 2nd ser., 10 (1819): 177–89.

Fowler, D. H. *The Mathematics of Plato's Academy: A New Reconstruction.* 2nd ed. Oxford: Oxford University Press, 1999.

Frazer, Sir James G. *The Golden Bough.* Abridged ed. New York: Macmillan, 1922.

Frege, Gottlob. *Die Grundlagen der Arithmetik.* Breslau: Wilhelm Koebner, 1884.

Frege, Gottlob. *The Foundations of Arithmetic.* Translated by J. L. Austin. Evanston, IL: Northwestern University Press, 1978.

Frege, Gottlob. "Über Sinn und Bedeutung." *Zeitschrift für Philosophie und philosophische Kritik,* n.s., 100 (1892): 25–50.

French, Steven, and Décio Krause. *Identity in Physics: A Historical, Philosophical, and Formal Analysis.* Oxford: Oxford University Press, 2006.

Freud, Sigmund. *The Standard Edition of the Complete Psychological Works of Sigmund Freud.* Translated and edited by James Strachey and Anna Freud. 24 vols. London: The Hogarth Press and The Institute of Psycho-analysis, 1953–1974. Reprint, New York: Vintage, 1999.

Freudenthal, Hans. "The Main Trends in the Foundations of Geometry in the 19th Century." *Studies in Logic and the Foundations of Mathematics* 44 (1966): 613–21.

Friedländer, Saul. *Franz Kafka: The Poet of Shame and Guilt.* New Haven, CT: Yale University Press, 2013.

Friedman, Michael. *Kant and the Exact Sciences.* Cambridge, MA: Harvard University Press, 1992.

Friedman, Milton. "The Methodology of Positive Economics." In *Essays in Positive Economics,* 3–43. Chicago: University of Chicago Press, 1953.

Friedman, Milton. "The Methodology of Positive Economics." In *The Methodology of Positive Economics: Reflections on the Milton Friedman Legacy,* edited by Uskali Mäki, 3–42. Cambridge: Cambridge University Press, 2009.

Fuchs, C. A. and Asher Peres. "Quantum Theory Needs No 'Interpretation.'" *Physics Today* 53 (2000): 70–71.

Garff, Joakim. *Søren Kierkegaard: A Biography.* Translated by Bruce H. Kirmmse. Princeton, NJ: Princeton University Press, 2005.

Garrett, Don. *Cognition and Commitment in Hume's Philosophy.* New York: Oxford University Press, 1997.

Geach, Peter. "Truth and God." *Proceedings of the Aristotelean Society,* Supplement 56 (1982): 83–97.

Gentzen, G. "The Consistency of Arithmetic." In *The Collected Papers of Gerhard Gentzen,* edited by M. E. Szabo, 132–213. Amsterdam: North-Holland, 1969.

Gentzen, G. "Die Widerspruchfreiheit der reinen Zahlentheorie." *Mathematische Annalen* 112 (1936): 493–565.

Ghazālī, Abū Ḥāmid al-. *The Incoherence of the Philosophers*. Translated by Michael E. Marmura. Provo, UT: Brigham Young University Press, 2000.

Gibbs, Josiah. *Elementary Principles in Statistical Mechanics*. Mineola, NY: Dover, 1960.

Gill, Christopher. "The Body's Fault? Plato's *Timaeus* on Psychic Illness." In *Reason and Necessity: Essays on Plato's "Timaeus,"* edited by M. R. Wright, 59-84. London: Duckworth, 2000.

Gilson, Etienne. *L'Être et l'essence*. Paris: Vrin, 1948.

Ginzburg, Carlo. *Wooden Eyes: Nine Reflections on Distance*. New York: Columbia University Press, 2001.

Glynn, Michael. *Vladimir Nabokov: Bergsonian and Russian Formalist Influences in His Novels* New York: Palgrave Macmillan, 2007.

Gödel, Kurt. "On Intuitionistic Arithmetic and Number Theory." In *The Undecidable: Basic Papers on Undecidable Propositions, Unsolvable Problems and Computable Functions*, edited by Martin Davis, 75-81. 1965. Reprint, Mineola, NY: Dover, 1993.

Gödel, Kurt. "What Is Cantor's Continuum Problem." In *Collected Works*. Vol. 2, *Publications 1938-1974*, edited by Solomon Feferman, 254-70. Oxford: Oxford University Press, 1990.

Gödel, Kurt. "Zur intuitionistischen Arithmetik und Zahlentheorie." In *Ergebnisse eines mathematischen Kolloquiums*, edited by Karl Menger, 4:34-38. Leipzig: F. Deuticke, 1933.

Goethe, Johann Wolfgang von. *Maximen und Reflexionen*. Edited by Max Hecker. Weimar: Goethe-Gesellschaft, 1907.

Goldstine, Herman H. *A History of the Calculus of Variations from the 17th through the 19th Century*. New York: Springer, 1980.

Gordon, Peter. *Continental Divide: Heidegger, Cassirer, Davos*. Cambridge, MA: Harvard University Press, 2010.

Graham, Daniel W. *Explaining the Cosmos: The Ionian Tradition of Scientific Philosophy*. Princeton, NJ: Princeton University Press, 2006.

Graham, Daniel W. "Once More unto the Stream." In *Doctrine and Doxography: Studies on Heraclitus and Pythagoras*, edited by David Sider and Dirk Obbink, 303-20. Berlin: De Gruyter, 2013.

Grattan-Guinness, Ivor. "Numbers, Magnitudes, Ratios, and Proportions in Euclid's Elements: How Did He Handle Them?" *Historia Mathematica* 23 (1996): 355-75.

Gray, Jeremy, ed. *The Symbolic Universe: Geometry and Physics 1890-1930*. Oxford: Oxford University Press, 1999.

Greenblatt, Stephen. *The Swerve: How the World Became Modern*. New York: W. W. Norton, 2011.

Gregory, Andrew. *The Presocratics and the Supernatural: Magic, Philosophy and Science in Early Greece*. London: Bloomsbury, 2013.

Griffel, Frank. *Al-Ghazālī's Philosophical Theology*. New York: Oxford, 2009.

Guthrie, W. K. *History of Greek Philosophy*. Vol. 2, *The Presocratic Tradition from Parmenides to Democritus*. Cambridge: Cambridge University Press, 1965.

Haase, G. Wolfgang. "Untersuchungen zu Nikomachos von Gerasa." PhD diss., University of Tübingen, 1974.

Hacking, Ian. "Biopower and the Avalanche of Printed Numbers." *Humanities in Society* 5 (1982): 279–95.

Hacking, Ian. *The Taming of Chance*. Cambridge: Cambridge University Press, 1990.

Hacking, Ian. "What Mathematics Has Done to Some and Only Some Philosophers." In *Mathematics and Necessity: Essays in the History of Philosophy*, edited by Timothy Smiley, 83–138. Oxford: Oxford University Press on behalf of the British Academy, 2000.

Hadot, Pierre. *The Inner Citadel: The Meditations of Marcus Aurelius*. Cambridge, MA: Harvard University Press, 1998.

Halévy, Daniel. *Essai sur l'accélération de l'histoire*. Paris: Éditions Self/Les Îles d'Or, 1948.

Hankinson, R. J. "Stoic Epistemology." In *The Cambridge Companion to the Stoics*, edited by Brad Inwood, 59–84. Cambridge: Cambridge University Press, 2003.

Hansen, Morkore Stigel. "The Spirit of Europe: Heidegger and Valéry on the 'End of Spirit.'" PhD diss., London School of Economics and Political Science, 2017.

Harrison, Carol. *Rethinking Augustine's Early Theology: An Argument for Continuity*. Oxford: Oxford University Press, 2006.

Hatfield, Gary. "The Cognitive Faculties." In *The Cambridge History of Seventeenth-Century Philosophy*, edited by Daniel Garber and Michael Ayers, 953–1002. Cambridge: Cambridge University Press, 1998.

Hausman, Daniel. "John Stuart Mill's Philosophy of Economics." *Philosophy of Science* 48 (1981): 363–85.

Hausman, Daniel. *The Separate and Inexact Science of Economics*. Cambridge: Cambridge University Press, 1991.

Hawking, Stephen. "Unified Theory—Stephen Hawking at European Zeitgeist 2011." Posted by Google Zeitgeist, May 18, 2011, video, 26:04. https://www.youtube.com /watch?v=r4TOliLZmcw.

Hayek, F. A. "The Use of Knowledge in Society." *American Economic Review* 35 (1945): 519–30.

Heath, Thomas L. *A History of Greek Mathematics*. 2 vols. Minneola, NY: Dover, 1981.

Hecht, Anthony. *The Transparent Man: Poems*. New York: Knopf, 1990.

Heckman, James. "Causal Parameters and Policy Analysis in Economics: A Twentieth Century Retrospective." *Quarterly Journal of Economics* 115, no. 1 (February 2000): 45–97.

Heckman, James A. "Randomization and Social Policy Evaluation." In *Evaluating Welfare and Training Programs*, edited by Charles F. Manski and Irwin Garfinkel, 201–30. Cambridge, MA: Harvard University Press, 1992.

Heidegger, Martin. *Basic Problems of Phenomenology*. Translated by Albert Hofstadter. Indianapolis: Indiana University Press, 1982.

Heidegger, Martin. *Being and Time*. Translated by John Macquarrie and Edward Robinson. New York: Harper Perennial, 2008.

Heidegger, Martin. *Being and Truth*. Translated by Gregory Fried and Richard Polt. Bloomington: Indiana University Press, 2010.

Heidegger, Martin. *Gesamtausgabe*. Vol. 4, *Erläuterung zu Hölderlins Dichtung*, edited by F.-W. von Herrmann. Vol. 16, *Reden und andere Zeugnisse eines Lebensweges*, edited by H. Heidegger. Frankfurt am Main: Vittorio Klostermann, 1996, 2000.

Heidegger, Martin. *Holzwege*. Frankfurt am Main: Vittorio Klostermann, 1950.

Heidegger, Martin. *Identity and Difference*. Translated by Joan Stambaugh. Chicago: University of Chicago Press, 1969.

Heidegger, Martin. *Kant and the Problem of Metaphysics*. Translated by Richard Taft. Bloomington: Indiana University Press, 1997.

Heidegger, Martin. *Poetry, Language, Thought*. Translated by Albert Hofstadter. New York: Harper Collins, 1971.

Heisenberg, Werner. *Der Teil und das Ganze: Gespräche im Umkreis der Atomphysik*. Munich: R. Piper, 1969.

Heller-Roazen, Daniel. *The Fifth Hammer: Pythagoras and the Disharmony of the World*. Cambridge, MA: MIT Press, 2011.

Hempel, Carl. "Geometry and Empirical Science." *American Mathematical Monthly* 52 (1945): 7–17.

Hempel, Carl. "On the Nature of Mathematical Truth." *American Mathematical Monthly* 52 (1945): 543–56.

Hemsterhuis, François. *Lettre sur la sculpture*. Amsterdam: Marc Michel Rey, 1769.

Hentschel, Klaus, ed., and Ann M. Hentschel, trans. *Physics and National Socialism: An Anthology of Primary Sources*. Basel: Birkhäuser, 1996.

Herakleitos and Diogenes. *Herakleitos and Diogenes*. Translated by Guy Davenport. Bolinas, CA: Grey Fox Press, 1979.

Herder, Johann Gottfried von. *Philosophical Writings*. Edited by M. N. Forster. Cambridge: Cambridge University Press, 2002.

Hernández, José. *El gaucho Martín Fierro y La vuelta de Martín Fierro*. Bueno Aires: El Ateneo, 1950.

Hilbert, David. *Foundations of Geometry*. Chicago: Open Court, 1971.

Hobbes, Thomas. *Leviathan*. Edited by Edwin Curley. Indianapolis, IN: Hackett, 1994.

Hobbes, Thomas. *The Metaphysical System of Hobbes: In Twelve Chapters from "Elements of Philosophy Concerning Body," Together with Briefer Extracts from "Human Nature" and "Leviathan."* Edited by Mary Whiton Calkins. Chicago: Open Court, 1913.

Hoefer, Carl. "Causal Determinism." In *Stanford Encyclopedia of Philosophy*. Updated January 21, 2016. https://plato.stanford.edu/entries/determinism-causal/.

Høgel, Christian. "An Early Anonymous Greek Translation of the Qur'ān: The Fragments from Niketas Byzantios' Refutatio and the Anonymous Abjuratio." *Collectanea Christiana Orientalia* 7 (2010): 65–119.

Horn, Christoph. "Augustins Philosophie der Zahlen." *Revue des Études Augustiniennes* 40 (1994): 389–415.

Høyrup, Jens. *Influences of Institutionalized Mathematics Teaching on the Development and Organization of Mathematical Thought in the Pre-Modern Period: Investigations into an Aspect of the Anthropology of Mathematics*. Materialien und Studien: Institut für Didaktik der Mathematik der Universität Bielefeld, vol. 20. Roskilde: Roskilde University Center, 1980.

Høyrup, Jens. *Lengths, Widths, Surfaces: A Portrait of Old Babylonian Algebra and Its Kin*. New York: Springer, 2002.

Huffman, C. A. "Limité et illimité chez les premiers philosophes grecs." In *La Fêlure du Plaisir: Ètudes sur le Philèbe de Platon*, vol. 2, *Contextes*, edited by M. Dixsaut, 11–31. Paris: Vrin, 1999.

Huffman, C. A. *Philolaus of Croton: Pythagorean and Presocratic; A Commentary on the Fragments and Testimonia with Interpretive Essays.* Cambridge: Cambridge University Press, 1993.

Huffman, C. A. "The Pythagorean Tradition." In *The Cambridge Companion to Early Greek Philosophy*, edited by A. A. Long, 66–87. Cambridge: Cambridge University Press, 1999.

Huffman, C. A. "The Role of Number in Philolaus' Philosophy." *Phronesis* 33 (1988): 1–30.

Hume, David. *Enquiries Concerning Human Understanding and Concerning the Principles of Morals.* 3rd ed. Edited by L. A. Selby-Bigge and revised by P. H. Nidditch. Oxford: Clarendon Press, 1975.

Hume, David. *A Treatise of Human Nature.* 2nd ed. Edited by L. A. Selby-Bigge and revised by P. H. Nidditch. Oxford: Clarendon Press, 1975.

Hurwicz, Leonid. "On the Structural Form of Interdependent Systems." In *Logic, Methodology, and Philosophy of Science: Proceedings of the 1960 International Congress*, edited by Ernest Nagel, Patrick Suppes, and Alfred Tarski, 232–39. Stanford, CA: Stanford University Press, 1962.

Husserl, Edmund. *The Crisis of European Sciences and Transcendental Phenomenology.* Translated by D. Carr. Evanston, IL: Northwestern University Press, 1970.

Husserl, Edmund. *Gesammelte Werke*, edited by Walter Biemel. Vol. 6, *Die Krisis der europäischen Wissenschaften und die transzendentale Phänomenologie: Eine Einleitung in die phänomenologische Philosophie.* The Hague: Martinus Nijhoff, 1954.

Husserl, Edmund. *Logical Investigations.* 2 vols. Translated by J. N. Findlay from the 2nd German ed. (1913). London: Routledge, 1970.

Husserl, Edmund. *Phenomenology and the Crisis of Philosophy: Philosophy as Rigorous Science and Philosophy and the Crisis of European Man.* Translated by Quentin Lauer. New York: Harper & Row, 1965.

Ibn Ṭufayl, Muḥammad ibn ʿAbd al-Malik. *Hayy ben Yaqdhān: Roman philosophique d'Ibn Thofaïl.* 2nd ed. Translated by Léon Gauthier. Beirut: Imprimerie catholique, 1936.

Ibn Ṭufayl, Muḥammad ibn ʿAbd al-Malik. *Ibn Tufayl's Hayy Ibn Yaqzan: A Philosophical Tale.* Translated by Lenn Evan Goodman. Chicago: University of Chicago Press, 2009.

Ibn Ṭufayl, Muḥammad ibn ʿAbd al-Malik. *Philosophus autodidacticus.* Translated by Edward Pococke. Oxford, 1671.

Ichigō, Masamichi. *Madhyamakālaṃkāra of Śāntarakṣita with His Own Commentary or Vṛtti, and with the Subcommentary or Pañjikā of Kamalaśīla.* 2 vols. Kyoto: Buneido, 1985.

Imhausen, Annette. *Mathematics in Ancient Egypt: A Contextual History.* Princeton, NJ: Princeton University Press, 2016.

Isaac, Joel. "Tool Shock: Technique and Epistemology in the Postwar Social Sciences." In "The Unsocial Science? Economics and Neighboring Disciplines since 1945," edited by Roger E. Backhouse and Philippe Fontaine. Supplement, *History of Political Economy* 42, suppl. 1 (2010): 133–64.

Jacquette, Dale. *David Hume's Critique of Infinity*. Leiden: Brill, 2001.

James, William. *Pragmatism: A New Name for Some Old Ways of Thinking*. London: Longmans, Green, 1907.

James, William. *The Principles of Psychology*. 2 vols. Mineola, NY: Dover, 1950. First published 1890 by Henry Holt (New York).

Janos, D. "The Greek and Arabic Proclus and al-Fārābī's Theory of Celestial Intellection and Its Relation to Creation." *Documenti e studi sulla tradizione filosofica medievale* 19 (2010): 19–44.

Janos, D. *Method, Structure, and Development in al-Fārābī's Cosmology*. Leiden: Brill, 2012.

Jaspers, Karl. *The Way to Wisdom: An Introduction to Philosophy*. New Haven, CT: Yale University Press, 2003.

Jauernig, Anja. "The Modal Strength of Leibniz's Principle of the Identity of Indiscernibles." *Oxford Studies in Early Modern Philosophy* 4 (2008): 191–225.

Jaynes, Edwin T. "The Gibbs Paradox." In *Maximum Entropy and Bayesian Methods*, edited by C. R. Smith, G. J. Erickson, and P. O. Neudorfer, 1–22. Dordrecht: Kluwer Academic, 1992.

Jaynes, Edwin T. "Gibbs vs Boltzmann Entropies." *American Journal of Physics* 33 (1965): 391–98.

Jevons, William Stanley. "Brief Account of a General Mathematical Theory of Political Economy." *Journal of the Royal Statistical Society* 29 (1866): 282–87.

Joas, Hans. "The Axial Age Debate in Religious Discourse." In *The Axial Age and Its Consequences*, edited by Robert N. Bellah and Hans Joas, 9–29. Cambridge, MA: Harvard University Press, 2012.

Johansen, T. K. *Plato's Natural Philosophy: A Study of the* Timaeus-Critias. Cambridge: Cambridge University Press, 2004.

Johnson, Samuel. *The History of Rasselas, Prince of Abyssinia*. Oxford World's Classics. Oxford: Oxford University Press, 2009.

Jonas, Hans. *The Imperative of Responsibility: In Search of an Ethics for the Technological Age*. Chicago: University of Chicago Press, 1984.

Jones, Matthew L. *The Good Life in the Scientific Revolution: Descartes, Pascal, Leibniz, and the Cultivation of Virtue*. Chicago: University of Chicago Press, 2006.

Jones, Matthew L. *Reckoning with Matter: Calculating Machines, Innovation, and Thinking about Thinking from Pascal to Babbage*. Chicago: University of Chicago Press, 2016.

Jones, Matthew L. "Space, Evidence, and the Authority of Mathematics." In *Routledge Companion to Eighteenth Century Philosophy*, edited by Aaron Garrett, 203–31. London: Routledge, 2014.

Josephson-Storm, Jason Ā. *The Myth of Disenchantment: Magic, Modernity, and the Birth of the Human Sciences*. Chicago: University of Chicago Press, 2017.

Jullien, François. *The Great Image Has No Form, or On the Nonobject through Painting*. Translated by Jane Marie Todd. Chicago: University of Chicago Press, 2009.

Jung, Carl G., and Wolfgang Pauli. *Atom and Archetype: The Pauli/Jung Letters*. Edited by C. A. Meier. Princeton, NJ: Princeton University Press, 2001.

Jung, Carl G., and Wolfgang Pauli. *The Interpretation of Nature and the Psyche (1932–1958)*. Princeton, NJ: Princeton University Press, 1969.

Kafka, Franz. *Aphorisms*. Translated by W. and E. Muir and M. Hofmann. New York: Schocken Books, 2015.

Kafka, Franz. *Die Erzählungen*. Edited by Roger Hermes. Frankfurt am Main: Fischer, 1996.

Kafka, Franz. *Der Prozess*. Berlin: Die Schmiede, 1925.

Kafka, Franz. *Das Urteil und andere Erzählungen*. Berlin: Fischer Taschenbücher, 1952.

Kagel J. H. and A. E. Roth, eds. *Handbook of Experimental Economics*. Princeton, NJ: Princeton University Press, 1995.

Kahn, Charles H. *Anaximander and the Origins of Greek Cosmogony*. New York: Columbia University Press, 1960.

Kahn, Charles H. *The Art and Thought of Heraclitus*. Cambridge: Cambridge University Press, 1979.

Kahneman, Daniel. *Thinking, Fast and Slow*. New York: Farrar, Straus and Giroux, 2011.

Kalvesmaki, Joel. *The Theology of Arithmetic: Number Symbolism in Platonism and Early Christianity*. Hellenic Studies 59. Cambridge, MA: Harvard University Press, 2013.

Kamenica, Emir. "Contextual Inference in Markets: On the Informational Content of Product Lines." *American Economic Review* 98 (2008): 2127–49.

Kanamori, Akihiro. *The Higher Infinite*. Berlin: Springer, 1994.

Kant, Immanuel. *Critique of Pure Reason*. Translated by Paul Guyer and Allen W. Wood. Cambridge: Cambridge University Press, 1998.

Kant, Immanuel. *Critique of Practical Reason*. Translated by Mary Gregor. Cambridge: Cambridge University Press, 2015.

Kant, Immanuel. *Gesammelte Schriften*. Edited by Akademie der Wissenschaften. Vol. 5, *Kritik der praktischen Vernunft, Kritik der Urtheilskraft*. Berlin: De Gruyter, 1913.

Kant, Immanuel. *Prolegomena to Any Future Metaphysics*. Translated by James W. Ellington. Indianapolis, IN: Hackett, 1977.

Kapstein, Matthew. "Mereological Considerations in Vasunandhu's 'Proof of Idealism.'" In *Reason's Traces: Identity and Interpretation in Indian and Tibetan Buddhist Thought*, 181–204. Boston: Wisdom Publications, 2001.

Kaufmann, W. *Discovering the Mind: Goethe, Kant, and Hegel*. New York: McGraw Hill, 1980.

Kavka, Gregory S. "The Toxin Puzzle." *Analysis* 43 (1983): 33–36.

Kay, Leslie M. "Two Minds about Odors." *Proceedings of the National Academy of Sciences* 101 (2004): 17569–70.

Kay, Leslie M. "Two Species of Gamma Oscillations in the Olfactory Bulb: Dependence on Behavioral State and Synaptic Interactions." *Journal of Integrative Neuroscience* 2 (2003): 31–44.

Kay, Leslie M., M. Krysiak, L. Barlas, and G. B. Edgerton. "Grading Odor Similarities in a Go/No-Go Task." *Physiology and Behavior* 88 (2006): 339–46.

Kay, Leslie M., and S. M. Sherman. "Argument for an Olfactory Thalamus." *Trends in Neurosciences* 30 (2006): 47–53.

Keats, John. *The Complete Poetical Works and Letters of John Keats, Cambridge Edition.* London: Houghton Mifflin, 1899.

Kenney, John Peter. *Contemplation and Classical Christianity: A Study in Augustine.* Oxford: Oxford University Press, 2013.

Kierkegaard, Søren. *The Concept of Anxiety: A Simple Psychologically Orienting Deliberation on the Dogmatic Issue of Hereditary Sin.* Edited and translated by Reidar Thomte and Albert B. Anderson. Princeton, NJ: Princeton University Press, 1980.

Kierkegaard, Søren. *The Concept of Irony.* Edited and translated by H. V. Hong and E. H. Hong. Princeton, NJ: Princeton University Press, 1989.

Kierkegaard, Søren. *Concluding Unscientific Postscript to "Philosophical Fragments."* Edited and translated by H. V. Hong and E. H. Hong. 2 vols. Princeton, NJ: Princeton University Press, 1992.

Kierkegaard, Søren. *Either/Or Part II.* Edited and translated by H. V. Hong and E. H. Hong. Princeton, NJ: Princeton University Press, 1987.

Kierkegaard, Søren. *Fear and Trembling/Repetition.* Edited and translated by H. V. Hong and E. H. Hong. Princeton, NJ: Princeton University Press, 1983.

Kierkegaard, Søren. *The Sickness unto Death: A Christian Psychological Exposition for Upbuilding and Awakening.* Edited and translated by H. V Hong and E. H. Hong. Princeton, NJ: Princeton University Press, 1980.

Kim, J. and E. Sosa, eds. *Metaphysics: An Anthology.* London: Blackwell, 1999.

al-Kindī. *Al-Kindî's Metaphysics.* Translated by Alfred Ivry. Albany: State University of New York Press, 1974.

al-Kindī. *Rasâ'il al-Kindî al-Falsafîya.* Edited by Muḥammad 'Abd-al-Hādī Abû Rîdah. 2 vols. Cairo: Dâr al-Fikr al-'Arabî, 1950–1953.

Kirkeboen, Lars J., Edwin Leuven, and Magne Mogstad. "Field of Study, Earnings, and Self-Selection." *Quarterly Journal of Economics* 131 (2016): 1057–111.

Kitcher, Philip. *Kant's Transcendental Psychology.* New York: Oxford, 1990.

Kitcher, Philip. "The Plight of the Platonist." *Noûs* 12 (1978): 119–36.

Klein, Jacob. *Greek Mathematical Thought and the Origin of Algebra.* Translated by Eva Brann. New York: Dover, 1992.

Knorr, W. *The Evolution of the Euclidean Elements: A Study of the Theory of Incommensurable Magnitudes and Its Significance for Early Greek Geometry.* Dordrecht: D. Reidel, 1975.

Knorr, W. "On the Early History of Axiomatics: The Interaction of Mathematics and Philosophy in Greek Antiquity." In *Theory Change, Ancient Axiomatics, and Galileo's Methodology: Proceedings of the 1978 Pisa Conference on the History and Philosophy of Science,* edited by J. Hintika, D. Gruender, and E. Agazzi, 1:145–86. Dordrecht: Reidel, 1980.

Korsgaard, Christine. *The Sources of Normativity.* Edited by Onora O'Neill. Cambridge: Cambridge University Press, 1996.

Kraemer, Joel. *Philosophy in the Renaissance of Islam: Abū Sulaymān Al-Sijistānī and His Circle.* Leiden: Brill, 1986.

Kripke, Saul. *Naming and Necessity.* Cambridge MA: Harvard University Press, 1980.

Kunen, Kenneth. *Set Theory: An Introduction to Independence Proofs.* Amsterdam: North-Holland, 1980.

Lagrange, Joseph-Louis. *Œuvres*. Vols. 11, 12. Paris: Gauthier-Villars et Fils, 1888, 1889.

Laks, André, and Glenn W. Most, eds. and trans. *Early Greek Philosophy*. 9 vols. Loeb Classical Library. Cambridge, MA: Harvard University Press, 2016.

Lamarra, Antonio. "Leibniz e la περιχώρησις." *Lexicon Philosophicum* 1 (1985). http:// scholarlysource.daphnet.org/index.php/DDL/article/view/21/10.

Lebedev, Andrei V. "Idealism (Mentalism) in Early Greek Metaphysics and Philosophical Theology: Pythagoras, Parmenides, Heraclitus, Xenophanes and Others (With Some Remarks on the *Gigantomachia about Being* in Plato's *Sophist*." *Indo-European Linguistics and Classical Philology* 23 (2019): 651–704.

Lécuyer, Bernard-Pierre. "Probability in Vital and Social Statistics: Quetelet, Farr, and the Bertillons." In *The Probabilistic Revolution*, vol. 1, *Ideas in History*, edited by Lorenz Krüger, Lorraine Daston, and Michael Heidelberger, 317–35. Cambridge, MA: MIT Press, 1987.

Leibniz, G. W. *De summa rerum: Metaphysical Papers, 1675–1676*. Edited and translated by G. H. R. Parkinson. New Haven, CT: Yale University Press, 1992.

Leibniz, G. W. *Die philosophischen Schriften*. Edited by C. I. Gerhardt. 7 vols. 1875–1890. Reprint, Hildesheim: Olms, 1960–1961.

Leibniz, G. W. *G. W. Leibniz: Philosophical Essays*. Edited and translated by Roger Ariew and Dan Garber. Indianapolis, IN: Hackett, 1989.

Leibniz, G. W. *The Labyrinth of the Continuum: Writings on the Continuum Problem, 1672–1686*. Edited and translated by Richard T. W. Arthur. New Haven, CT: Yale University Press, 2002.

Leibniz, G. W. *The Leibniz-Des Bosses Correspondence*. Edited and translated by Donald Rutherford and Brandon Look. New Haven, CT: Yale University Press, 2007.

Leibniz, G. W. *Leibniz: Philosophical Papers and Letters*. 2nd ed. Edited and translated by Leroy E. Loemker. Dordrecht: D. Reidel, 1989.

Leibniz, G. W. *Mathematische Schriften*. 7 vols. Edited by C. I. Gerhardt. Halle, 1849–1863. Reprint, Hildesheim: Georg Olms, 1963.

Leibniz, G. W. *New Essays on Human Understanding*. Translated by Peter Remnant and Jonathan Bennett. Cambridge: Cambridge University Press, 1981.

Leibniz, G. W. *Sämtliche Schriften und Briefe*. Edited by Deutsche Akademie der Wissenschaften zu Berlin. Darmstadt: O. Reichl; Leipzig: Koehler und Amelang; Berlin: Akademie, 1923–.

Leibniz, G. W. *Theodicy*. Chicago: Open Court Classics, 1985.

Leibniz, G. W., and Samuel Clarke. *A Collection of Papers Which Passed between the Late Learned Mr. Leibnitz, and Dr. Clarke, in the Years 1715 and 1716. Relating to the Principles of Natural Philosophy and Religion*. London: James Knapton, 1717.

Leibniz, G. W., and Samuel Clarke. *Correspondence*. Edited by Roger Ariew. Indianapolis, IN: Hackett, 2000.

Lenard, Philipp. *Deutsche Physik*. 4 vols. Munich: J. F. Lehmann, 1936.

Levenson, Jon D. *The Death and Resurrection of the Beloved Son: The Transformation of Child Sacrifice in Judaism and Christianity*. New Haven, CT: Yale University Press, 1993.

Levitt, Steven D. and John A. List. "What Do Laboratory Experiments Measuring Social Preferences Reveal about the Real World?" *Journal of Economic Perspectives* 21, no. 2 (2007): 153–74.

Lewis, David. "How Many Lives Has Schrödinger's Cat?" *Australasian Journal of Philosophy* 82 (2004): 3–22.

Lewis, David. "New Work for a Theory of Universals." *Australasian Journal of Philosophy* 61 (1984): 343–77.

Lewis, David. *On the Plurality of Worlds*. Oxford: Blackwell, 1986.

Lewis, David. *Parts of Classes*. Oxford: Blackwell, 1991.

Lewis, Wyndham. *Time and Western Man*. London: Chatto & Windus, 1927.

Lindberg, David, and Ronald Numbers. *When Science and Christianity Meet*. Chicago: University of Chicago Press, 2003.

Lisi, Leonardo F. "Rainer Maria Rilke: Unsatisfied Love and the Poetry of Living." In *Kierkegaard's Influence on Literature, Criticism and Art*, edited by Jon Stewart, 231–5. Kierkegaard Research, vol. 12. Abingdon: Routledge, 2016.

Lloyd, G. E. R. *Aristotle: The Growth and Structure of His Thought*. Cambridge: Cambridge University Press, 1968.

Lloyd, G. E. R. *Demystifying Mentalities*. Cambridge: Cambridge University Press, 1990.

Lloyd, G. E. R. "The Meno and the Mysteries of Mathematics." *Phronesis* 37 (1992): 166–83.

Lloyd, G. E. R. *Polarity and Analogy: Two types of Argumentation in Early Greek Thought*. Cambridge: Cambridge University Press, 1966.

Lloyd, G. E. R. "The Theories and Practices of Demonstration." In *Aristotelian Explorations*, edited by G. E. R. Lloyd, 7–37. Cambridge: Cambridge University Press, 1996.

Locke, John. *An Essay Concerning Human Understanding*. Edited by Peter H. Nidditch. Oxford: Clarendon Press, 1975.

Locke, John. *Of the Conduct of the Understanding*. Oxford: Clarendon, 1901.

Loeb, Louis. "The Cartesian Circle." In *The Cambridge Companion to Descartes*, edited by J. Cottingham, 200–235. Cambridge University Press, 1992.

Loeb, Louis. "The Priority of Reason in Descartes." *Philosophical Review* 99 (1990): 3–43.

Lucretius. *On the Nature of Things*. Translated by Martin Ferguson Smith. Indianapolis, IN: Hackett, 2001.

Luis de León. *Poesías de Fray Luis de León*. Edited by A. C. Vega. Madrid: Saeta 1955.

Ma, X.-S. et al. "Quantum Erasure with Causally Disconnected Choice." *Proceedings of the National Academy of Sciences* 110 (2013): 1221–26.

Ma, X.-S., J. Kofler, and A. Zellinger. "Delayed-Choice Gedanken Experiments and Their Realization." *Reviews of Modern Physics* 88 (2016), article 015005. https://doi.org/10.1103/RevModPhys.88.015005.

Manilius. *Astronomica*. Translated by G. P. Goold. Loeb Classical Library. Cambridge, MA: Harvard University Press, 1977.

Mann, N., M. Gibson, and C. Nally, eds. *W. B. Yeats's* A Vision: *Explications and Contexts*. Oxford: Oxford University Press, 2012.

Marius Victorinus. *Le Livre des XXIV philosophes: Résurgence d'un texte du IVe siècle*. Edited and translated by Françoise Hudry. Paris: Vrin, 2009.

al-Marrākushī, ʿAbd al-Wāḥid. *Al-Muʿjib fī talkhīṣ akhbār al-Maghrib*. Beirut: al-Maktaba al-ʿAṣriyya, 2006.

Marshall, Alfred. *Principles of Economics*. 9th variorum ed. London: Macmillan, 1961.

Marx, Karl. *Capital: A Critique of Political Economy*. 3 vols. 1867. New York: International, 2003.

Matte Blanco, Ignacio. *The Unconscious As Infinite Sets: An Essay in Bi-logic*. London: Duckworth, 1975.

Maturana, Humberto R., and Francisco J. Varela. *Autopoesis and Cognition: The Realization of the Living*. Dordrecht: D. Reidel, 1980.

Matz, Aaron. "The Years of Hating Proust." *Comparative Literature* 60 (2008): 355–69.

Maupertuis, Pierre Louis Moreau de. "Les Loix de mouvement et du repos, déduites d'un principe de métaphysique (lu le 15. Avril 1744, dans l'Assemblée publique de l'Académie Royale des Sciences de Paris)." In *Histoire de l'Académie Royale des Sciences et des Belles Lettres*, 267–94. Paris: Académie Royale des Sciences et des Belles Lettres, 1746.

Maxwell, James Clerk. *Theory of Heat*. London: Longmans, Green, 1872.

Mazur, Barry. "When Is One Thing Equal to Some Other Thing?" In *Proof and Other Dilemmas: Mathematics and Philosophy*, edited by B. Gold and R. Simons, 221–42. Washington, DC: Mathematical Association of America, 2008.

McGucken, W. *Nineteenth Century Spectroscopy: Development of the Understanding of Spectra, 1802–1897*. Baltimore: Johns Hopkins University Press, 1969.

McIntosh, Donald. *The Foundations of Human Society*. Chicago: University of Chicago Press, 1969.

McTaggart, John M. E. "The Unreality of Time." *Mind*, n.s., 68 (1908): 457–84.

Mehrtens, Herbert. "Irresponsible Purity: On the Political and Moral Structure of the Mathematical Sciences in the National Socialist State." In *Scientists, Engineers, and National Socialism*, edited by Monika Renneberg and Mark Walker, 324–38. Cambridge: Cambridge University Press, 1994.

Mehrtens, Herbert. "Ludwig Bieberbach and 'Deutsche Mathematik.'" In *Studies in the History of Mathematics*, edited by Esther R. Phillips, 195–241. Washington DC: Mathematical Association of America, 1987.

Mehrtens, Herbert. "Mathématiques, sciences de la nature et national-socialisme: Quelles questions poser?" In *La Science sous le Troisième Reich: Victime ou alliée du nazisme?*, edited by Josiane Olff-Nathan, 33–49. Paris: Seuil 1993.

Mehrtens, Herbert. *Moderne—Sprache—Mathematik: Eine Geschichte des Streits um die Grundlagen der Disziplin und des Subjekts formaler Systeme*. Frankfurt am Main: Suhrkamp, 1990.

Mehrtens, Herbert. "Modernism vs. Counter-modernism, Nationalism vs. Internationalism: Style and Politics in Mathematics, 1900–1950." In *L'Europe mathématique: Histoires, mythes, identités*, edited by Catherine Goldstein, Jeremy Gray, and Jim Ritter, 518–29. Paris: Éditions de la Maison de l'homme, 1996.

Melamed, Y., and M. Lin. "Principle of Sufficient Reason." In *The Stanford Encyclopedia of Philosophy*. Updated September 7, 2016. https://plato.stanford.edu/archives/spr2018/entries/sufficient-reason/.

Mencken, H. L. *A Mencken Chrestomathy*. New York: Vintage, 1982.

Mercer, Christia. "Leibniz, Aristotle, and Ethical Knowledge." In *The Impact of Aristotelianism on Modern Philosophy*, edited by Riccardo Pozzo, 113–47. Washington, DC: Catholic University of America Press, 2004.

Mercer, Christia. "Leibniz on Mathematics, Methodology, and the Good: A Reconsideration of the Place of Mathematics in Leibniz's Philosophy." *Early Science and Medicine* 11 (2006): 424–54.

Mercer, Christia. *Leibniz's Metaphysics: Its Origins and Development*. Cambridge: Cambridge University Press, 2001.

Meyerson, Émile. *Identité et réalité*. Paris: F. Alcan, 1908.

Meyerson, Émile. *Identity and Reality*. Translated by Kate Loewenberg. London: George Allen & Unwin, 1930.

Meyerson, Émile. *Lettres françaises*. Edited by Bernadette Bensaude-Vincent and Eva Telkes-Klein. Paris: CNRS, 2009.

Mill, John Stuart. *The Collected Works of John Stuart Mill*. Vol. 1, *Autobiography and Literary Essays*. Edited by John M. Robson and Jack Stillinger with an introduction by Lord Robbins. Toronto: University of Toronto Press, 1981.

Miłosz, Czesław. *Emperor of the Earth: Modes of Eccentric Vision*. Berkeley: University of California Press, 1977.

Minkowski, Hermann. "Raum und Zeit." *Jahresbericht der Deutschen Mathematiker-Vereinigung* 18 (1909): 75–88.

Minkowski, Hermann. "Space and Time." In *The Principle of Relativity*, translated by W. Perrett and G. B. Jeffery, 73–91. New York: Dover, 1952.

Mirowski, Philip. *More Heat Than Light: Economics as Social Physics, Physics as Nature's Economics*. Cambridge: Cambridge University Press, 1989.

Mises, Ludwig von. *Epistemological Problems of Economics*. Auburn, AL: Ludwig von Mises Institute, 1960.

Mohr, R., K. Sanders, and B. Sattler, eds. *One Book, the Whole Universe: Plato's "Timaeus" Today*. Las Vegas, NV: Parmenides, 2010.

Momigliano, Arnaldo. *Alien Wisdom: The Limits of Hellenization*. Cambridge: Cambridge University Press, 1975.

Mondolfo, Rodolfo. *Heráclito: Textos y problemas de su interpretación*. Mexico City: Siglo 21, 1966.

Morgenstern, Oskar. "Logistik und Sozialwissenschaften." *Zeitschrift für Nationalökonomie* 7 (1936): 1–24.

Morgenstern, Oskar. *Wirtschaftsprognose*. Vienna: Springer, 1928.

Mueller, Ian. "Mathematics and Education: Some Notes on the Platonist Programme." In *"Peri Tōn Mathēmaton": Essays on Greek Mathematics and Its Later Development*, edited by Ian Mueller. Special issue, *Apeiron* 24 (1991): 85–104.

Mueller, Ian. "On the Notion of a Mathematical Starting Point in Plato, Aristotle and Euclid." In *Science and Philosophy in Classical Greece*, edited by A. C. Bowen, 59–97. New York: Garland, 1991.

Muller, Richard A. *Now: The Physics of Time*. New York: W. W. Norton, 2016.

Muñoz, Jerónimo. *Libro del nuevo cometa*. Valencia: Pedro de Huete, 1573.

Muñoz, Jerónimo. *Libro del nuevo cometa, Valencia, Pedro de Huete, 1573; Littera ad Bartholomaeum Reisacherum, 1574; Summa del prognostico del cometa: Valencia,* [. . .] *1578.* Valencia: Hispaniae Scientia, 1981.

al-Muqammas, Dawud. *Twenty Chapters.* Edited and translated by Sarah Stroumsa. Provo, UT: Brigham Young University Press, 2015.

Musil, Robert. "Der mathematische Mensch." 1913. Projekt Gutenberg. https://www.projekt-gutenberg.org/musil/essays/essays.html.

Musil, Robert. *The Man without Qualities.* Translated by Sophie Wilkins and Burton Pike. New York: Vintage International, 1996.

Musil, Robert. *Precision and Soul: Essays and Addresses.* Translated by Burton Pike and David S. Luft. Chicago: University of Chicago Press, 1995.

Musil, Robert. *Tagebücher, Aphorismen, Essays und Reden.* Edited by Adolf Frisé. Hamburg: Rowohlt, 1955.

Nabokov, Vladimir. *Insomniac Dreams: Experiments with Time.* Edited by Gennady Barabtarlo. Princeton, NJ: Princeton University Press, 2018.

Natenzon, Paulo. "Random Choice and Learning." *Journal of Political Economy* 127 (2019): 419–57.

Nelböck, Johann. "Die Bedeutung der Logik im Empirismus und Positivismus." PhD diss., University of Vienna, 1930.

Netz, Reviel. *The Shaping of Deduction in Greek Mathematics: A Study in Cognitive History.* Cambridge: Cambridge University Press, 1999.

Neumann, John von. "Die Axiomatisierung der Mengenlehre." *Mathematische Zeitschrift* 27, no. 1 (1928): 669–752.

Neumann, John von. "Eine Axiomatisierung der Mengenlehre." *Journal für die Reine und Angewandte Mathematik* 154 (1925): 219–24.

Neumann, John von. *Mathematical Foundations of Quantum Mechanics.* Translated by Robert T. Beyer. Princeton, NJ: Princeton University Press, 1955, revised 2018.

Neumann, John von. "The Mathematician." In *Works of the Mind,* edited by Robert B. Heywood, 180–96. Chicago: University of Chicago Press, 1947.

Neumann, John von. *Mathematische Grundlagen der Quantenmechanik.* Berlin: Springer, 1932.

Neumann, John von. "Zur Theorie der Gesellschaftsspiele." *Mathematische Annalen* 100 (1928): 295–320.

Neumann, John von, and Oskar Morgenstern. *A Theory of Games and Economic Behavior.* 1944. Reprint, Princeton, NJ: Princeton University Press, 2004.

Neurath, P. *Gesammelte philosophische und methodologische Schriften.* Edited by R. Haller and H. Rutte. Vienna: Hölder-Pichler-Tempsky, 1981.

Neurath, P. and E. Nemeth, eds. *Otto Neurath oder die Einheit von Wissenschaft und Gesellschaft.* Vienna: Böhlau, 1993.

Newton, Isaac. *The Principia: Mathematical Principles of Natural Philosophy; A New Translation.* Translated I. B. Cohen and Anne Whitman. Berkeley: University of California Press, 1999.

Nicomachus of Gerasa. *Introductio arithmetica libri I.* Edited by Richard Gottfried Hoche. Leipzig: Teubner, 1866.

Nicomachus of Gerasa. *Nicomachus of Gerasa: Introduction to Arithmetic*. Translated by Martin Luther D'Ooge, Frank E. Robbins, and Louis C. Karpinsky. New York: Macmillan, 1926.

Nicomachus of Gerasa. *Theologoumena arithmeticae*. Edited by Victor De Falco. Stuttgart: Teubner, 1975. First published 1922 by Teubner (Leipzig).

Niehoff, Maren. *Philo of Alexandria: An Intellectual Biography*. New Haven, CT: Yale University Press, 2018.

Nietzsche, Friedrich. *Beyond Good and Evil*. Translated by Walter Kaufmann. New York: Vintage 1966.

Nietzsche, Friedrich. *The Gay Science*. Translated by Walter Kaufmann. New York: Vintage, 1974.

Nietzsche, Friedrich. *Human, All Too Human: A Book for Free Spirits*. Translated by R. J. Hollingdale. Cambridge: Cambridge University Press, 1996.

Nietzsche, Friedrich. *On the Genealogy of Morals*. Translated by Walter Kaufmann and R. J. Hollingdale. New York: Vintage, 1967.

Nietzsche, Friedrich. *Die Philosophie im tragischen Zeitalter der Griechen*. In *Werke in drei Bänden*, vol. 3, edited by Karl Schlechta. Munich: Hanser, 1954.

Nietzsche, Friedrich. *Untimely Meditations*. Translated by R. J. Hollingdale. Cambridge: Cambridge University Press, 1983.

Nirenberg, David. *Anti-Judaism: The Western Tradition*. New York: W. W. Norton, 2013.

Nirenberg, Ricardo L., and David Nirenberg. "Badiou's Number: A Critique of Set Theory as Ontology." *Critical Inquiry* 37 (2011): 583–614.

Nirenberg, Ricardo L., and David Nirenberg. "Critical Response." *Critical Inquiry* 38 (2012): 362–87.

O'Brien, E. "Enjoy It Again: Repeat Experiences Are Less Repetitive Than People Think." *Journal of Personality and Social Psychology* 116, no. 4 (2019): 519–40. https://doi.org/10.1037/pspa0000147.

O'Neil, Cathy. *Weapons of Math Destruction: How Big Data Increases Inequality and Threatens Democracy*. New York: Broadway Books, 2016.

Origen. *Homilies on Genesis and Exodus*. Translated by Ronald E. Heine. Washington, DC: The Catholic University of America Press, 1982.

Ormsby, Eric Linn. *Theodicy in Islamic Thought: The Dispute over al-Ghazali's Best of All Possible Worlds*. Princeton NJ: Princeton University Press, 1984.

Owen, David. *Hume's Reason*. Oxford: Oxford University Press, 1999.

Palmer, Ada. *Reading Lucretius in the Renaissance*. Cambridge, MA: Harvard University Press, 2014.

Palmer, John. *Parmenides and Presocratic Philosophy*. Oxford: Oxford University Press, 2009.

Panofsky, Erwin. *Three Essays on Style*. Edited by William S. Heckscher. Cambridge, MA: MIT Press, 1997.

Pascal, Blaise. *Pensées*. Edited by Léon Brunschvicg. Paris: Hachette, 1897.

Pascal, Blaise. *Pensées and Other Writings*. Translated by Honor Levi. Oxford World's Classics Oxford: Oxford University Press, 1999.

Pascal, Blaise. *Traité du triangle arithmétique*. Paris: Guillaume Desprez, 1665.

Pauli, Wolfgang. "Quantentheorie." In *Handbuch der Physik*, vol. 23, *Quanten*. Berlin: Springer, 1926.

Pearson, Karl. *The History of Statistics in the 17th and 18th Centuries against the Changing Background of Intellectual, Scientific, and Religious Thought*. Edited by E. S. Pearson. London: Charles Griffin, 1978.

Peirce, C. S. "The Doctrine of Necessity Examined." *Monist* 2 (1892): 481–86.

Penrose, Roger. *The Emperor's New Mind: Concerning Computers, Minds, and Laws of Physics*. New York: Oxford University Press, 1989.

Penrose, Roger. *Shadows of the Mind: A Search for the Missing Science of Consciousness*. Oxford: Oxford University Press, 1994.

Périer, A. "Un traité de Yaḥyâ ben 'Adī: Défense du dogme de la trinité contre les objections d'al-Kindī." *Revue de l'Orient chrétien*, 3rd. ser., 2 (1920–1921): 3–21.

Perler, Dominik, and Ulrich Rudolph. *Occasionalismus: Theorien der Kausalität im arabisch-islamischen und im europäischen Denken*. Göttingen: Vandenhoek & Ruprecht, 2000.

Pešić, Peter. "The Principle of Identicality and the Foundations of Quantum Theory I. The Gibbs Paradox." *American Journal of Physics* 59 (1991): 971–74.

Pešić, Peter. "The Principle of Identicality and the Foundations of Quantum Theory II: The Role of Identicality in the Formation of Quantum Theory." *American Journal of Physics* 59 (1991): 975–978.

Pešić, Peter. *Seeing Double: Shared Identities in Physics, Philosophy, and Literature*. Cambridge, MA: MIT Press, 2002.

Pétrement, Simone. *Simone Weil: A Life*. 1976. New York: Schocken Books, 1988.

Phillips, J. F. "Neo-Platonic Exegeses of Plato's Cosmology." *Journal of the History of Philosophy* 35 (1997): 173–97.

Pierce, Charles Sanders. *Collected Papers of Charles Sanders Peirce. volume 3, Exact Logic (Published Papers). Volume 4, The Simplest Mathematics*. Edited by Charles Hartshorn and Paul Weiss. Cambridge, MA: The Belknap Press of Harvard University Press, 1933.

Pierce, Benjamin. "Address of Professor Benjamin Pierce, President of the American Association for the Year 1853." *Proceedings of the American Association for the Advancement of Science* 8 (1855): 1–17.

Pindar. *Pindar*. Translated by J. E. Sandys, Loeb Classical Library. Cambridge, MA: Harvard University Press, 1968.

Planck, Max. *Acht Vorlesungen über Theoretische Physik, gehalten an der Columbia University, NY*. Leipzig: S. Hirzel, 1910.

Planck, Max. "Dynamische und statistische Gesetzmässigkeit." In *Physikalische Abhandlungen und Vorträge*, vol. 3, edited by Verband Deutscher Physikalischer Gesellschaften und der Max-Planck-Gesellschaft zur Förderung der Wissenschaften e. V., 77–90. Braunschweig: F. Vieweg, 1958.

Planck, Max. *Eight Lectures on Theoretical Physics*. Translated by A. P. Wills. New York: Columbia University Press, 1915. Reprint, Mineola, NY: Dover, 1998.

Plato. *The Collected Dialogues of Plato*. Edited by Edith Hamilton and Huntington Cairns. Princeton, NJ: Princeton University Press, 1961.

Plato. *Plato: Cratylus, Parmenides, Greater Hippias, Lesser Hippias*. Translated by H. N. Fowler. Loeb Classical Library. Cambridge, MA: Harvard University Press, 2002.

Plato. *Plato: Phaedrus*. Translated by A. Nehamas and P. Woodruff. Indianapolis, IN: Hackett, 1995.

Plato. *Plato: Theaetetus, Sophist*. Translated by H. N. Fowler. Loeb Classical Library. Cambridge, MA: Harvard University Press, 2006.

Pliny. *Natural History*. Translated by H. Rackham, Loeb Classical Library. Cambridge, MA: Harvard University Press, 1938.

Poe, Edgar Allan. *Eureka: A Prose Poem*. Amherst, NY: Prometheus, 1997.

Poincaré, H. *Dernières pensées*. Paris: Flammarion, 1917.

Poincaré, H. "Jubilé scientifique de Camille Flammarion." *L'Astronomie* 26 (1912): 97–153.

Poincaré, H. *La Science et l'hypothèse*. Paris: Flammarion, 1917.

Poincaré, H. "L'Espace et la géométrie." *Revue de métaphysique et de morale* 3 (1895): 631–46.

Poincaré, H. *Science and Method*. Translated by Francis Maitland. New York: Dover: 1952.

Popper, Karl. "The Nature of Philosophical Problems and Their Roots in Science." In *Conjectures and Refutations: The Growth of Scientific Knowledge*, 66–96. London: Routledge and Kegan Paul, 1963.

Popper, Karl. *The Open Society and Its Enemies*. Vol. 1, *The Spell of Plato*. 5th ed. London: Routledge and Kegan Paul, 1945.

Porta, Horacio. "In Memory of Jacob Schwartz." *Notices of the AMS* 62, no. 5 (2015): 488.

Porter, Theodore M. *Karl Pearson: The Scientific Life in a Statistical Age*. Princeton, NJ: Princeton University Press, 2004.

Porter, Theodore M. *The Rise of Statistical Thinking, 1820–1900*. Princeton, NJ: Princeton University Press, 1986.

Poulet, Georges. *Études sur le temps humain*. Vol. 2, *De l'instant éphémère à l'instant éternelle*. 1964. Paris: Librairie Plon, 2017.

Proust, Marcel. *Du côté de chez Swann*. Paris: Bibliothèque de la Pléiade, 1966.

Proust, Marcel. *Le Temps retrouvé*. Paris: Bibliothèque de la Pléiade, 1966.

Ptolemy. *Lettre à Flora*. 2nd ed. Edited by Gilles Quispel. Paris: Cerf, 1966.

Putnam, Hilary. "Time and Physical Geometry." *Journal of Philosophy* 64 (1967): 240–47.

Quine, W. v. O. "Two Dogmas of Empiricism." *Philosophical Review* 60 (1951): 20–43.

Rand, Ayn. *Atlas Shrugged: 35th Anniversary Edition*. New York: Signet, 1992.

al-Rāzī, Abū Bakr Muhammad ibn Zakariyya ibn Yahya. *Opera philosophica*. Edited by Paul Kraus. Cairo, 1939.

Read, Kay Almere. *Time and Sacrifice in the Aztec Cosmos*. Bloomington: Indiana University Press, 1998.

Reed, Robert C. "Decreation as Substitution: Reading Simone Weil through Levinas." *Journal of Religion* 93 (2013): 25–40.

Renger, A.-B., and A. Stavru, eds. *Pythagorean Knowledge from the Ancient to Modern World: Askesis, Religion, Science*. Wiesbaden: Harrassowitz, 2016.

Repgow, Eike von. *Der Sachsenspiegel*. Edited by August Lüppen and Friedrich Kurt von Alten. Amsterdam: Rodopi, 1970.

Rickless, Samuel, "Plato's *Parmenides*." *Stanford Encyclopedia of Philosophy*. Updated January 14, 2020. https://plato.stanford.edu/archives/spr2020/entries/plato-parmenides/.

Rietdijk, C. W. "A Rigorous Proof of Determinism Derived from the Special Theory of Relativity." *Philosophy of Science* 33 (1966): 341–44.

Rietdijk, C. W. "Special Relativity and Determinism." *Philosophy of Science* 43 (1976): 598–609.

Rilke, Rainer Maria. *Briefe aus Muzot, 1921–1926*. Leipzig: Insel, 1936.

Rilke, Rainer Maria. *Rainer Maria Rilke, Lou Andreas-Salomé: Briefwechsel*. Leipzig: Insel, 1952.

Rilke, Rainer Maria. *Les Elégies de Duino, Sonnets à Orphée*. Translated by J.-F. Angelloz. Paris: Aubier, 1943.

Rilke, Rainer Maria. *Letters of Rainer Maria Rilke*. Vol. 2. Translated by Jane Bannard Greene and M. D. Herter Norton. New Haven, CT: Yale University Press, 1947.

Rilke, Rainer Maria. *Rainer Maria Rilke Briefe*. 2 vols. Edited by Ruth Sieber-Rilke and Karl Altheim. Berlin: Insel, 1950.

Rilke, Rainer Maria, and André Gide. *Correspondances 1909–1926*. Paris: Corrêa, 1952.

Robb, Richard. *Willful: How We Choose What We Do*. New Haven, CT: Yale University Press, 2019.

Robbins, Frank Egleston. "Arithmetic in Philo Judaeus." *Classical Philology* 26 (1931): 345–61.

Rochberg, Francesca. *Before Nature: Cuneiform Knowledge and the History of Science*. Chicago: University of Chicago Press, 2016.

Rockmore, Tom. *Art and Truth after Plato*. Chicago: University of Chicago Press, 2013.

Rodriguez-Pereyra, R. *Leibniz's Principle of Identity of Indiscernibles*. Oxford: Oxford University Press, 2014.

Rorty, Richard. *Contingency, Irony, and Solidarity*. Cambridge: Cambridge University Press, 1989.

Rorty, Richard. *Philosophy as Poetry*. Charlottesville: University of Virginia Press, 2016.

Rorty, Richard. "Pragmatism and Romanticism." In *Philosophy as Cultural Politics*, 4:105–19. Cambridge: Cambridge University Press, 2007.

Rosato, L., and G. Álvarez, eds. *Borges, libros y lecturas: Catálogo de la colección Jorge Luis Borges en la Biblioteca Nacional*. Buenos Aires: Biblioteca Nacional, 2010.

Rosenberg, Alexander. *Economics: Mathematical Politics, or Science of Diminishing Returns?* Chicago: University of Chicago Press, 1992.

Rota, Gian-Carlo. "*Fundierung* as a Logical Concept." *Monist* 72 (1989): 70–77.

Rousseau, Jean-Jacques, *Œuvres complètes*. 4 vols. Paris: Gallimard, 1964–1969.

Rovelli, Carlo. *The First Scientist: Anaximander and His Legacy*. Translated by M. L. Rosenberg. Yardley, PA: Westholme, 2011.

Rowett, Catherine. "Philosophy's Numerical Turn: Why the Pythagoreans' Interest in Number Is Truly Awesome." In *Doctrine and Doxography: Studies on Heraclitus and Pythagoras*, edited by David Sider and Dirk Obbink, 3–31. Berlin: De Gruyter, 2013.

Roy, Subroto. *Philosophy of Economics: On the Scope of Reason in Economic Inquiry*. London: Routledge, 1989.

Rubin, Uri. "Al-Ṣamad and the High God: An Interpretation of Sūra CXII." *Der Islam* 61 (1984): 197–217.

Runia, David. *Philo in Early Christian Literature: A Survey*. Assen: Van Gorcum, 1993.

Runia, David. *Philo of Alexandria and the Timaeus of Plato*. Philosophia Antiqua 44. Leiden: Brill, 1986.

Runia, David. "Why Does Clement of Alexandria Call Philo 'The Pythagorean?'" *Vigiliae Christianae* 49 (1995): 1–22.

Ruskin, John. *Unto This Last, and Other Writings*. New York: Penguin, 1986.

Rusnock, Paul. "Was Kant's Philosophy of Mathematics Right for Its Time?" *Kant-Studien* 95 (2004): 426–42.

Russell, Bertrand. *Logic and Knowledge: Essays 1901–1950*. Edited by Robert Charles Marsh. 1956. Reprint, Nottingham: Spokesman, 2007.

Russell, Bertrand. *My Philosophical Development*. 1959. Reprint, London: Routledge, 1993.

Russell, Bertrand. *Mysticism and Logic, and Other Essays*. London: Longmans, Green, 1919.

Russell, Bertrand. "On Denoting." *Mind* 14, no. 56 (1905): 479–93.

Russell, Bertrand. *Our Knowledge of the External World as a Field for Scientific Method in Philosophy*. Chicago: Open Court, 1914.

Russell, Bertrand. *The Problems of Philosophy*. London: Williams & Norgate, 1912.

Saint-Hélier, Monique. *À Rilke pour Noël*. Berne: Éditions du Chandelier, 1927.

Salinas, Pedro. *Poesías Completas*. Edited by J. Marichal. Madrid: Aguilar, 1961.

Sant'Anna, A. S. "Individuality, Quasi-Sets and the Double-Slit Experiment." *Quantum Studies: Mathematical Foundations* (2019). https://link.springer.com/article/10.1007/s40509-019-00209-2.

Santayana, G. *Winds of Doctrine*. New York: Scribner's, 1913.

Sarraute, Nathalie. *The Age of Suspicion*. Translated by M. Jolas. New York: G. Braziller, 1963.

Sarraute, Nathalie. *L'Ère du soupçon*. Paris: Gallimard, 1956.

Saunders, Simon. "Indistinguishability." In *Oxford Handbook in Philosophy of Physics*, edited by R. Batterman, 340–80. Oxford: Oxford University Press, 2013.

Scheibe, Erhard. "Albert Einstein: Theorie, Erfahrung, Wirklichkeit." *Heidelberger Jahrbücher* 36 (1992): 121–38.

Scheler, Max. *Die Stellung des Menschen im Kosmos*. Darmstadt: Reichl, 1928.

Scheler, Max. *The Human Place in the Cosmos*. Translated by Manfred S. Frings. Evanston, IL: Northwestern University Press, 2009.

Schelling, F. W. J. *The Grounding of Positive Philosophy: The Berlin Lectures*. Translated by Bruce Matthews. Albany: State University of New York Press, 2007.

Schelling, Thomas. "Egonomics, or the Art of Self-Management." *American Economic Review* 68 (1978): 290–94.

Schlick, Moritz. "Naturphilosophische Betrachtungen über das Kausalprinzip." *Naturwissenschaft* 8 (1920): 461–47.

Schliesser, Eric. "Newton and Newtonianism." In *The Routledge Companion to Eigh-*

teenth Century Philosophy, edited by Aaron Garrett, 62–90. New York: Routledge, 2014.

Schmidt, Lothar, J. Lower, T. Jahnke, S. Schößler, M. S. Schöffler, A. Menssen, C. Lévêque, N. Sisourat, R. Taïeb, H. Schmidt-Böcking, and R. Dörner. "Momentum Transfer to a Free-Floating Double Slit: Realization of a Thought Experiment from the Einstein-Bohr Debates." *Physical Review Letters* 111 (2013): 103201-1-5.

Schmitt, Alois. "Mathematik und Zahlenmystik." In *Aurelius Augustinus: Die Festschrift der Görres-Gesellschaft zum 1500; Todestage des Heiligen Augustinus*, edited by Martin Grabmann and Joseph Mausbach, 353–66. Cologne: J. P. Bachem, 1930.

Schöck, Cornelia. "The Controversy between al-Kindī and Yaḥyā b. ʿAdī on the Trinity, Part One: A Revival of the Controversy between Eunomius and the Cappadocian Fathers." *Oriens* 40 (2012): 1–50.

Schöck, Cornelia. "The Controversy between al-Kindī and Yaḥyā b. ʿAdī on the Trinity, Part Two: Gregory of Nyssa's and Ibn ʿAdī's Refutation of Eunomius' and al-Kindī's 'Error.'" *Oriens* 42 (2014): 220–53.

Schoenemann, P. Thomas. "Syntax as an Emergent Property of Semantic Complexity." *Minds and Machines* 9 (1999): 309–46.

Schopenhauer, Arthur. *On the Fourfold Root of the Principle of Sufficient Reason and On the Will in Nature*. Translated by Karl Hillebrand. London: George Bell and Sons, 1903.

Schrödinger, Erwin. *The Interpretation of Quantum Mechanics: Dublin Seminars (1949–1955) and Other Unpublished Essays*. Edited by Michel Bitbol. Woodbridge, CT: Oxbow Press, 1995.

Schrödinger, Erwin. *Mind and Matter*. Cambridge: Cambridge University Press, 1958.

Schrödinger, Erwin. *My View of the World*. Cambridge: Cambridge University Press, 1964.

Schrödinger, Erwin. "Quantisierung als Eigenwertproblem (Zweite Mitteilung)." *Annalen der Physik* 79 (April 1926): 489–527.

Schrödinger, Erwin. "Was ist ein Naturgesetz?" *Naturwissenschaft* 17 (1929), 9–11.

Schrödinger, Erwin. "What Is a Law of Nature?" In *Science, Theory, and Man*, 133–47. New York: Dover, 1957.

Schwartz, Benjamin I. "The Age of Transcendence." *Daedalus* 104 (1975): 1–7.

Segal, Sanford. *Mathematicians under the Nazis*. Princeton, NJ: Princeton University Press, 2014.

Serres, Michel. *The Birth of Physics*. Manchester: Clinamen Press, 2000.

Sewell, Elizabeth. *Paul Valéry: The Mind in the Mirror*. Cambridge: Bowes & Bowes, 1952.

Sigmund, Karl. *Exact Thinking in Demented Times*. New York: Basic Books, 2017.

Simelidis, Christos. "The Byzantine Understanding of the Qur'anic Term al-Ṣamad and the Greek Translation of the Qur'an." *Speculum* 86 (2011): 887–913.

Simon, Herbert A. "Review of Theory of Games and Economic Behavior." *American Journal of Sociology* 50, no. 6 (1945): 558–60.

Sorabji, Richard. "The Origins of Occasionalism." In *Time, Creation and the Continuum: Theories in Antiquity and the Early Middle Ages*, 297–306. London: Duckworth, 1983.

Sorabji, Richard. *The Philosophy of the Commentators, 200-600 AD*. Vol. 2, *Physics*. Ithaca, NY: Cornell University Press, 2005.

Sorabji, Richard. *Time, Creation, and the Continuum: Theories in Antiquity and the Early Middle Ages*. Chicago: University of Chicago Press, 1983.

Spencer-Brown, George. *Laws of Form*. Leipzig: Bohmeier, 2014.

Spengler, Oswald. *The Decline of the West: Form and Actuality*. 2 vols. Translated by Charles Francis Atkinson. New York: Alfred A. Knopf, 1926-1928.

Spengler, Oswald. *Der Untergang des Abendlandes: Umrisse einer Morphologie der Weltgeschichte*. 2 vols. 1923. Reprint in one volume, Munich: C. H. Beck, 1990.

Spengler, Oswald. *Spengler ohne Ende: Ein Rezeptionsphänomen im internationalen Kontext*. Schriften zur politischen Kultur der Weimarer Republik, vol. 16. Edited by Gilbert Merlio and Daniel Meyer. Frankfurt am Main: Lang, 2014.

Spinoza, Baruch. *Spinoza: Complete Works*. Edited by Michael L. Morgan. Translated by Samuel Shirley. Indianapolis, IN: Hackett, 2002.

Stadler, Friedrich. "Documentation: The Murder of Moritz Schlick." In *The Vienna Circle: Studies in the Origins, Development, and Influence of Logical Empiricism*, edited by Friedrich Stadler, 866-909. Vienna: Springer, 2001.

Staehle, Karl, *Die Zahlenmystik bei Philon von Alexandreia*. Leipzig: Teubner, 1931.

Stamelman, Richard. *Lost Beyond Telling: Representations of Death and Absence in Modern French Poetry*. Ithaca, NY: Cornell University Press, 1990.

Stark, Johannes. "The Pragmatic and the Dogmatic Spirit in Physics." *Nature* 141 (1938): 770-72.

Stendhal. *Souvenirs d'égotisme*. Paris: Le Divan, 1927.

Stigler, Stephen M. "Richard Price, the First Bayesian." *Statistical Science* 33 (2018): 117-25.

Stolzenberg, Gabriel. "Can an Inquiry into the Foundations of Mathematics Tell Us Anything Interesting about Mind?" In *The Invented Reality: How Do We Know What We Believe We Know? Contributions to Constructivism*, edited by Paul Watzlawick, 257-308. New York: W. W. Norton, 1984.

Stone, Martin. "Interpretation and Pluralism in Law and Literature: Some Grammatical Remarks." In *Reason and Reasonableness*, edited by Riccardo Dottori, 129-158. Münster: Lit Verlag Münster, 2005.

Strassburg, Gottfried von. *Tristan and Isolde*. Translated by Francis G. Gentry. New York: Continuum, 2003.

Strobach, Niko. *The Moment of Change: A Systematic History in the Philosophy of Space and Time*. Dordrecht: Springer, 1998.

Stroumsa, Guy G. "From Abraham's Religion to the Abrahamic Religions." *Historia Religionum* 3 (2011): 11-22.

Sturm, Thomas. "Kant on Empirical Psychology." In *Kant and the Sciences*, edited by Eric Watkins, 163-80. Oxford: Oxford University Press, 2001.

Swift, Jonathan. "Abstract of Collins's 'Discourse on Free Thinking.'" In *The Prose Works of Jonathan Swift*, edited by Temple Scott, 3:163-92. London: George Bell & Sons, 1898.

Swift, Jonathan. *The Benefit of Farting Explain'd*. Exeter: Old Abbey Press, 1996.

Szabó, Árpád. *The Beginnings of Greek Mathematics*. Dordrecht: D. Reidel, 1978.

al-Ṭabarī, Jarīr. *The Commentary on the Qur'ān by Abū Ja'far Muḥammad b. Jarīr al-Ṭabarī.* Vol. 1. Translated by J. Cooper. Oxford: Oxford University Press, 1987.

Taine, Hyppolite. *De l'intelligence.* 1870. Reprint, Paris: L'Harmattan, 2005.

Tait, William. "Gödel on Intuition and on Hilbert's Finitism." In *Kurt Gödel: Essays for His Centennial,* edited by Solomon Feferman, Charles Parsons, and Stephen Simpson, 88–108. Cambridge: Cambridge University Press, 2010.

Tarán, L. *Academica: Plato, Philip of Opus, and the Pseudo-Platonic "Epinomis."* Philadelphia: American Philosophical Society, 1975.

Tarán, L. "Amicus Plato sed magis amica veritas: From Plato and Aristotle to Cervantes." *Antike und Abendland* 30 (1984): 93–124.

Tarán, L. "Nicomachus of Gerasa." In *Collected Papers (1962–1999),* 544–48. Leiden: Brill, 2001.

Tarán, L., ed. and trans. *Parmenides.* Princeton, NJ: Princeton University Press, 1965.

al-Tawḥīdī, Abū Ḥayyān. *Al-Muqābasāt.* Edited by Ḥassan al-Sandūbī. Kuwait: Dār Su'ād al-Ṣabāḥ, 1992.

Taylor, Richard. "The Agent Intellect as 'Form for Us' and Averroes's Critique of al-Fârâbî." *Proceedings of the Society for Medieval Logic and Metaphysics* 5 (2005): 18–32.

Thaler, Richard, and H. M. Shefrin. "An Economic Theory of Self Control." *Journal of Political Economy* 89 (1981): 392–406.

Thiele, R. "Georg Cantor, 1845–1918." In *Mathematics and the Divine: A Historical Study,* edited by Teun Koetsier and Luc Bergmans, 523–48. Amsterdam: Elsevier, 2004.

Tokarczuk, Olga. *Flight.* Translated by Jennifer Croft. London: Fitzcarraldo, 2017.

Toulmin, Stephen. *Return to Reason.* Cambridge: Harvard University Press, 2001.

Tversky, Amos, and Daniel Kahneman. "Rational Choice and the Framing of Decisions." In *Rational Choice: The Contrast between Economics and Psychology,* edited by R. Hogarth and M. Reder, 67–94. Chicago: University of Chicago Press, 1986.

Tversky, Amos. "Features of Similarity." *Psychological Review* 84 (1977): 327–52.

Tzu, Lao. *Tao Te Ching: A Book about the Way and the Power of the Way.* Translation by Ursula K. Le Guin. Boston: Shambhala, 1997.

Vailati, Ezio. *Leibniz and Clarke: A Study of Their Correspondence.* New York: Oxford University Press, 1997.

Valéry, Paul. *Charmes.* Paris: NRF, 1922.

Valéry, Paul. "La Crise de l'esprit, première lettre." In *Œuvres,* vol. 1, 988–94. Paris: Gallimard, 1957.

Valéry, Paul. *History and Politics: Collected Works of Paul Valéry.* Edited by J. Mathews and translated by D. Folliot. Vol. 10. Princeton, NJ: Princeton University Press, 1971.

Valéry, Paul. *Œuvres.* 2 vols. Paris: Pléiade, 1960.

Van Kooten, Geurt Hendrik. *Paul's Anthropology in Context: The Image of God, Assimilation to God, and Tripartite Man in Ancient Judaism, Ancient Philosophy, and Early Christianity.* Tübingen: Mohr Siebeck, 2008.

Vasari, Giorgio. *Le Vite de più eccellenti pittori scultori ed architettori scritte da Giorgio Vasari.* 9 vols. Florence: Sanson, 1906.

Vernant, Jean-Pierre. *Myth and Thought among the Greeks.* London: Routledge, 1983.

Vico, Giambattista. *On the Most Ancient Wisdom of the Italians.* Translated by L. M. Palmer. Ithaca, NY: Cornell University Press, 1988.

Viereck G. S. *Glimpses of the Great*. New York: Macaulay, 1930.

Wais, Karin. *Studien zu Rilkes Valéry-Übertragungen*. Tübingen: Max Niemayer, 1967.

Wazeck, Milena. *Einsteins Gegner: Die öffentliche Kontroverse um die Relativitätstheorie in den 1920er Jahren*. Frankfurt: Campus, 2009.

Weatherson, Brian. "David Lewis." In *Stanford Encyclopedia of Philosophy*. Updated September 19, 2014. https://plato.stanford.edu/entries/david-lewis/.

Weber, Heinrich. "Leopold Kronecker." *Jahresberichte der Deutschen Mathematiker-Vereinigung* 2 (1891/1892): 5–31. Reprinted in *Mathematische Annalen* 43 (1893): 1–25.

Weber, Max. *Max Weber: Essays in Sociology*. Edited and translated by H. H. Gerth and C. Wright Mills. New York: Oxford University Press, 1946.

Weber, Max. *Max Weber Gesamtausgabe*. Edited by W. J. Mommsen, W. Schluchter, and B. Morgenbrod. 17 vols. Tübingen: J. C. B. Mohr, 1992.

Weber, Max. "Wissenschaft als Beruf." In *Schriften 1894–1922*, edited by Dirk Kaesler, 474–511. Stutgart: Kröner 2002.

Weil, Simone. *First and Last Notebooks*. Translated by R. Rees. Oxford: Oxford University Press, 1970.

Weil, Simone. *Œuvres*. Paris: Gallimard, 1999.

Weil, Simone. *Simone Weil*. Paris: Cahiers de l'Herne, 2014.

Weizsäcker, C. F. von. "Zur Deutung der Quantenmechanik." *Zeitschrift für Physik* 118 (1941): 489–509.

Weyl, Hermann. *Gruppentheorie und Quantenmechanik*. Leipzig: S. Hirzel, 1928.

Weyl, Hermann. *Mind and Nature*. Princeton, NJ: Princeton University Press, 2009.

Weyl, Hermann. *Philosophy of Mathematics and Natural Science*. New York: Atheneum, 1963.

Weyl, Hermann. *The Theory of Groups and Quantum Mechanics*. Translated by H. P. Robertson. Mineola, NY: Dover, 1950.

Wheeler, J. A. "The 'Past' and the 'Delayed-Choice' Double-Slit Experiment." In *Mathematical Foundations of Quantum Theory*, edited by A. R. Marlow, 9–48. Cambridge, MA: Academic Press, 1978.

Wheeler, J. A. *Quantum Theory and Measurement*. Princeton, NJ: Princeton University Press, 1983.

Whitehead, Alfred North. *An Introduction to Mathematics*. New York: Henry Holt, 1911.

Wigand, Johann. *De neutralibus et mediis libellus*. 1562.

Wigner, Eugene P. *Philosophical Reflections and Syntheses*. Edited by Jagdish Mehra and Arthur S. Wightman. Berlin: Springer, 1995.

Wigner, Eugene P. "The Unreasonable Effectiveness of Mathematics in the Natural Sciences." *Communications on Pure and Applied Mathematics* 13 (1960): 1–14.

Wilson, F. P. *The Oxford Dictionary of English Proverbs*. 3rd ed. Oxford: Clarendon Press, 1970.

Winter, Alison. *Memory: Fragments of a Modern History*. Chicago: University of Chicago Press, 2012.

Wittgenstein, Ludwig. *Blue and Brown Books*. New York: Harper, 1965.

Wittgenstein, Ludwig. *Philosophical Investigations*. Edited by G. E. M. Anscombe and R. Rhees and translated by G. E .M. Anscombe. 3rd ed. Oxford: Blackwell, 1958.

Wordsworth, William. *The Prelude, or Growth of a Poet's Mind*. Oxford: Oxford University Press, 1933.

Wright, M. R. *Cosmology in Antiquity*. London: Routledge and Kegan Paul, 1995.

Wyller, Egil A. "Zur Geschichte der platonischen Henologie: Ihre Entfaltung bis zu Plethon/Bessarion und Cusanus." In *Greek and Latin Studies in Memory of Cajus Fabricius*, edited by Sven-Tage Teodorsson, 239–65. Göteborg: University of Göteborg, 1990.

Yaḥyā b. ʿAdī. *Maqālah fī'l-Tawḥīd*. Edited by Samīr Khalil. Beirut: Al-Maktaba al-Būlusiyya, 1980.

Young, Charles M. "Plato and Computer Dating." *Oxford Studies in Ancient Philosophy* 12 (1994): 227–50. Reprinted in Nicholas D. Smith, ed., *Plato: Critical Assessments*. Vol. 1, *General Issues of Interpretation*. London: Routledge, 1998.

Young, Edward. *Night Thoughts*. London: C. Whittingham, 1798.

Zhmud, Leonid. *Pythagoras and the Early Pythagoreans*. Oxford: Oxford University Press, 2012.

Zhmud, Leonid. "Pythagorean Communities: From Individuals to Collective Portraits." In *Doctrine and Doxography: Studies on Heraclitus and Pythagoras*, edited by David Sider and Dirk Obbink, 33–52. Berlin: De Gruyter, 2013.

Zhmud, Leonid. *Wissenschaft, Philosophie und Religion im frühen Pythagoreismus*. Berlin: Akademie, 1997.

Zhou, Yang, and David J. Freedman. "Posterior Parietal Cortex Plays a Causal Role in Perceptual and Categorical Decisions." *Science* 365, no. 6449 (July 12, 2019): 180–85.

Zhuang Zhou. *Zhuangzi, The Complete Writings*. Translated by Brook Ziporyn. Indianapolis, IN: Hackett, 2020.

Ziporyn, Brook. *Beyond Oneness and Difference: Li and Coherence in Chinese Buddhist Thought and Its Antecedents*. Albany: State University of New York Press, 2013.

Ziporyn, Brook. *Ironies of Oneness and Difference: Coherence in Early Chinese Thought; Prolegomena to the Study of Li*. Albany: State University of New York Press, 2013.

INDEX

Printed and bound by CPI Group (UK) Ltd, Croydon, CR0 4YY

27/10/2024